Dogfight

Dogfight

The Supermarine Spitfire and the Messerschmitt Bf109

David Owen

Pen & Sword
AVIATION

First published in Great Britain in 2015 by
Pen & Sword Aviation
an imprint of
Pen & Sword Books Ltd
47 Church Street
Barnsley
South Yorkshire
S70 2AS

ISBN 978 1 47382 806 3

Typeset in Ehrhardt by
Mac Style Ltd, Bridlington, East Yorkshire
Printed and bound in the UK by CPI Group (UK) Ltd,
Croydon, CRO 4YY

Pen & Sword Books Ltd incorporates the imprints of Pen & Sword
Archaeology, Atlas, Aviation, Battleground, Discovery, Family History,
History, Maritime, Military, Naval, Politics, Railways, Select, Transport,
True Crime, and Fiction, Frontline Books, Leo Cooper, Praetorian Press,
Seaforth Publishing and Wharncliffe.

For a complete list of Pen & Sword titles please contact
PEN & SWORD BOOKS LIMITED
47 Church Street, Barnsley, South Yorkshire, S70 2AS, England
E-mail: enquiries@pen-and-sword.co.uk
Website: www.pen-and-sword.co.uk

Contents

Acknowledgments

Writing a novel may still be a solitary occupation, provided the author's inspiration and talent are up to creating memorable and credible characters, and putting them through a series of challenging situations and recording how they respond and develop. This can never be true of non-fiction, especially when dealing with the past. A writer venturing into this arena needs a varied and capable team of helpers and advisers to bring a project to completion; and thanking all concerned for their assistance and support is one of the most pleasant of duties in preparing a book for publication. So here goes.

In no particular order, except perhaps the chronological, I'd like to set the record straight by thanking everyone who shared my belief that the story was worth the telling, and needed a measure of perseverance to turn it into reality. My own inspiration on this, the 75th anniversary of the start of the Battle of Britain, stemmed originally from my eldest brother Norman, killed in a training flight at 19, on his way to become an RAF pilot in 1940. Old college friends like Jim Doxey and Johnnie Mudge, Graham McLean, Phil Patience and Martin Taylor, helped keep the nose pressed to the grindstone by frequent enquiries about progress. Sources of information like the anonymous, but ever-helpful staff of the Liverpool and Manchester reference libraries, located books, websites and unpublished but essential documents, bearing on the subject of these two remarkable aircraft, how they were designed and built, and the stories of those who flew and fought in them.

Illustrations could have been difficult to find, had not Pete Bunting and Sean Penn, and their colleagues at the RAF Museums of Hendon and Cosford, made it possible to photograph wartime aircraft. My elder son, James, and his family, took us to Washington DC so that we could visit the Smithsonian Air and Space Museum to photo aircraft and artefacts available nowhere else, and younger son Will, a pilot himself, brought his own photographic expertise to bear on the Hendon collections. Stephen Vizard and Chris Michell of Airframe Assemblies showed me Spitfires and 109s being rebuilt and restored at their Isle of Wight workshops and produced unique wartime drawings and photos of Spitfire and 109 production.

Wikipedia made available hundreds of official wartime pictures, and through their agreement with the German Bundesarchiv, an equally wide range of German photographs, while US sources like the National Archives, the US Navy Historic Centre, and most especially the USAF Museum at Dayton in Ohio, enabled high-quality downloads to be sourced with the minimum of delay or expense. Alfred

Price, perhaps the ultimate authority on the Spitfire and on wartime air operations, and a valued contributor on many other projects, provided pictures and background information. Finally, Captain Eric 'Winkle' Brown gave freely of his time and unrivalled experience to fill in the background to these formidable fighters and assess their relative strengths and weaknesses. My grateful thanks to you all – I only hope you feel the finished work is worth it !

David Owen
Chester, UK, May 2015

Foreword

Captain Eric 'Winkle' Brown CBE DSC AFC RN (Retd)

In a field where most wartime, fighter pilots tended to prefer the types they flew in combat over those that fought against them, 'Winkle' Brown stands out for two main reasons: his incredible breadth of experience having flown more types of aircraft than anyone else since the invention of flight, and his objectivity having been a trained military test pilot charged with evaluating the behaviour and performance of Allied and Axis warplanes in detail. Here is his evaluation of the Spitfire's two main wartime opponents:

Throughout World War Two, Germany's fighter defence relied mainly on the piston engined Messerschmitt 109 and the Focke-Wulf 190. Like Britain's Hurricane and Spitfire, one preceded the other and both kept up with the hunt by aerodynamic modifications, increases in engine power and in firepower, as well as weapon carrying capacity. Hence there are many marks or models of both aircraft, but it is most interesting to compare what pilots believe were the classic examples of each type, namely the Bf109F and the FW190A-4.

Let me say from the outset that I would not like to be the pilot of a Bf109F in the western operational area of World War Two, where the opposition was of a high standard in both aircraft and pilot quality. I feel the drawbacks of the 109 were the claustrophobic cockpit and the poor harmony of control. On the plus side it had very good firepower, especially when equipped with the Mauser cannon firing through the airscrew hub, and it had a fast dive to evade trouble.

On the other hand I would have been very happy to be the pilot of a Fw190 A-4 with its fine performance, superb rate of roll, wonderful view from the cockpit, reliable radial engine and formidable firepower. With this package I would have been ready to take on all-comers. Indeed in 1943 I had the opportunity to lock horns with such an aircraft while flying a Spitfire IX over France at about 25,000 feet, and after some eight minutes of exhilarating thrust-and-parry we both mutually agreed the combat situation had reached stalemate and parted company before we both ran out of fuel.

With reference to the 109, many will point to the 352 kills made by Erich Hartmann, but when I interviewed him in 1958 he pointed out that all of his successes were on the eastern front against the Russians who were tactically naive in air operations, and he swore by the effectiveness of the airscrew hub firing Mauser cannon …

Eric Brown, July 2014

Chapter One

Bombers over Breakfast

The most crippling blow to Britain's pre-war air defences began emerging early one morning in February 1934, on the main up departure platform at Bristol's Temple Meads Station. Amid the passengers boarding the Great Western Railway express to London Paddington was Robert Lewis, editor of the *Bristol Evening World*, part of the Rothermere group of national and provincial newspapers. He was attending a meeting with fellow editors, chaired by Lord Rothermere, head of the group. Rothermere had been a member of Lloyd George's Air Council during the First World War, and was extremely aviation-minded. On the agenda was the need for a high-speed aircraft to match the twelve-seat Lockheed monoplane recently ordered by Lord Beaverbrook, head of the rival Express group of papers.

On the crowded platform Lewis spotted a face he recognized, Roy Fedden, chief engine designer of the Bristol Aircraft Company. The men had met at the Bristol Yacht Club, where Lewis was a member and Fedden had given a lecture at the Spa Hotel in Clifton. The subject of that lecture and their chance meeting on Temple Meads Station would have massive consequences. One was part of the lethal threat presented by a new generation of enemy bombers. Another, and even more serious, would be the sudden crippling of all Britain's defending fighters.

Fedden's lecture described his company's latest project, the Bristol 135 twin-engined monoplane. Lewis knew this was something Rothermere would want to know about, and when he and Fedden entered the restaurant car for breakfast, he invited him to share his table. They spent some minutes chatting about their common interest in fishing, before Lewis switched the subject to the 135, and asked Fedden for more information that he could present to Rothermere at the London meeting.

Two Bristol Mercury radial engines provided the power to enable the new Bristol aircraft to outstrip all existing first-line fighters. *(via Wikipedia)*

Fedden was delighted to oblige. The early 1930s had been difficult years for Bristol. The company had flourished during the First World War, when their successful two-seat Bristol Fighter biplane became a mainstay of the Royal Flying Corps, and more than 5000 had been produced by 1918. But the return of peace meant business had slowed, orders had dried up and Bristol only remained afloat commercially by turning out competitors' designs under licence. The Bristol 135 was a brave attempt at a radical new design to kick the company's market back into life. Sadly, aviation was slow to accept new ideas and radical designs. At the time Fedden and his team began work on the 135 project, the Bristol Fighter was still in front-line military service in New Zealand!

This made Rothermere's response crucial from Fedden's point of view. By the time both men had finished their breakfast and the train was pulling into Paddington Station, Lewis had all the facts about the 135. He had a welcome idea for the meeting, but remained completely unaware about the threat the project would pose to the stability of military aviation and to Britain's ability to defend itself from air attack.

Neither man realized the imminence of a drastic change of course. Only sixteen years before, the most terrible war in history had smouldered to an end with an appalling casualty bill. During the conflict, military aircraft had developed into powerful weapons, but the process had been interrupted by the Armistice. The public view of military aviation was still inspired by stories of daring single combat between fighter aces, high above the mud, blood and squalor of the trenches. Unfortunately, this comforting and inspiring epic was a small part of the truth, and concealed a darker shadow. During the final months of the war, large, long-range multi-engined biplane bombers had been able to reach enemy territory for the first time, raining high explosive and incendiaries down on a helpless civilian population. The Armistice had prevented a full demonstration of the bomber's power, but it remained a massive threat for the future.

At first, aircraft engineers were happy to develop civilian airliners. Thanks to wartime improvements these could now fly further, faster and more reliably than before. New engines delivered more power. New construction techniques produced stronger and lighter airframes, and better streamlining reduced drag. Wooden skeletons of spars, ribs and stringers were replaced by steel tube frameworks, still covered by linen fabric, coated with dope to stretch it tight. All too often they still needed bracing wires, which hampered performance and made improvements slow and sporadic.

Against this familiar and unthreatening background, the new 135 was an ambitious step further. Fedden had concentrated on the main problem limiting aero-engine development. To produce more power, an engine had to turn over more quickly. This shortened the time available for the fuel-air mixture to be forced into the cylinders and the exhaust gases ejected on each cycle. Engines therefore tended to run short of breath, especially as the aircraft climbed and air pressure fell with increasing altitude.

One way around this obstacle was to increase the area of the valves providing access to each cylinder. For a given design, four smaller valves gave a larger area than two bigger ones. Unfortunately this complicated the valve gear, and Fedden decided to use sleeve valves instead of the normal poppet valves. These promised greater reliability especially at higher speeds and power settings, but the first sleeve valve engine was only just taking shape. This was the Bristol Aquila, a smaller six-cylinder radial based on the existing Bristol Mercury, and the 135 would have two of them.

Steel and fabric would be used in small amounts. The engines would be carried on steel frames, and steel would be used for the flanges of the wing-spar webs. The control surfaces, rudders, elevators and ailerons would be covered with fabric, but that was all. The rest of the airframe – fuselage, wings and tail plane – would use light alloy, including the stressed skin covering. This neat and carefully streamlined twin-engine monoplane was an extremely promising design, except for one drawback. Just before the meeting at the Spa Hotel, the Bristol board had cancelled the project as too expensive. One reason why Fedden kept his appointment at the Yacht Club was to drum up some interest and possibly an order for the new aircraft.

Later that day, Lewis told Rothermere about the 135. Designed by Frank Barnwell, originally a naval architect and shipbuilding draughtsman, it was a small feeder airliner, connecting local airfields with major airports where passengers could switch to long-distance international flights. The slim fuselage held eight seats: two in the nose for the crew, and three rows of double seats for half-a-dozen passengers, and it promised to be light, fast and economical. Rothermere's wartime Air Council experience convinced him he could recognize something good when he heard about it, and ordered Lewis to go back to Fedden and his colleagues for more information.

When he did, Barnwell and Fedden showed him the 135 drawings. Later, on 26 March 1934, Rothermere ordered the 135 for a total price of £18,500 on condition the aircraft was delivered within a year. In fact, he was looking for a rather different type of plane; not an efficient mini-airliner, but a high-speed transport for reporters and photographers to cover breaking news stories anywhere in Europe. Also on his personal agenda was a chance to outdo Beaverbrook's US-built Lockheed 12, and to do this with a British aircraft.

Nevertheless, meeting this order would be a challenge for Bristol. The Aquila engines would not be ready for bench testing for another six months, and their power output was still relatively modest. Instead, Rothermere's insistence on speed before economy opened up a different possibility. His priorities persuaded Bristol to replace the Aquilas with the larger and more powerful supercharged Bristol Mercury engines. These were already performing well, and the engineers predicted a spectacular top speed for the time of 240 mph. This revised design was renamed the Bristol 142 and work began on the necessary changes.

By now, the order was attracting criticism. Ironically the source was Rothermere's own aviation adviser, an ex-Royal Flying Corps officer and former Secretary General of the Air League of the British Empire, Brigadier General P.R.C. Groves. He was appalled at the new design's predicted performance. He insisted it would be lethally fast and predicted it would cause the death of anyone foolish enough to fly in it. He tried to convince Rothermere the spectacular failure of the project would make him a laughing stock. At first Rothermere was deterred by these lurid warnings, but after listening to the engineers' detailed arguments, he backed their judgment and the order went ahead.

Thereafter, progress proved surprisingly straightforward. Fitting the more powerful engines meant a great deal of extra work, but the prototype was ready less than three weeks after the original deadline. On 12 April 1935, Bristol test pilot Cyril Uwins opened up the 142's twin Mercury engines and took off from the company airfield at Filton. The flight went well. Uwins reported the handling was excellent, with no real vices and decided all was in order for the handover to the new owner. In the meantime, Roy Fedden had one valuable suggestion to make. The 142 had flown with conventional four-bladed, fixed–pitch wooden propellers. Fedden recommended making the most of the low drag factor by using the far more efficient three-bladed variable-pitch airscrews made by the American Hamilton-Standard Corporation.

These allowed the pilot to vary the blade angles between fine pitch for landing and take-off, and coarse pitch for cruising. The improvement provided by the new propellers was truly spectacular. During performance trials, the 142 reached a top

As new bombers were flying faster, new fighter designs remained trapped in a time warp, and the RAF searched for a replacement for its slow and inadequately armed Bristol Bulldog. *(Author photo)*

speed of 307 mph. To Rothermere's delight, this was more than 100 mph faster than Beaverbrook's Lockheed. The Air Ministry on the other hand, was appalled. Rothermere's new plane was faster than all the RAF's current fighters by an even wider margin.

Completely unaware of this problem, Rothermere saw his purchase as a triumphant vindication of his hopes that Britain could hold its own against the latest aviation designs from overseas. Consequently, simply calling it the Bristol 142 did not do justice to the message the aircraft carried. Instead he had the words 'Britain First' painted on the nose. Others, aware of the more sinister implications of such a fast aircraft, apparently immune to fighter attack, called it the 'Rothermere Bomber.'

This was a symbol of the death's head at the peacetime feast. During the early 1930s, the euphoria of the 1918 victory had given way to growing disquiet at the increasing threat of the long-range bomber. Prime Minister Stanley Baldwin told the House of Commons in November 1932 that it would be well for the 'man in the street' to realize 'there is no power on earth that can protect him from being bombed', and it was clear the 'man in the street' was all too well aware of it. Experts predicted that massed bombing raids would cause a terrible death toll, with every ton of bombs dropped in a future war causing at least seventy-two casualties. Claims were made that thousands of people would be killed by devastating raids within hours of hostilities breaking out, claims providing fertile soil for German propaganda.

This placed the Air Ministry on the horns of a terrible dilemma. If 'Britain First' could be developed into a fast, modern medium bomber, able to outstrip enemy fighters, then Britain should have this capability and have it as quickly as possible. But the achievement of Barnwell and Fedden could be copied by others. It was all very well for the RAF to have a bomber which could attack enemy countries without fear of defending fighters, but what could be done about enemy bombers intruding into British airspace to do the same?

'Britain First' or the 'Rothermere Bomber' wearing RAF markings for service acceptance trials at Martlesham Heath. *(Author drawing)*

First, they needed to take a closer look at what 'Britain First' could do, and Rothermere agreed that the aircraft could go to the Aeroplane and Armament Experimental Establishment at Martlesham Heath, near Ipswich, for RAF experts to examine and test it. In June 1935, 'Britain First' joined a series of aircraft waiting for Service evaluation. Just how dramatic an improvement it represented over current military designs was underlined by the plane immediately ahead of it in the test line-up. This was the latest bomber to be accepted for RAF service, the Boulton Paul Overstrand, a large and cumbersome biplane with an enormous fixed undercarriage, and a massive network of bracing struts and wires. Though its engines delivered similar power to those of the Bristol 142, its huge drag penalty limited its top speed to 148 mph, less than half that of the Bristol monoplane! Given that this slowest of targets was the threat RAF fighters were designed to combat, meeting something like the Bristol 142 would render them completely useless.

Although the performance of the Bristol 142 had surprised the Air Ministry as much as its new owner, Barnwell and Fedden kept them fully informed. With a depressed aircraft market, the last thing they wanted was to offend their main military customer by making a modern machine for a civilian owner without telling them, and without updating them on engine progress. They then built a second prototype, fitted with the original Aquila sleeve-valve engines, called the Bristol 143. Even with engines smaller than those of Rothermere's machine, the plane's top speed was an encouraging 250 mph.

Nonetheless, Rothermere's plane seemed the best bet for future British bomber development, and Rothermere himself announced he would hand it over to the RAF if this would help provide it with the best possible machines. Certainly the Service pilots found it much to their liking, though more complex than the slow and undemanding biplanes they were used to. On 17 July 1935, one Martlesham Heath pilot was preparing to land the 142 when he tried to lower the undercarriage. It failed to lock into position and when he touched down, it collapsed and severely damaged the aircraft. This was profoundly embarrassing for the RAF, and they insisted on paying back Rothermere's initial purchase price for 'Britain First' by way of compensation. His flamboyant and popular gesture actually cost him nothing at all.

By this time, Barnwell and Fedden had produced a revised version of the 142, the 142M, better suited to military requirements. During August 1935 they raised the wing to a mid position to provide room for an enclosed bomb bay beneath, with stowage for four 250lb bombs. The tailplane was also raised to avoid being blanked by the higher wing, and space inside the glazed nose was provided for a bomb-aimer. Armament was limited to just two machine guns; one fixed in the port wing firing forward under the pilot's control, and another in a power-operated gun-turret mounted on the upper side of the fuselage behind the wing.

The Air Ministry moved equally quickly. They drew up an official medium bomber specification, 28/35, with the 142M in mind. Within a month they had

The RAF was not alone in depending on biplane fighters; Italy's Fiat CR42 saw widespread service in World War Two, and suffered accordingly. *(Author photo)*

ordered 150 of the new bombers, almost a year before the 142M prototype actually flew. This emerged from the Bristol plant in June 1936 with revised carburettor air intakes and cooling gills around the engine cowlings, controllable by the pilot to ensure the engines ran at optimum temperature under different loads. The retractable tail wheel of the prototypes was replaced by a fixed wheel, and another 450 Blenheim Mark Is, as the design was known in the RAF, were ordered to bring the initial total to 600.

For the RAF's thriving bomber lobby, these were the happiest of times. In the final months of the First World War, public opinion had demanded heavier bombing raids on Germany, in response to German attacks on Allied cities by Zeppelin airships and large Gotha biplanes. The RAF was founded on 1 April 1918 by combining the Army's Royal Flying Corps and the Royal Naval Air Service, and ordered to develop better air defences for the British Isles and a strategic bombing offensive against Germany. The Independent Force of heavy Handley Page biplane bombers had been formed in June 1918 to do this. It was commanded by Hugh Trenchard, later the Chief of the Air Staff and a fervent advocate of the value of air power to win wars without the slaughter of the trenches.

Trenchard was convinced this thinking was essential for future funding and continuing independence for the RAF. For the most devoted supporters of this 'Trenchard doctrine' the power of the bomber made fighters irrelevant. Winning future wars meant more and more bombers delivering heavier bomb loads to defeat and demoralise the enemy. At its most fanatic, the bomber lobby insisted that spending money on fighters was completely wasted, diverting resources from the bombers, which alone could bring victory. It seemed that the yawning performance gap between bombers and fighters was winning the argument for them.

Unfortunately, the situation was already worse than the Air Ministry feared. While Rothermere's plane was being completed, the newly established Nazi regime in Germany had begun massive rearmament, geared to another world war, but hidden for the time being under a strict security blanket. At first, Rothermere himself welcomed what he saw as Hitler's forward thinking on aviation matters and he sympathised with the Nazi leader's warnings of the Bolshevik menace of Stalin's Russia. Consequently, as soon as Hitler felt secure enough to lift the curtain on Germany's rising military strength, he sent Rothermere details of the formation of the new Luftwaffe on 8 March 1935 to ensure maximum publicity in national newspapers like Rothermere's Daily Mail.

Even before this, the Nazis had commissioned designs for future military aircraft under various pretences. Dornier had been ordered to design and build a civilian airliner for the national airline *Deutsche Luft Hansa* and a 'high-speed mail plane' for the German State Railways. The resulting Dornier 17 flew for the first time in November 1934, just eight months after Rothermere's order for the Bristol 135.

As an airliner, the Dornier was unconvincing. Its extremely narrow fuselage would justify its future nickname of the 'Flying Pencil'. Officially it would carry six passengers, like the Bristol 135, but its cramped interior ruled out the chance of a commercial passenger payload. The idea of a mail plane was slightly more credible, but it was increasingly obvious this was a bomber in all but name. It provided room for an operational crew of pilot, navigator, radio operator, bomb aimer and gunners in the forward section of the aircraft. German military psychology grouped crews of tanks and bombers closely together for mutual support.

Unfortunately for Hitler's plans, Germany lacked one vital requirement; high power aircraft engines. Aero-engine development had only recently been allowed under the Versailles Treaty, so for the time being performance was disappointing. The Dornier 17 prototype used two BMW engines delivering 750 bhp apiece, giving a top speed of 225 mph. This would outstrip most biplane fighters, but improvements would soon ensue. Flight tests showed a lack of yaw stability, cured by replacing the single fin and rudder with a twin tail. Switching to more powerful engines eventually pushed the top speed up to a useful 259 mph.

Perhaps the most convincing confirmation that Barnwell and Fedden, and Dornier for that matter, were on the right track came from Heinkel. The German firm had decided to build the fastest monoplanes in the world even before the Nazi takeover. First, they studied the high-speed American monoplanes built by Lockheed. The single-engined Lockheed 9 Orion appeared in 1931. It was the first airliner with retractable undercarriage, based on a mainly wooden airframe, with a

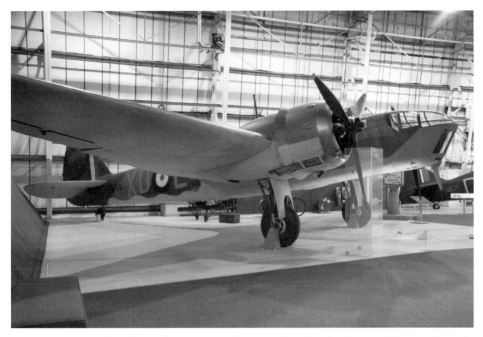

The Mark IV Blenheim had a longer nose with room for a bomb aimer and became the main version of the design. *(Author photo)*

220 mph top speed. Its compact fuselage could carry half-a-dozen passengers, with the pilot sitting in a cockpit above, and it promised to capture a huge share of the airliner market.

Unfortunately, in October 1934 the US civil aviation authority insisted that safety demanded that passenger-carrying aircraft on major routes had to have two pilots and two engines. By this time, Heinkel had already been catching up on Lockheed with the first flight of the Heinkel He 70 monoplane on 1 December 1932. Like the Orion, it was a compact and carefully streamlined monoplane with retractable undercarriage. It was powered by a single BMW V12 engine and carried a pilot with a radio operator behind him, and four passengers sitting in pairs facing one another with their backs to the cabin sides. Unlike the Orion, it had a stressed magnesium alloy skin to save weight, but its performance was almost identical.

Fortunately Lockheed were ready for the new requirements with their twin-engined Electra, their first all-metal stressed-skin monoplane. This had first flown in February 1934, eight months before the laws changed. Powered by two Pratt and Whitney Wasp radials, it carried ten passengers at a top speed of just over 200 mph. Other US aircraft builders were working along similar lines, with Boeing's 247 flying for the first time in 1933 with similar capacity and performance, and the Douglas DC2 which first flew on 11 May 1934 and carried fourteen passengers at a top speed of 210 mph.

Unlike their German equivalents, these American designs were not simply stepping stones towards bombers, but genuine attempts to meet a rapidly increasing market for safe, long distance transport for passengers and freight across their vast country. When the Americans built bombers of their own as war approached, these would be completely different designs. In Nazi Germany though, the military tail wagged the aviation dog, and Heinkel's attempt to follow Lockheed's example in switching from a single-engine monoplane to a twin would have a more sinister purpose.

Like the Dornier 17, the Heinkel 111 was commissioned as a potential airliner, but here too its design only made sense as a bomber. It had similarly graceful elliptical wing and tail surfaces to the Heinkel 70, but scaled up to twin-engine size it was officially capable of carrying ten passengers at a top speed of 225 mph. The passenger cabin was split into two small sections, the after part carrying six non-smokers, but the smaller smoking section had four seats set lower in the fuselage, a bomb bay in all but name. Indeed, for all save the few genuine civilian variants, this section would hold up to two tons of bombs. The Heinkel 111 made its maiden flight on 24 February 1935 and would be followed by the next bomber design, the Junkers 88 appearing as a military aircraft from the start. The aviation world was changing fast, with performance climbing from month to month.

Once the Luftwaffe had been officially revealed, and with it the powerlessness of the western democracies to frustrate his intentions, Hitler's policy switched from hiding the new force to emphasising its power, its strength and its readiness. By now, the regime's underlying ruthlessness had been revealed by the *Night of the Long Knives*, the cold-blooded murders of the leaders of the *SturmAbteilung* or *SA*, a violent band of thugs who had helped Hitler seize power. Now he saw them as a threat to the consolidation of that power, and for many inside and outside Germany, the killings of his former collaborators were conclusive proof that Hitler was a dangerous megalomaniac. Unfortunately, it was also becoming clear he was a megalomaniac with an increasingly powerful air force.

So how had it been possible for these new bomber designs to turn the tables so dramatically on the fighters they would face in combat? Why could fighter designers not effect a similar transformation? The truth was that every one of these new bombers made current fighters more incapable. By the time the Blenheim finally entered service in the summer of 1937, the RAF's current fighter biplanes were about to be replaced by a whole new generation – of fighter biplanes. Front-line fighter squadrons, like those of France, Germany and the USA, still depended on planes all too familiar to pilots of 1918. They were happiest with an open cockpit for the best possible view of the skies around them. They wanted manoeuvrability to dodge enemy fire and bring their own guns to bear on a fleeing opponent, which meant the low wing loading of a biplane. They wanted the supreme simplicity and reliability of a fixed undercarriage, and they wanted the familiar armament of a pair of machine guns mounted within easy reach on the engine cowling in front of them. Their tendency to jam in combat could be tackled by a hammer carried in the cockpit.

This was the prescription underlying British fighters of the time like the Bristol Bulldog and the Hawker Fury, French fighters like the Nieuport Type 62, Italian fighters like the Fiat CR42, American fighters like the Boeing F4B and Grumman FF-1, and German fighters like the Heinkel 51 and the Arado 68. Unfortunately, it also brought with it enormous amounts of drag, limiting top speed to between 150 mph and 250 mph at a time when bombers were capable of between 225 and 275 mph, even with wartime loads. This was the reason why the new metal-skinned monoplane bombers were so immune to fighter interception.

The situation was unavoidable. Aero-engines of the early 1930s delivered a maximum power of around 600 bhp. With a traditional fighter design and all the accompanying drag, this enabled a top speed of around 150 mph, far too slow to catch and shoot down one of the new bombers. Even if designers switched to all metal stressed skin construction and retractable undercarriage, their planes would still be lucky to exceed 250 mph – still insufficient to catch a twin-engine bomber.

This was a matter of simple mathematics. Putting two modern engines, even at 600 bhp each, into a streamlined monoplane produced enough performance to outrun fighters. Fitting a single, similar engine into the most modern fighter airframe would never overcome enough weight and drag to give a high enough top speed and rate of climb. To make matters worse, any fighter able to catch one of the new bombers would be unlikely to hold it in its gunsights for longer than a second or two. Delivering a powerful enough punch within that timeframe demanded much heavier armament.

Fighter pilots' preferences made failure inevitable. The drag caused by an open cockpit meant the only view it was likely to give was of a monoplane bomber disappearing into the distance. Pilots insisted the inherent manoeuvrability of a biplane was essential in fighter-to-fighter combat. Their large wing area enabled biplanes to roll more quickly and turn more tightly in a dogfight. So long as fighters were evenly matched, designers had little incentive to explore new ideas, but now the game was well and truly up.

With faster twin-engine monoplanes like 'Britain First', manoeuvrability suddenly mattered less. How could a fighter shoot down a bomber if it could not even catch it? Against other fighters, a faster opponent could choose the conditions of combat, rush in and deliver a quick blow, shooting down his adversary without manoeuvring at all. Only five years before the second war it was clear the fast monoplane fighter must be the shape of the future. But without much more powerful engines, fighter performance seemed doomed to remain unattainable.

Half-measures would be useless. The French Air Force, the *Armée de l'Air*, introduced a stressed-skin monoplane fighter in 1935 that seemed as modern as any in the business. The Dewoitine D.500 was so vital that production was licensed to two other manufacturers in addition to the parent firm. But the details were significant. It still had fixed undercarriage, with spatted wheels carried on stout struts, and the pilot sat in an open cockpit. It was armed with two machine guns on the engine cowling with the option of two more in the wings.

There was a commendable lack of bracing wires, and drag was probably lower than most of its contemporaries. But the 600 bhp of its Hispano–Suiza V12 engine limited its top speed to 223 mph, still slower than the fastest biplane fighters of the day. Even the improved D.501 with a 690 bhp Hispano engine could only make 250 mph. Its armament was boosted by a cannon between the engine cylinder banks, firing through the propeller hub, but it would never get close enough to an opponent to prove a threat.

Its limited performance meant that when war began, it was fit only for training duties. In the end the most powerful lesson resulting from Lewis and Fedden's breakfast time conversation on that train to London was that only if the engine manufacturers could boost power output beyond 1000 bhp could the balance between fighter and bomber, so upset by the Bristol 142 and its contemporaries, be restored. Given the lack of funds for fighter development, this meant starting from scratch once such engines were available. Suddenly, the aviation future which appeared so promising across a breakfast table on the Bristol to London express in 1934 now seemed darker and more threatening. Meeting the challenge it presented would call for two of the most brilliant and original creations in the whole history of aviation.

Chapter Two

Whistling Toilets and Flying Radiators

For the first time since the invention of the aeroplane it seemed that modern bombers could do anything they wanted to their helpless victims, for the lack of engines powerful enough for fighters to catch them, beat them off, or shoot them down. Without them there would be no chance of rebuilding air defences and restoring the balance between bomber and fighter. Yet truth is often said to be stranger than fiction. There were in fact already engines powerful enough for high performance fighters and there had been for several years. Unfortunately, they belonged to the specialized world of international air racing. They ran on exotic fuel mixtures. They had to be dismantled and rebuilt after every few hours' running. Their spectacular power produced vast amounts of torque, making aircraft almost impossible to fly. They poured out so much heat that if a plane did fly, it could not do so for long. Until these failings were mastered, they would never cope with military service.

On 23 October 1925 at Baltimore, on the shore of Chesapeake Bay on the US East Coast, competitors had gathered on the eve of the latest Schneider Trophy race. One entrant was a small British monoplane. Its slim fuselage carried a massive 24-litre Napier Lion engine with three banks of four cylinders each, arranged in a broad-arrow configuration, and developing 700 horsepower. Its single wing was mounted high on the fuselage to reduce drag, but its massive floats greatly increased it again. This was the Supermarine S4, brainchild of a young designer named Reginald Mitchell. Born in 1895, son of a Staffordshire schoolmaster and printer, he began as an apprentice in a Potteries company making industrial steam locomotives. His indentures completed, he travelled to Woolston in Southampton to be interviewed for the job of assistant to Supermarine boss Hubert Scott-Paine. So eager was he to begin work designing aircraft that on being offered the job, he asked his family to send on his luggage rather than returning home first to collect it and say his farewells.

At first, the conventional workhorse qualities of his steam locomotives were echoed in the ponderous but ruggedly reliable biplane flying boats he later made for Supermarine. No Mitchell design ever fell apart in the air. The latest of these flying boats, the Stranraer, was chiefly famous for its whistling toilet, because the on-board lavatory opened to the outside air, and lifting the lid produced a piercing shriek long remembered by users. Far less memorable was its performance, with a top speed of 165 mph, but its endurance was truly remarkable. The final civilian boat operated in Canada until 1962!

The 1925 Schneider Trophy race was won by the US Navy's team of Curtiss biplanes. Standing on the float is Jimmy Doolittle, who later led the first long range bombing attack on Tokyo. *(NASA picture)*

Ponderous but reliable, the Supermarine Stranraer seemed to have more in common with the industrial steam locomotives which began R. J. Mitchell's engineering career than the temperamental racing seaplanes of the Schneider contests. *(Author photo)*

However, for the Schneider Trophy races, speed was paramount. The Trophy was a massive baroque sculpture, depicting the female Spirit of Flight diving to kiss the crest of a breaking wave. It was presented by Jacques Schneider, son of a French armaments manufacturer, who assumed that flying's future depended on seaplanes and flying boats. In spite of enormous drag, they could land on any expanse of water close to their destination.

Entrants were limited to flying boats with boat-like hulls, or seaplanes, supported on a pair of floats. Each race involved a series of high-speed laps at low altitude around a set of markers. There were no limits to engine size or power, and the winning country hosted the following year's race. Should a country win three races in succession, it would keep the Trophy, though this simple target proved highly elusive.

Mitchell had already won the 1922 race with another biplane flying boat, the Supermarine Sea Lion II. Powered by an earlier 450 hp version of the Napier Lion, this had robbed the Italians of a third win and permanent possession of the Trophy. The 1923 race was won by two US Navy Curtiss seaplanes. Now Mitchell was trying again with a monoplane, designed and built in six months. He used a cantilever wing and tailplane, connected to a wooden monocoque fuselage and supported by a pair of floats.

As its pilot, Henri Biard, took off on a pre-race test, increased power made his machine vicious to handle. As he eased into the first turn, the S4 flicked into a spin and crashed into the sea. Biard insisted violent wing flutter was responsible, though later analysis suggested the real cause was aileron flutter. The evidence remained on the seabed, beyond the reach of current diving equipment. The Americans finished first with a British Gloster III in second place.

In 1926, the Italians beat the Americans, preventing them keeping the Trophy. The Italian dictator, Benito Mussolini, announced Italy would win the 1927 event 'at all costs.' Two more wins would prove the triumph of Fascism to all. Sadly, engine power was now reaching the point where the laws of physics trumped engineering ingenuity. Newton's Third Law of Motion stipulates any action must produce an equal and opposite reaction. So a powerful aero-engine turning a propeller in one direction creates a reaction to twist an aircraft in the opposite direction, pushing one float deeper and swinging the aircraft off course. Under full power a seaplane would swing through more than a right angle before leaving the water.

The Italians found the torque reactions of their Macchi M39 seaplanes were almost great enough to turn them over on take-off. To combat this, the starboard float was loaded with extra fuel and the port wing was made slightly wider to make left turns simpler. Nevertheless, engines overheated and one caught fire on a test flight. Another broke a connecting rod and had to be rebuilt. The same happened to its replacement.

Meanwhile Reginald Mitchell's 1927 entrant, the Supermarine S5, would join the Italians in this problem area. The latest Napier Lion now delivered almost 900 hp. The fuselage was made from aluminium alloy with wooden wings built around twin spars and covered in plywood. Frontal area was cut to the minimum, with each aircraft tailored around its pilot. The Supermarine team now had the backing of the RAF High Speed Flight to ensure the S5s were flown by vastly experienced pilots. It was just as well, as they were lethally difficult to control.

To counter the swing to port from torque reaction, Mitchell re-aligned the starboard float eight inches further out from the centre line to help keep the aircraft straight. Limited space in the tiny fuselage meant fuel had to fit in the starboard float, so extra weight helped counter torque reaction. Unfortunately, as both floats dipped downwards under power, the resulting bow wave blinded the pilot with spray until increasing speed lifted them clear of the water.

Still convinced the S4 was lost through wing flutter, Mitchell strengthened the S5 wing. To avoid a heavier structure and a thicker aerofoil, he fitted bracing wires between floats, wings and fuselage. Fortunately, cutting weight and removing the transverse struts between floats raised the top speed by 5 mph. Shaping the nose tightly around the engine cut frontal area by more than one-third, and modifying the floats saved another 14 percent. Extra power caused extra heat, so flat copper radiators were fitted to the wing surfaces. He also switched to a low-wing configuration, with the cockpit moved forward to improve visibility.

The 1927 race was held off Venice. Three Supermarine S5s and three Gloster IVB biplanes, all using the uprated Napier Lion engine, were either shipped direct or carried to Venice aboard the aircraft carrier HMS *Eagle*. After the French and Americans withdrew, the event was a straightforward contest between the British and the Italians.

Each of the latest Macchi M52 seaplanes now had the powerful 1000 hp Fiat AS3 under a large streamlined cowling, producing an oddly humpbacked appearance. The wings were slightly swept back, and the cruciform tail had a fin and rudder extending above and below the tailplane. Overall it was smaller but more powerful than its predecessor and weighed slightly less, so it should have been faster.

The M52s began the race well, but the S5s outran them. Two Italians retired with engine failures, followed by Mussolini's last hope, Guayetti, out of the race with a fuel leak. The surviving Gloster IV suffered a jammed propeller, and at the end only two S5s remained. Flight Lieutenant Webster set a record average of 281.66 mph with Flight Lieutenant Worsley close behind. It was a complete triumph for Mitchell's design, marred only by the Italian journal *Aeronautica*. Backed by the French, the Italians insisted Mitchell had copied the M52. This was nonsense. The designs were developed at the same time under the tightest secrecy, and there were many detail differences between them.

Jacques Schneider died in 1928 so the race was postponed to 1929, with future races held in alternate years to give more time for new designs. Mitchell realized the Lion engine was reaching the end of its racing life, but a new power source was about to replace it: Rolls-Royce. Meanwhile, following the British win at Venice, the 1929 race would be held off England's south coast.

Macchi replaced the Fiat engine and the humpback M39 with the much more beautiful and aerodynamic M67, powered by an awe-inspiring 57-litre Isotta-Fraschini engine with no less than eighteen cylinders. These were arranged as a 19-litre in-line bank of six with two more banks arranged at 60 degrees in a broad-arrow formation similar to the Napier Lion, but delivering 1800 hp. Even to cover the relatively short distance race, extra fuel had to be carried in the floats.

The makers built twenty-seven engines. They would need them, as several blew up during tests. Three M67s were entered, but during trials on Lake Garda in August 1929, team pilot Giuseppe Motta was killed when he crashed at 362 mph. The Italians wanted time to solve the problems, but the Royal Aero Club refused as the race had already been postponed by a year for this reason. The Italian team entered two M67s and one M52 from the previous race.

Increased power brought more problems. Lieutenant Remo Caldringher was blinded and choked by smoke and fumes, and his M67 flicked into a spin after the first turn. He recovered but landed again as smoke blocked his view. His teammate, Lieutenant Giovanni Monti, covered his opening lap at 301 mph before a radiator burst, drenching him with steam and boiling water, and he was rushed to hospital. Their only consolation was second place for the older M52, between the two Supermarine machines.

Meanwhile, what of Mitchell and Supermarine? Following the 1923 US victory, the Air Ministry ordered Rolls-Royce and Napier to build similar engines for British racing planes. After a stormy boardroom battle, Rolls-Royce decided their traditional road car reliability should be proved in this more demanding field. It reflected Mitchell's approach with flying boats. Having seen Italian hopes blighted by power without reliability, he and Rolls-Royce would take greater care.

The Curtiss D12 engine had a light alloy casting for each bank of cylinders, rather than separately cast cylinders screwed into the crankcase. Experience showed these cast-block engines were lighter and more reliable. Arthur Rowledge, designer of the Napier Lion, had moved to Rolls-Royce in 1921 to design the new V12 Kestrel, supercharged at all altitudes to increase power output at the risk of greater stress.

The Kestrel's two cylinder banks were arranged in a 60-degree Vee, the basic arrangement for all classic Rolls-Royce aero engines up to the end of the war and beyond. With 5-inch cylinder bores, a 5.5-inch stroke and 21.25 litre capacity, this delivered 450 hp in unblown form. Reliability was targeted from the start, with four valves per cylinder and a single overhead camshaft for each bank, and a pressurized cooling system to maintain temperature at 150 degrees C. Later versions with higher blower speeds and boost pressures would deliver 700 hp by 1940.

However, the Kestrel was too small for Trophy racers. Simply increasing boost risked unreliability, so with no engine size limits, Rolls-Royce scaled up the engine as a safer way of boosting power. They had already built an engine called the Buzzard, with 6-inch bores and a 6.6-inch stroke for a 36.7 litre capacity, delivering 955 hp in supercharged form. Only 100 were made, but it provided the inspiration for the Rolls-Royce 'R' racing engines for the last and most successful of Mitchell's seaplanes, the Supermarine S6 and S6B.

The Air Ministry agreement of February 1929 left six months of development time before the next race in September. This was a dauntingly tight schedule. Though the basic configuration of the 'R' was inherited from the Buzzard, most details were different, to promote reliability. The crank case, cylinder blocks and reshaped housings for propeller reduction gears were cast in a new Rolls-Royce aluminium alloy. Connecting rods were strengthened and the engine's size trimmed. Each cylinder was set against its equivalent in the opposite bank, so connecting rods of opposed cylinders had to share the same big-end bearing which was difficult to arrange.

Rocker and camshaft covers were reshaped to match aircraft nose contours. To prevent spray entering the air intake, it was moved between the cylinder banks. Parts normally finished in bronze and steel were replaced by aluminium forgings to last the short duration of the race, and the engine was fed by a new single-stage, double-sized compressor. Every cylinder had twin sparking plugs, each connected to one of two magnetos mounted at the rear of the engine, so magneto failure would not affect combustion or smooth running.

Increased power produced even more heat, so the S5's surface radiators were extended. Engine coolant passed between inner and outer layers of double skinned panels in the surfaces of wings and floats. Engine oil was cooled by double skins on the fuselage sides, the fin and the tailplane. Even then, the danger of overheating was always present. The engine ran on an exotic fuel mixture of benzole and aviation spirit with 5cc of tetraethyl lead per gallon, set to a richer than normal fuel-air ratio to increase reliability and reduce temperature.

The 'R' engine delivered enormous power from the start. The blower ran at eight times engine speed, delivering 18lbs of boost, three times that of the Kestrel. This caused spark plug failures until a replacement to withstand the harsh conditions was developed. But overall reliability remained a concern. The prototype engine delivered 1545 horsepower for a fifteen-minute bench test and then collapsed under the strain. One by one, weaknesses were revealed, identified and eliminated. Burned out valves were replaced by sodium cooled versions to withstand hot gases. Pistons, connecting rods and crankshaft main bearings were redesigned and strengthened, along with smaller components like valve springs. When the fourteenth prototype engine was tested, it delivered 1850 horsepower for an hour and forty minutes.

Bench testing was noisy and stressful. Simulating the cooling effect of the slipstream at full racing speeds needed four Kestrels: one to create the slipstream, one to cool the 'R' crankcase, another to ventilate the test area and a fourth to drive the supercharger test rig. Four unsilenced Kestrels running at full power with the 'R' under full boost produced epic noise levels and inflicted intermittent deafness on the testers. Sixty gallons of warm castor oil sprayed out of the exhaust valves on each test, treating engineers to a massive dose of laxative. To remedy this, they followed the prescription of Great War fighter pilots sprayed with castor oil while flying rotary-engined biplanes: brandy and condensed milk.

Because of strict running limits for the 'R', engines needed dismantling and rebuilding, even when fitted to the seaplanes. Rolls-Royce removed the rear bodywork from a Phantom and installed an open load bed for a high-speed pick-up to rush engines between Calshot and Derby. Racing through the dark it was inevitably named 'The Night Phantom'.

Fortunately, hard work produced spectacular power increases. Starting from 1400 hp, higher boost pressures pushed the 'R' to more than 2000 hp, with even more to come. For the moment, they decided further power increases might overstrain the airframe and increase fuel consumption beyond the capacity of the floats.

To exploit this power Mitchell designed another supremely elegant seaplane, similar to the S5, but larger to carry the 'R' engine, with an all metal structure for the cooling panels to fit into the skin rather than on top of it. The leading float struts were moved further forward to support the heavier engine, and both floats were now used for fuel. Two S6s, N247 and N248 were completed and the engines fitted by early August.

Unfortunately, the first test did not go according to plan. Increased power caused a violent swing to port as the float dipped under torque reaction, and the S6 refused to take off in a straight line. Mitchell loaded more fuel into the starboard float to even up the loads. Finally the changes worked and the S6 took to the air.

Even so it was tricky. Team leader, Squadron Leader Orlebar, first fought for twenty minutes to control the machine, deluged with spray and struggling to gain speed. Later that afternoon he tried again. A 12 mph breeze kicked up a chop on the water, which did the trick. He prevented the nose dropping and took off normally, with just two problems. The seaplane flew left wing low, with severe rudder vibration. Mitchell had predicted the port wing drop on take-off, so Orlebar was ready. He also predicted a top speed of 340 mph, and on a high speed run, Orlebar's instruments showed 345 mph.

Changing the fuel mixture eased vibration, but caused overheating, which meant still larger radiator panels. These leaked as wings flexed under the loads of tight turns around the markers. With no time for radical redesign, Mitchell asked for ideas. One pilot cured his Austin Seven's leaking radiator with a compound called 'Neverleak' from his local garage. Pouring it through the radiator cap heated it up and sealed the leaks. Mitchell used it on the seaplane engines, where it worked first time.

On 1 September 1929, Flight Lieutenant Richard Waghorn took off in the S6 on his seventh attempt. Adjusting propeller pitch raised engine speed and power but created more heat, so pilots had to reduce power wherever possible. Problems persisted up to 6 September, the night before the race around seven laps of a thirty-two mile course between the Isle of Wight and the mainland. Weather was critical, as a fresh breeze could raise enough swell for flying to be abandoned, and the smallest floating objects could cause damage. That night the seaplane moorings were protected by sentries. A sergeant from the High Speed Flight realized one of the S6s – N248 – was listing with a leaking float. Overnight repairs were banned, but the next morning they hauled the sinking plane ashore to plug the leak.

The second problem was less obvious but more serious. While changing plugs on the engine of N247, a mechanic noticed cylinder damage, showing the piston had seized. Racing rules barred changing the engine. They could fit a new cylinder block, but only on the aircraft itself. Switching pistons was simple, but the block was too heavy to lift by hand, and it had to be lowered extremely accurately.

It was past midnight and the maintenance crew had worked a full twenty-four hours. Their only help was an Experimental Department team that had come down for the race. After an evening party, they were asleep at the Crown Hotel. Their boss, Ernest Hives, went to fetch them, and to find which rooms they occupied he knocked on doors with men's shoes left outside for cleaning. Unfortunately a practical joker had switched them with shoes of ladies in other rooms, but embarrassment gave way to determination once the problem was explained.

The job was done by early morning. Piston failure must have occurred after slow running the previous day. This had let fuel drain into the supercharger casing, where it remained until the engine was started. Before the fuel vaporized, the fan threw enough of it into the cylinder to wash the walls clear of lubricating oil and leave them vulnerable to friction.

There were now three British entrants – two S6s and an S5 brought in to replace the Gloster IV, which suffered persistent engine problems, and three Italians. Waghorn's S6 would take off first, assuming the engine held together, followed by the first M67, then the S5, the second M67, then the second S6 and finally the M52. However, the race was no real contest. Waghorn accelerated through a cloud of spray, and spectators cheered when his floats left the water, his engine bellowing with mechanical health. By the end of the lap, at an average of 324 mph, he closed in on the back marker, De Molins' M52. By the time they reached the Cowes marker, Waghorn climbed around the beacon before diving back to sea level and passing the Italian machine.

By this time both M67s' engines had failed and the second S6 had been disqualified for rounding a marker on the wrong side. Since the pilot was out of contact, he completed the full seven laps, and set the fastest average of 337 mph. Now everything depended on Waghorn and the repaired engine. Would it break down, overheat or run out of fuel as the port float had been drained to straighten the take-off run?

On Waghorn's final lap the engine misfired. As it seemed to recover, he climbed, hoping to glide across the finishing line. Then, as he rounded the Cowes marker, it stopped altogether and his race was over. He landed on the water and waited for the launch. To his amazement, the crew were smiling and cheering. He had miscounted the laps and had won the race.

His winning average was 328.63 mph, with the M52 in second place and the S5 third. It was a convincing victory, and one more would ensure the Schneider Trophy remained in British hands permanently, though there were ominous rumblings. By the end of 1930, both France and Italy planned to contest the 1931 race, but British participation was in doubt. Costs for developing two new racing seaplanes were estimated at £100,000, too high a political price. The axe fell on 15 January 1931. The Air Ministry refused permission for the 1929 seaplanes to be used or for RAF pilots to fly them. A week later, the Royal Aero Club offered to raise £100,000 if the Government relaxed its ban. It refused, allegedly because the Chancellor, Philip Snowden, was bitterly opposed to expensive international competitions.

Help came from 73-year-old Lady Lucy Houston, widow of a ship-owner MP and a multi-millionairess in her own right. Sharply critical of Government policies, she used the Schneider Trophy to attack its lack of patriotism. With resources in plenty she signed a cheque for £100,000. Even the Prime Minister realized the way public opinion was moving, yearning for good news to counter economic gloom. The offer was accepted and preparations began with eight months to the race.

This was not enough time for Mitchell to design, build and develop new machines, so he improved the S6 instead. Rolls-Royce increased engine and blower speeds to boost the 'R' to 2350 horsepower, but this meant more heat and more torque. Mitchell lengthened the floats to carry more fuel and fitted larger radiators. The ultimate Mitchell seaplane was renamed the S6B, while two of the S6s were given the larger floats and the more powerful engine, and designated the S6A as training and reserve aircraft. Finally, the exotic fuel cocktail used in 1929 had acetone added to eliminate intermittent misfiring.

These changes could not cope with all the extra power. Even though almost every square foot of the seaplanes' skin now carried cooling panels, justifying Mitchell's description of the aircraft as a flying radiator, using full power caused engine temperatures to rise dangerously high. Pilots had to watch coolant temperatures rather than air speed and hope the extra power would keep them ahead of the Italians. By the time the S6 was upgraded to the S6A, its handling on the water was greatly improved, but pilots still met severe vibration in the climb. Orlebar had to interrupt one high-speed run and glide to a dead-stick landing in the Beaulieu estuary. While waiting for the launch, he noticed the fuselage was buckled in front of the tail.

Both S6As had their control surfaces modified and on Orlebar's next flight on 26 June, the vibration was gone, but the engine spluttered badly enough to force him down onto the water. Dirt was blocking the petrol filters and recurred after nine minutes on a later flight. The filters were choked with a fibrous compound from the fuel mixture, attacking the chemical sealant in the tanks and pipes. Mitchell responded tersely, 'You'll just bloody well have to fly them until all that stuff comes off.' Fortunately, the compound was stripped away by the fuel, and after refitting the pipes and purging the system, all was well.

Next it was the turn of the S6Bs to suffer. Orlebar was twice beaten by the huge swing on opening the throttle. First he nearly hit the team's base at Calshot Castle. Then he hit a barge and damaged his wing. Take-off was a nightmare, even for experienced Service pilots. The delay between opening up the throttle and rudder control becoming effective allowed the swing to worsen.

In the words of Flying Officer Snaith of the High Speed Flight[1], 'the start of the take-off had to be at least 45 degrees to the right of the wind, and full right rudder had to be applied and held. If all the preliminaries had been co-ordinated nicely, the aircraft accelerated to a speed at which the airflow over the rudder was just sufficient for directional control by the time the aircraft had been dragged round into the wind.'

There were more problems. The noses of the floats dug into the water causing a persistent porpoising movement, resolved by holding the stick back throughout the take-off. The final unsticking from the water happened suddenly as the machine leapt off the water in a partially stalled condition. Normally the automatic reflex was to push the stick forward to trade height for speed, but this proved fatal – literally.

Another High Speed Flight pilot, Lieutenant Jerry Brinton of the Fleet Air Arm on his first take-off in an S6A, failed to gain enough height and pushed the nose down to avoid a stall. His S6A fell back on the water, leapt back into the air, fell back and bounced again to a height of 30 feet before plunging nose first into the sea, and poor Brinton died from a broken neck.

Eventually the take-off was studied with an S6A. Mitchell suggested an older nine-foot propeller, and immediately the aircraft responded properly. The pilots then complained the S6B was nose-heavy, so Mitchell added flexible metal strips to the trailing edges of the elevators, bent downwards by 1 degree, to perform as trim tabs and avoid pilots having to keep pulling back on the stick. But if the British pilots faced deadly problems in handling such high-powered aircraft, so did the Italians. Instead of a new engine, they simply fitted two existing ones together. The Macchi MC72 had a fuselage consisting of a steel-tube framework fixed to a wooden monocoque by four massive bolts. An oil tank was housed in the nose, and the all-metal wing had flat radiators faired into its surface. In addition each float carried three radiators, the leading one for water cooling and the others for oil cooling. Extra radiators were hidden in the float struts and during hot weather an extra radiator could be slung below the fuselage.

Two Fiat AS5 V12s were squeezed, one behind the other, to minimize frontal area and drag. The massive engine which resulted was 11 feet long, with twenty-four cylinders in four banks of six apiece and a capacity of 50 litres. The rear section drove one propeller through a reduction gearbox between the engines and a drive shaft running between the cylinder banks of the forward section. The

---

1. (Appendix 8, *R J Mitchell, Schooldays to Spitfire*, Gordon Mitchell, pp. 337–346)

Mitchell's supreme racer – the Supermarine S6B at Calshot Castle. *(ex-Crown Copyright, via Wikipedia)*

forward drive shaft enclosed the rear shaft and this drove a second propeller. This ensured the two propellers turned in opposite directions. With a massive rear-mounted supercharger this doubled the output to a formidable 3000 horsepower, but the huge torque reactions worked in opposite directions and cancelled one another out.

The earlier S6A in the Solent Sky Museum in Southampton. *(Nimbus 227 via Wikipedia)*

The Macchi-Castoldi MC72 eliminated the torque reactions which made these racing machines so difficult to handle by fitting two engines in the fuselage, driving propellers which turned in opposite directions. *(via Wikipedia)*

Sadly, this ingenious idea failed due to stupendous bouts of backfiring, which lost power and imposed additional stresses. The first French entry for eight years withdrew when one of its pilots died in a training crash, and the Italian team followed suit after two similar disasters. The British victory was therefore a formality, when on 13 September 1931, Flight Lieutenant Boothman completed the seven lap race without the 'R' engine missing a beat, crossing the finish line forty-seven minutes later to win the Schneider Trophy outright at an average speed of just over 340 mph.

Unfortunately, these super-powerful engines still lacked the day-to-day reliability needed for interceptor fighters. The Rolls-Royce 'R' and its competitors delivered their power under closely controlled conditions and at too high a practical price. They had to be made tractable, dependable and reliable, not for an hour or two at a time, but on a constant basis under wartime restrictions and requirements. Compared with this, winning the Schneider Trophy had been an aeronautical walk in the park.

So the persistent myth that the Schneider Trophy seaplanes were direct ancestors of the Spitfire could hardly be further from the truth. Solving problems of high power and high speed would teach all kinds of lessons, some of them entirely misleading. Taming torque reaction, eliminating wing flutter and coping with excess heat emissions would between them blight the performance and the prospects of the first Spitfire to the point where it became a complete dead end in fighter development.

Chapter Three

Messerschmitt's Deadly Designs

Sunday 26 February 1928 should have been a red-letter day for Willy Messerschmitt. While Reginald Mitchell was hard at work preparing the Supermarine S6 to fly at more than five miles a minute, Messerschmitt had much more modest objectives in mind. He was about to watch the first flight of his M20a, a single engine, ten-seat high-wing monoplane with a top speed of 127 mph. He hoped to sell it to the German state airline *Deutsche Luft Hansa* to build a secure future for his fledgling company. Reliability, economy and load capacity were the qualities he was offering, rather than speed and power.

To reach this point, Messerschmitt had followed a radically different route from Mitchell, despite similarities at the beginning. Mitchell, born in 1895, was three years older than Messerschmitt. Both belonged to fairly large families. Mitchell's father was a headmaster who later switched careers to run a prosperous printing business. Messerschmitt senior trained as an engineer, only to take over the family's wine shop on *Langestrasse* in the small Franconian town of Bamberg after his elder brother emigrated to the USA.

Both boys were fascinated by flying. They built model aircraft from frameworks of wooden strips covered with doped tissue paper and powered by twisted rubber bands driving wooden propellers. Although the younger, Messerschmitt was first to move to full-size aircraft, building gliders with friends during the summer holidays. This taught him the paramount importance of lightness and simplicity for any of his designs to stay airborne for more than seconds at a time. Even though he and his chief collaborator, Friedrich Harth, were called up for army service in the First World War, both were lucky enough to survive.

After the war they returned to gliders. It was now their only choice, as the Versailles Treaty prevented Germany from building a new air force, and a crippling combination of inflation and recession ruled out aircraft manufacturing from scratch. So Messerschmitt studied engineering at Munich's Technical University. This presented him with the steepest of learning curves and frequent crashes set an uncomfortable precedent for the future.

One of the crashes wrecked the partnership between Messerschmitt and Harth. On 13 September 1921, Harth took off in Messerschmitt's S8 sailplane in a high wind over the *Wasserkuppe*, a high and exposed plateau in the Rhön mountains, in the German state of Hessen which was a popular site for endurance flights. He soared upwards to more than 200 feet and stayed aloft for more than twenty minutes, which broke the existing record, when the machine broke too. A control failure made it plunge into the ground. Harth spent six months in hospital before making a partial and long delayed recovery.

Meanwhile Messerschmitt moved on. In 1923 he set up a company in the family home in Bamberg and began making a much-improved sailplane, the S14, which won a series of events in the Rhön mountains on 30 August, once again in a 40 mph wind. This time, the pilot, Hans Hackmack, reached a record height of almost 1000 feet and set a new endurance record in conditions that killed one competitor and caused two others to crash. He then landed safely, perhaps an even greater achievement.

In 1924, Germany was finally allowed to build powered aircraft once again. Messerschmitt began with powered sailplanes, but with more failures and crash landings along the way. His first powered glider, the S15, failed to leave the ground at all, crashing during its take-off run. He built two variants of his S16 sailplane; the single-seat S16a, and his first two-seater, the S16b. Fitted with ailerons instead of wing warping, and aided by elevators and a rudder, these were much closer to proper aircraft.

Sadly, at a time when Mitchell's flying boats were showing how reliable aircraft could be, this was a quality Messerschmitt had yet to master. Both S16s were entered in the 1924 Rhön competition, the first to admit powered gliders. The single-seater took off first, and climbed steadily into the wind until the propeller reduction gear collapsed. The pilot turned around and glided back towards the field. When he reached it, he crashed into a line of trees and wrecked the aircraft.

The two-seater fared little better. Again a promising take-off and climb ended in failure when the drive chain connecting the engine to the propeller gearbox broke. The pilot made a successful forced landing in a nearby field, where the chain was mended and he took off again, until water in the carburettor caused the engine to fail. Again he made an emergency landing, but on ground so rough it was impossible to take off again once the problem was remedied. The only success for the company was gained by the old engineless S14, which performed impeccably in the practised hands of a former fighter pilot from the *Richthofen Geschwader*.

From then on Messerschmitt turned his back on gliders. This was a shrewd decision as the 1924 contest was won by a former fighter pilot and popular sports flier, Ernst Udet. His personal machine, the *Colibri*, was much more a conventional powered aeroplane than a powered sailplane, giving a clear indication of future development.

Messerschmitt continued his numerical sequence of designs, following the S16 with the M17, with the new initial signifying 'M' for 'motor' rather than 'S' for '*Segelflugzeug*' or 'Sailplane'. His designs broadened from powered sailplanes to light aircraft and sports planes, and finally to small commercial aircraft, feeder airliners to connect local fields with major airports and long distance flights by major operators.

Messerschmitt's M17 in the Deutsche Museum clearly shows its sailplane ancestry through high aspect-ratio wings and slab-sided fuselage. *('RuthAS' via Wikipedia)*

His M17 was a two-seat light aircraft, virtually the S16b with a new label. Powered by the same 799cc Douglas motor-cycle engine and chain-driven propeller, it had the same fuselage profile, with the passenger tucked into a cutaway space in front of the pilot between the engine and the high wing for maximum exposure to the slipstream. Entered for the 1925 *Zugspitz* contest, it fared little better than before. Its engine lacked the power to reach the specified height of 3000 metres to enter the contest, and it was rejected on safety grounds.

Messerschmitt M17 in flight. *('Kogo' via Wikipedia)*

Finally Messerschmitt learned his lesson. All seven subsequent production M17s used more powerful engines like the 24 horsepower ABC Scorpion and the 25 horsepower Bristol Cherub, and the fuselage was reshaped to better protect both pilot and passenger. To make room for production machines, Messerschmitt's team moved to a former munitions factory in the *Hauptmoorswald* industrial estate outside Bamberg, with space to build a series of M17s.

Each was made almost entirely from wood, including both framing and skin panelling. Their shape would be followed by most of Messerschmitt's early commercial planes. The long, narrow wing showed its sailplane ancestry, with straight leading and trailing edges tapering to sharply rounded wingtips. Four massive bolts fixed it on top of the slab-sided fuselage. This had a deep nose, which gave the aircraft the breast of a portly pigeon rather than the slimness of a hawk. As the fuselage was a closed box, pilot and passenger could only see through curved side cut-outs in the panelling, while the pilot had to lift a hinged panel set into the wing trailing edge to climb in and out of his seat.

The cantilever wing was built around a single box spar and a plywood leading edge with stretched and doped linen fabric covering the framework from the spar back to the trailing edge. Ailerons were built up in the same way, and elevators formed the rear of a movable tailplane whose angle of incidence could be adjusted on the ground. All control surfaces were moved by cables apart from the inner sections of the aileron linkages, which involved rods running through the wing centre sections. The engine support was fabricated from welded steel, and the streamlined cowling was beaten to shape and riveted to the fuselage structure, with the six-gallon fuel tank linked to the engine through a gravity feed.

The Messerschmitt M17 was no aeronautical beauty. Its lines suggested little of grace or performance. It looked like what it was; a simple, workmanlike design, with economy and efficiency top priorities. Its top speed was some 85 mph and it could cruise at 75 mph with a ceiling of 13,000 feet and a range of 360 miles, but its most remarkable features were a payload almost exactly equal to its weight, and radically low fuel consumption.

These machines chalked up more successful results than the original M17. In the spring of 1925, one won a series of performance prizes at the *Oberfrankenflug* competition in Bamberg. In the autumn, another M17 entered the International Flying Competition at *Schleißheim* where it won first prize overall in the speed and height categories, and fifth place in the relay event. It was flown by a former combat flyer on the Turkish front in the First World War, Karl Croneiss, who worked for Messerschmitt as a part-time test pilot.

In May 1925, contestants were assembling at the Tempelhof airfield in central Berlin for the ten-day *Deutsche Rundflug* contest, involving return flights to different cities within the *Reich* borders, from Stettin and Breslau in the east to Munster and Augsburg in the west. Two M17s were entered, one flown by Hans Hackmack and the other by Heinz Seywald, a First World War fighter pilot earning a growing reputation as a sports flier.

As the start date approached, Press interest in the two Messerschmitts grew more intense, but both failed to appear. The reason was revealed in a letter from Willy Messerschmitt to *Flight* magazine two weeks afterwards. In it he explained that, 'During a test flight in the first M17 on May 14, I got caught in the overhead cables of an electric high-tension power line. The machine was badly damaged, and I was myself rather seriously injured, which is the reason I am writing from this sanatorium, where I have been since the accident. It took so long to effect repairs that the machine could not be ready for the *BZ* [*Berliner Zeitung* – the newspaper sponsoring the *Rundflug*] competition. The second machine was not finished until two days before the start, and was compelled to make a forced landing in the *Thuringer Wald* on account of plug trouble. The machine was only slightly damaged, but repairs could not be completed in time.'[1]

Messerschmitt would not in fact have flown the plane himself. Records show he only qualified for his private pilot's licence four years later. Even then, he was a reluctant solo pilot, always preferring the company of a more experienced flier whenever possible. In view of what would happen with several of his later designs, he was wise to avoid the experience until reliability improved.

Even when his aircraft did turn up, results were mixed. In September 1925 Karl Croneiss took first place in the three-day Munich International Flying Competition, but in the 1926 *Suddeutschlandflug* contest, two M17s finished tenth and thirteenth in the flying competitions, though one won first prize on technical performance. In September 1926 the same machine achieved the first ever crossing of the Alps in a light aircraft, with Eberhard von Conta and Dr Werner von Langsdorff sharing the flying to finish a 1000 mile journey from Bamberg to Rome in a total flying time of fourteen hours and twenty minutes.

These were heady days for German fliers. Fewer restrictions created a boom in civil aviation, with the total route mileage of German airlines being greater than their American equivalents. Most were provided by the heavily subsidized national airline *Deutsche Luft Hansa*, but former wartime flyers soon realized the opportunities of filling the gaps in the *Luft Hansa* long-distance route network. One of these was Theo Croneiss, younger brother of Karl. He had served in the Turkish campaign as leader of the so-called '*Dardanellen Staffel*' ['Dardanelles Squadron'] and was credited with five confirmed victories.

Croneiss was a burly and intimidating figure with the build of a brawler, but nevertheless a skilled and able pilot. Working with other war veterans, he opened a flying school in 1924 called the *Sportflug GmbH für Mittelfranken und Oberpfalz* (Sports Flying Limited for Central Franconia and the Upper Palatinate, or *Sportflug*

---

1. Flight, 18 June 1925, p.373.

for short) for both civil and military pilots. Its first two courses alone turned out twenty-two novice pilots and sixty-three ex-military pilots. With around 180 flights a day it was soon the second busiest training school in Germany.

Inevitably, hard times meant Government aid for the school would soon dry up, so in 1925 Croneiss set up his own local airline to connect with *Luft Hansa* long-distance flights under the name *Nordbayerische Verkehrsflug GmbH Fürth*, (North Bavarian Transport Company Ltd) with headquarters at Fürth near Nuremberg, some thirty miles south of Messerschmitt's base at Bamberg.

For the business to succeed he had to undercut normal fares with more economical aircraft. As a first step he bought the Messerschmitt M17 his brother had flown to victory at Munich. What he next wanted from Messerschmitt was a larger four-seat version of the M17, with greater performance and economy. What he got was a pair of three-seater M18s. Croneiss specified four seats in terms of fare-paying passengers, but Messerschmitt assumed the four seats included one for the pilot. Croneiss also wanted an all-metal fuselage, but was happy enough for the wings to be made from wood in the conventional way. Here Messerschmitt was happy to go further and worked out that he could supply metal wings as well, at a total purchase price of 25,000 *Reichsmarks*, around a third of the normal price for an aircraft of this type!

The finished design had a similar slab-sided, pigeon-breasted profile to the M17. The pilot sat in front in an open cockpit, reached through an entry door on the starboard side, while the three passengers sat in a closed compartment with glassed in windows on the port side. With an air cooled 80 horsepower Siemens-Halske radial engine and a flying weight of around 1000kg, the M18 could run feeder services at around one-third to one-half the costs of other airlines. It first flew on the rainy and windy morning of 15 June 1926, with Theo Croneiss at the controls. He was delighted with the performance, but agreed with Messerschmitt that the remaining M18s would be modified to carry four passengers to reduce costs per passenger still further.

Messerschmitt built three of these five-seat M18bs, and then began another three for Croneiss's airline, but he was already busy on another project which would push his ideas on lightness and simplicity an important stage further. The M19 was a low-wing, single-seat sports plane, made almost entirely from wood and cleverly designed around the technical competition rules for the 1927 *Sachsenflug* to make it unbeatable. For technical evaluation, entrants were assessed on the ratio of the empty weight of the aircraft to its full flying weight. So he cut the empty weight of the M19 to less than 300kg, ensured it could take off in less than 200 metres and kept its loaded weight the same for all sections of the competition.

The M19 did exactly what it was supposed to do. Its low wing perpetuated the long and slender sailplane plan, and the fin and tailplane had the same simple straight edges as those of the M17 and M18. The one-piece wing carried the tubular framework for the narrow track landing gear, and four men could easily assemble the fuselage to the wing when required. Powered by a Bristol Cherub air cooled two-cylinder engine developing between 28 and 36 horsepower, its performance was good enough to take first and second places in the competition even before the technical assessments were made.

Theo Croneiss won the event in the first M19 to be completed, and the extra technical assessment points meant he took most of the prize money, in spite of running out of fuel and failing to complete the course. Eberhard von Conta flew the second machine, but crashed and wrecked the aircraft. He was still awarded second place and took most of the remaining prize money, so extra funds had to be found for other entrants to receive anything at all.

By now Messerschmitt's company had moved to its final centre of operations at Augsburg, where M18 production for Croneiss's airline resumed. As airline competition grew tougher, Messerschmitt realized the lower operating costs of his designs would appeal to a wider range of customers, but there remained a single cloud on the commercial horizon; on 22 September 1927 one of the first production 18as crashed at Schwarza near Weimar, killing the pilot and two passengers. As yet there was no criticism of the design, and mishaps were still an inseparable part of flying; normally one flight in ten ended in an emergency landing, and one in every 150 involved injury or worse to one or more passengers.

Messerschmitt decided his new M20 should be a larger M18, for even greater economy. He stretched and widened the fuselage to carry ten passengers, with two pilots sitting side by side in a separate cabin. The cantilever high-wing was also scaled up from the M18 and the aircraft featured all-metal construction with

BFW M20 at Berlin Tempelhof Airport. *(Dutch National Archives via Wikipedia)*

a 500 horsepower BMW V12 engine. Unfortunately Messerschmitt's obsession to reduce weight and drag to the minimum was reducing its safety margins.

He cut the wing area, to trim weight and drag. To stiffen the wing, he combined the single main spar with the metal leading edge skin to create a D-section box, a technique later used by Mitchell for the Spitfire wing. He covered it with thicker aluminium alloy panels to cut the weight of the inner framework. Only the trailing edge and the control surfaces were fabric covered, a feature which would have a tragic outcome. He replaced elaborate fuselage formers by stiffening the skin plates through rolling their edges where these were fitted to adjacent panels. Finally he lengthened the fuselage to increase the effect of the fin, rudder and elevators, so their size could also be reduced to cut drag.

This careful formula promised much for operators struggling with higher operating costs, and the German Transport Ministry instructed *Luft Hansa* to order two M20s, leaving Messerschmitt on the doorstep of the big league at last. Everything depended on the plane standing on the Messerschmitt airfield at Augsburg on that February Sunday morning being able to prove its performance.

The flight would bring together four people with pivotal parts in the Messerschmitt story. Apart from Messerschmitt and Croneiss, these were the pilot and the boss of *Deutsche Luft Hansa*. This last was another wartime officer named Erhard Milch, a forceful and efficient administrator who began his peacetime career as a manager for Junkers Airways. Out of thirty-eight airlines founded in Germany since 1918, only Junkers and Aero-Lloyd survived long enough for State subsidy. Late in 1925 they were merged to create *Luft Hansa* and the German Government insisted Milch should be a director.

He soon became head of operations, but lacked reliable engines and aircraft. Nevertheless he increased the route network and the miles flown, and transformed the company's safety record. He opened flying schools for pilots, introduced regular checks and servicing routines, and vastly improved safety and punctuality. In 1927 he flew on the inaugural service from Berlin to Vienna and back. On the return journey the plane failed to clear a mountain while approaching Dresden and crash-landed in a snowbound pine forest. There he persuaded passengers to sing to keep themselves warm until help arrived, and then he took them and their baggage to the city to share a bottle of 1921 Rhine wine.

Finally, the pilot that February day was the same Hans Hackmack who flew Messerschmitt's S14 sailplane and was later offered the chance to fly one of the M17s in the *Deutscher Rundflug*. A qualified engineer, in 1921 he was technical manager of a secret Junkers plant set up outside Moscow to develop warplanes away from Allied surveillance. The venture foundered after the German Government subsidy of ten million gold *Reichsmarks* over four years was used to pay Junkers workers at its Dessau plant instead. The plant was closed and the subsidy cancelled, and Junkers went into almost terminal decline.

Hackmack was then appointed chief test pilot at the *Deutsche Versuchsanstalt für Luftfahrt* (German Experimental Institute for Aviation) at the old Johannisthal airfield in Berlin. He had also been working for *Luft Hansa* and become a close friend of the formidable Milch, which made him ideal to deliver the verdict on the M20. When Messerschmitt asked him to make the flight, he cheerfully accepted the offer. It would lose him his life.

Hackmack warmed up the engine and began his take-off run. As the M20 reached flying speed and left the ground to begin its climb, it carried the hopes of Messerschmitt and his team. However, the first signs of disaster appeared within minutes. Keen eyed spectators could see the fabric covering the rear section of the port wing beginning to tear away in the slipstream. There was still no real threat to the aircraft structure, but Hackmack would not wait to find out. Some weeks earlier he had been forced to take to his parachute when testing a Heinkel, which caught fire in mid-air over Berlin. Now facing an equally dire emergency, he decided to do the same. It was a disastrous mistake.

The M20 was only some 250 feet above the ground when Hackmack jumped. This was too low for his parachute to open properly and he died on hitting the ground. Milch was appalled by what he saw as Messerschmitt's erosion of safety margins, and his apparent indifference to the victims of his flawed designs, in this case an old and valued friend. Though other pilots felt that had Hackmack stayed aboard, the flight could have been successful, Milch cancelled the *Luft Hansa* order.

The ever-dependable Theo Croneiss took over the test flying and on 3 August flew the second prototype M20 with no problems at all. Indeed, the performance showed that Messerschmitt had a winner on his hands. The M20 weighed 1800kg empty, but could carry a 2800kg load, a far better performance than other commercial aircraft of the time. Its maximum speed was 127 mph and it could cruise economically at 105 mph for a range of more than 600 miles at heights up to 16,000 feet. With figures like these, even Milch could be persuaded to change his mind and confirm the order for the two initial production M20as, accepted by *Luft Hansa* in the summer of 1929. By then Messerschmitt was turning out an improved M20b, with increased wing dihedral for better yaw stability, and *Luft Hansa* ordered ten more.

All seemed well until more accidents changed the picture, forcing Messerschmitt into bankruptcy and throwing his future as an aircraft maker into doubt. First to crash was one of the M20bs. On a *Luft Hansa* flight from Berlin to Vienna on 6 October 1930, the aircraft hit a strong downdraught on final approach to an intermediate stop at Dresden. It crashed onto an Army firing range short of the airfield and everyone aboard was killed. One of them was the wife of the *Luft Hansa* agent in Sofia, increasing Milch's hostility. On 4 April 1931 a second M20b flying a group of *Reichswehr* officers, crashed on a charter flight from Muskau to Görlitz, close to the Polish frontier. The pilot, the radio operator and eight officers were killed. Milch was furious at these threats to his airline's credibility. He cancelled orders for the ten more M20s and began a lawsuit to reclaim the payments already made.

Against all the odds, Messerschmitt eventually turned the tables on Milch and *Luft Hansa*. Thanks to fancy financial footwork and wealthy and powerful backers, he rebuilt his company. The official crash enquiries cleared him of blame, since they found the specifications had been at fault, and he had met them in full. With his company secure and sufficient funds to meet Milch in court, he forced *Luft Hansa* to drop its lawsuit and take delivery of all the aircraft ordered.

It proved a Pyrrhic victory. In terms of winning highly placed contacts for future projects like the 109 fighter, it was catastrophic. Milch had lost to an engineer he considered incompetent, and his enmity would endure. Matters were made worse by Messerschmitt's links with Theo Croneiss, since Milch saw his small airline as a dangerous rival. Nevertheless, the M20s were built and paid for, and gave faithful service to *Luft Hansa* well into the wartime years, but there would be no more airline orders and when both men next met, Milch would hold all the trump cards.

Messerschmitt's priorities were now to find new markets and new customers. His next two orders were for biplanes, ignoring his own preference for monoplanes. The M21 was a two-seat training machine for commercial pilots ordered by the German Transport Ministry, but no more followed. The M22 for the German Defence Ministry was even further from Messerschmitt's own ideas. Officially

Albert Speer, Erhard Milch and Willy Messerschmitt. *(Bundesarchiv Photo via Wikipedia)*

Messerschmitt Bf108B-1 in *Luft Hansa* colours at Duxford, 2009. *('TSRL' via Wikipedia)*

described as a 'mail plane' to evade Allied restrictions, it was intended as an experimental night fighter and optional bomber, but was a large and cumbersome biplane. Its deep fuselage, open cockpit, massive undercarriage and twin radial engines slung between the wings, seemed to demonstrate as many sources of monumental drag as possible. It was also a killer.

On 14 October 1930, the Defence Ministry test pilot, a wartime ace with nine victories to his credit named Eberhard Mohnike, was putting the plane through its paces. Trusting its description as a 'fighter' and using his own wartime experience as a successful combat pilot, he decided to carry out a loop, even though this was not planned. All went well enough until one of the engines exceeded its safe speed, and its three bladed propeller flew to pieces. With only the remaining engine producing power, the asymmetric loads prevented Mohnike regaining control and he was killed when it crashed near the airfield.

Still the crashes continued. On 16 September 1931, one of the old and dependable M18s was conducting aerial photography for a survey in Sweden. While flying over thick cloud, the pilot tried to recover his view of the ground by diving through a gap in the overcast. The stresses caused the wing to fail and the pilot died in the crash. Even Messerschmitt's sporting planes were not immune. After the M19 single-seater, Messerschmitt began a two-seater equivalent, the M23. This took his twin themes of lightness and simplicity even further, with a pentagonal cross-section wooden fuselage with two tandem open cockpits, perched on a single spar wooden wing. The workers who built it called it the Mahogany Coffin. Fortunately it never flew as it infringed a Junkers patent, and its 25 horsepower Mercedes engine left it seriously underpowered. Instead, work switched to a revised M23a, with a more powerful engine, ranging from 34 to 60 horsepower, and one flown by Theo Croneiss, won the *Ostpreussenflug* competition in March 1929. This more positive news helped generate a small production run of nine aircraft.

The improved M23b had a more streamlined fuselage and still more powerful engines. One used the German Siemens-Halske air-cooled five cylinder radial, producing some 80 horsepower, a welcome sign that German engines were now catching up with international competition. This won the 1929 French *Challenge de Tourisme Internationale*, representing the company's greatest success so far. In all, seventy M23bs were built and sold by the end of 1931. In particular, Romania built fifteen M23bs under a production licence from Messerschmitt by the end of 1933, as well as eight machines bought directly.

Following Messerschmitt's recovery from bankruptcy, the skies seemed to be clearing when the German Transport Ministry commissioned him to design and build six aircraft for the German team contesting the 1932 Tourism Challenge. He produced his sleekest and most advanced sports plane yet, the M29. It had two tandem cockpits for pilot and flight mechanic, enclosed under Perspex canopies for the first time, with single cantilever undercarriage legs to minimize drag, a controllable tailplane rather than a separate elevator, braced by struts, and leading-edge anti-stall slots based on a Handley Page patent. Nearly all of these would appear in similar form in the later Bf108 and the 109 fighter.

With such an advanced specification and such a vital contract, Messerschmitt took no chances. The prototype endured more than a hundred hours' flight testing over three months, and each team aircraft completed thirty more flying hours of tests and checks. Surely nothing could go wrong after this meticulous preparation? But with only days to go, on 8 August 1932, one of the team planes crashed and killed its pilot, a flying instructor named Fridolin Kreuzkamp. This was bad, but the following day saw the death of flight mechanic Starchinsky when his aircraft broke up in flight, even though the pilot survived by using his parachute. Immediately all M29s were banned from the competition. The enquiry found both accidents were probably caused by tailplane failure, caused by fatigue cracking of the control surfaces.

Events went from bad to worse. Messerschmitt's company, the *Bayerische Flugzeugwerke AG* (BFW) finally emerged from bankruptcy on 1 May 1933, just three months after the Nazis had taken power in Germany. Their policies of massive rearmament would create huge possibilities for designers like Messerschmitt, but one factor resulting from Hitler's rise to power seemed to make any role for him impossible.

On the very same day, Hermann Göring was made Minister for Aviation. As his deputy, with the rank of Secretary of State, he appointed Messerschmitt's most bitter opponent, Erhard Milch, and gave him responsibility for all new contracts for civilian and military aircraft. Milch made his position plain from the start. He ordered one of his new staff, *Oberstleutnant* Wilhelm Wimmer, to inform Messerschmitt that on safety grounds, he would be awarded no contracts for his own designs. Truly, his position seemed desperate.

Chapter Four

First Steps: the Supermarine 224 and the Bf108

The darkest hour is said to come before the dawn. For both Messerschmitt and Mitchell this proved true in different ways; the darkness of personal enmity for one and technical disappointment for the other. For the German engineer, there was even a false dawn when an attractive job offer in early 1934 seemed to offer a fresh start in his professional life. When facing the unrelenting hostility of Air Ministry officials, the relative peace of an academic post must have seemed attractive. The professorship of Aeronautical Design at the University of Danzig offered prestige and financial security at a time when both were in short supply, and Messerschmitt looked for advice.

He began with the Air Ministry. He asked the Technical Office two very direct questions. Did they see any merit in his work at all, and should he continue to compete for official military aircraft contracts? The answer was uncompromising. *Oberstleutnant* Wimmer's responses were no and no. He said it would be better for all concerned if Messerschmitt took up the post. As a brutal rejection it seemed successful. Milch might even have arranged the job offer to get rid of his enemy. In the bizarre world of Nazi Germany, anything was possible.

Messerschmitt realized that against this enduring hostility, the quality of his designs seemed to matter not at all. Was there any point in hoping his proposals might be accepted? He asked his friends. One of them, *Oberst* Fritz Loeb, had recently been appointed to the Technical Office. His advice was as clear as Wimmer's, but more encouraging. He advised Messerschmitt to stay and keep up the pressure. Things would soon change for the better.

Loeb was right from optimism, clairvoyance, inside knowledge or perhaps all three. Wimmer was on his way out, as promotion would take him to the higher echelons of the Luftwaffe General Staff. However, before leaving the Technical Office he had time to join Milch in a tactical blunder which would hand Messerschmitt the contract he needed, and provide the Luftwaffe with one of the most outstanding fighters of all time.

Neither Messerschmitt nor Mitchell designed their fighters in a vacuum. Both had to meet an official specification from their respective Air Ministries. In a perfect world, these would be explicit enough to make design simple and straightforward, bearing in mind the most extraordinary fact that neither man had any experience of designing even an obsolete biplane fighter, let alone a modern monoplane. Any official guidance on what was wanted would have made their tasks easier.

However, because fighter design was changing so quickly, both specifications remained vague. The British version, F7/30, issued on 1 October 1931, set out to find a replacement for the RAF's obsolete Bristol Bulldog, a classic open cockpit air-cooled engine single-seat biplane fighter. Instead of listing the features required, it simply said that 'A satisfactory fighting view is essential and designers should consider the advantages offered in this respect by low-wing monoplane or pusher [designs].'

So monoplanes were apparently not essential. The top speed requirement (195 mph at 15,000 feet) did not eliminate biplanes. A closed cockpit was not specified; many pilots wanted the clear view of an open cockpit. There was no insistence on retractable undercarriage. The truth was that a new design could meet this specification in every respect without greatly improving on existing RAF fighters. However, it would be totally inadequate to challenge what the Germans were developing.

The German specification was much more radical. It was fourth in a series of *Rüstungsflugzeuge*, or aircraft specifications, drawn up in 1933 by the Air Ministry's Technical Office after detailed studies into future air combat. It specified an all-metal single-seat monoplane interceptor fighter with a top speed of at least 400 kph (250 mph) at under 20,000 feet, using the Junkers Jumo engine, then delivering some 700 hp at take-off. It had to fly for an hour and a half on full tanks, and maximum wing loading should be 100 kg/sq.m or some 20 lbs/sq.ft, suggesting a highly manoeuvrable aircraft. However, manoeuvrability was only third on the list of essential qualities, behind climbing speed and all-out level speed.

Wimmer's Technical Office had simply rewritten two proposals from aircraft makers Heinkel and Arado, then returned the specification to them, and also to Focke-Wulf, another well-favoured manufacturer which would later produce the 109's closest rival, the FW190 of the middle war years. All three companies had to produce three prototypes for competitive testing by late 1934. This gave Messerschmitt's rivals an inside track. By the time he was allowed to take part, his prototype was already overdue. His competitors had more than a year's start, and Milch was said to have insisted Messerschmitt would never be allowed to win the contract in any case. Overtaking his rivals should have been completely impossible for a small and financially strapped company.

Messerschmitt knew he had just one opportunity. Radical new ideas might help him win on performance, but he would have to avoid alerting Milch. For the present, commercial survival was top priority, and his M36 switched back to the small airliner market. This small six-seater was designed and built for a Romanian company, which also wanted a licence to build the M35, an elegant open tandem two-seater sports machine and trainer.

Willy Messerschmitt described in an interview with the German news magazine *Der Spiegel* in January 1964 what happened next: 'In 1933 I'd built an airplane for a Romanian airline company, and then the chief of the *Technisches Amt* [Major

Wimmer] sent me a letter with not altogether friendly undertones. Asking if I wasn't aware of the times, and how could I be having my staff of twenty-five or thirty people working on foreign projects.... I was asked to come to Berlin and was told, "We've already awarded three companies the contract for developing a modern fighter, but we would be willing to give you a contract too." I studied the requirements in detail and went back to Berlin, where I told the gentlemen that the conditions weren't to my liking. Such a fighter would never be able to bring down a bomber embodying the latest technological advances, and these would most assuredly be appearing shortly. What would then be the purpose of the fighter in the first place ...? *Generalstabschef* [Chief of the Luftwaffe General Staff] Wever, a very intelligent man, saw to it that I got the contract without any preconditions attached and that I was able to build an airplane as I imagined a fighter to be....'[1]

This partial but vital victory sent Messerschmitt's reputation with the Ministry even lower. Even then they loaded the dice against him as heavily as they could. What they did not realize was that he intended to use the M37, a little four-seat sports plane for the Tourism Challenge, as the basis for his single-seat fighter. At first glance, the two designs had little in common; the M37's engine delivered one-third the power of that planned for the fighter, but Messerschmitt could try out advanced design ideas. His obsessions with cutting weight and drag for better performance were as valid for a tourer as a fighter.

The M37's light and simple frame carried a stressed skin of thin light alloy panels to provide stiffness and strength with minimum weight. The slim fuselage was made from two symmetrical halves butt-jointed together at the top and bottom centre lines. Each half consisted of transverse bulkheads and longitudinal stringers with skin panels flush riveted to the framework. The undercarriage legs were attached to the fuselage frame to minimize stress and driven by worm gears to retract into the underside of the wing.

Messerschmitt designed a very thin wing to minimize the parasitic drag of high-speed monoplanes. Looking from above, the wing had a simple straight taper on leading and trailing edges from root to square-cut tip, with no bulges or discontinuities to mar its shape. To provide stiffness without bracing wires, designers often included two girder type spars in each wing. Messerschmitt used a single box section spar instead, to save weight and material. This caused his first problem. The smaller the wing for a given engine, the faster the aircraft must fly to stay in the air. Normally this would mean a higher stalling speed and an even higher landing speed to provide a safety margin. In this case, the top speed of the aircraft would be just under 200 mph. A wing designed for this kind of speed would not create enough low-speed lift for easy landings and take-offs. Fortunately, designing gliders had made Messerschmitt an excellent aerodynamicist and he knew that by changing the

---

1. *Der Spiegel*, 15 January 1964, pp. 35–6.

camber of the wing and increasing lift and drag for approach and landing, he could combine high overall performance with low landing speeds.

He did this in two ways. He fitted trailing edge flaps which extended backwards and downwards to increase the wing area and camber, and generate extra lift on the landing approach. This also created additional drag to slow the aircraft down, so the pilot could lose speed so the plane's landing speed dropped to less than half its cruising speed. In fact, the M37 wing would cope with a maximum-minimum speed ratio of five to one.

He also fitted leading edge slots. Messerschmitt met the British aircraft engineer Sir Frederick Handley Page in the early 1920s and was fascinated by his use of leading-edge slots to delay an approaching stall. As the airflow over a wing becomes more turbulent, the slots lift to allow airflow to pass between them and the wing leading edge. This reduces the angle of attack, smoothes out the airflow and cuts the stalling speed. Furthermore the noise and vibration of the slots extending warns the pilot of the approaching stall.

This combination of trailing edge flaps and leading edge slots was used on the Handley Page HP21 of 1923, a US Navy monoplane fighter where good low-speed flying characteristics were even more essential than on the land-based 109. It also enabled pilots to approach as close as possible to the stall without actually stalling the aircraft when turning tightly in combat as a matter of survival, so Messerschmitt chose the same combination for both the M37 and the 109.

The M37 had a closed four-seat cabin with dual controls. The tailplane had a single main spar and, perhaps to avoid the failures of some earlier Messerschmitt designs, it was braced to the fuselage by a strut on either side, which would also appear on the early versions of the fighter. All control surfaces were fabric covered, and the tailplane incidence could be adjusted by a cockpit control wheel.

The fourth annual International Touring Challenge would be held in August and September 1934. Work on building the first two machines began in October 1933 with their designation changed from the M37 to the Bf 108, one numeral short of the fighter, which was already entering the detailed design process. The first 108 flew in spring 1934, and five more followed during the next four months.

By the time all six 108s were ready, the company's future was brighter. New funding was available for fighter prototypes for *Rüstungsflugzeug IV,* [Aircraft Specification IV], but the opposition was keeping its powder dry. When the second Bf 108 crashed during a training flight for the Challenge team on 27 July 1934, team manager Theo Osterkamp, another First World War flyer and a future Luftwaffe General, insisted it was unsafe and tried to persuade other team members to refuse to fly it.

Fortunately, Willy Messerschmitt's friend Fritz Loeb was able to help once more. By then, he pointed out that the 108 had been tested at the Luftwaffe establishment at Rechlin and passed as safe by official test pilots. He then flew to Berlin and demanded that Milch replace Osterkamp as team manager. There is no proof that

Milch took any action, as Osterkamp stayed in place, but he and three other team pilots took part in the Challenge – all four flying 108s.

Results were mildly promising. The Germans entered fifteen aircraft of different types, but the Poles took first and second places overall on handicap. Third was a German Fieseler Fi97, but fifth and sixth places went to Osterkamp and Werner Junck flying 108As, powered by 250 horsepower Hirth HM 8U engines. A slightly less powerful variant using a 218 hp Argus, flown by Carl Francke, finished tenth, but the 108s were fastest overall, which boded well for the future.

In 1935 the more powerful 108B *Taifun* appeared. BFW received a contract for thirty-five 108Bs, using the Argus As 10e air-cooled in line 270 horsepower engine. Seven more emerged in 1936, and after the production line was moved from Augsburg to Regensburg, 175 were made during 1938. In all, 529 were produced by the end of 1942. Production was then shifted to occupied France, where 170 more were made by 1944. The French made more after the war, and production revived again in the early 1970s, proving the Bf108 was far ahead of its time.

However, it still had weaknesses. Poor skinning at the wing-to-fuselage joint in one batch impaired its strength in tight turns or when pulling out of a dive. When Hitler's deputy, Rudolf Hess, flew an aerobatic display near Berlin, he found the metal skin had failed at the fuselage/wing joint, almost crippling the machine. Remedial action eliminated the problem, but Messerschmitt's stress and weight saving policy demanded quality checks during production.

The 108B could cruise at 155 mph, a respectable top speed for a front line fighter of only a few years before. Its top speed was 190 mph, and its Handley Page slots normally extended between 63 and 69 mph. It was simple to fly and manoeuvrable, rolling readily and stalling gently. In a tight turn under power, an aircraft will often stall with the lower wing first, which can end in a snap roll or spin. An experienced pilot could hold the 108 with its lower wing slot extended and remain balanced in the turn without stalling at all.

All this was good for the fighter. In fact the greatest problems came where the 109 design diverged most from the 108, because of very different purposes. The single-seat fighter had a narrower fuselage and a narrower undercarriage track, so landing and taking off needed quick reactions and a delicate touch. One area where the specification had been strictest, the new fighter's armament, was hopelessly old-fashioned. The irony was that Messerschmitt met that requirement so well, it would haunt his fighter for its entire operational life.

By comparison, Mitchell's approaching disappointment was more of his own making. His Supermarine design team had taken part in the F7/30 competition from the beginning. With the British aircraft industry in recession, new work was welcome and his design joined a bizarre mixed bag of competitors. There were twelve initial entrants, six biplanes and six monoplanes. Armstrong Whitworth entered two; the AW21 monoplane and the AW35 Scimitar biplane. The Blackburn F3 was a biplane; a whole series of Bristol designs included the Type 123 biplane

The fighter which beat the Supermarine 224 into RAF service – the fixed-undercarriage Gloster Gladiator. *(Will Owen)*

and the Type 133 monoplane, and the Gloster SS37 became the Gladiator biplane. Hawker entered the private venture PV3, a more powerful version of its Fury biplane fighter, while Westland entered another biplane, the PV4, and a parasol wing monoplane.

The Armstrong Whitworth Scimitar was a conventional radial engine biplane like most of the RAF fighters of 1918. It had failed to meet specification F.21/26 as the AWXVI Starling. Now it was back, with a more powerful Armstrong Whitworth Panther VII radial engine delivering 568 hp. The fin and tailplane had been reshaped, and the cockpit placed higher up in a reshaped fuselage for a better view. The authorities were not impressed, so the company added their AW21 monoplane to the list. This had stubby wings and a portly fuselage with retractable undercarriage, an open cockpit, and either a Panther engine or the more unusual Armstrong Whitworth Hyena radial, with fifteen cylinders arranged in three rows of five to reduce frontal area at the cost of possible overheating.

One area where the specification was more modern than the German equivalent was its insistence on an armament of four machine guns. Two entrants grouped them together in the nose. The Boulton Paul P67 monoplane had twin 395 hp 16-cylinder in-line Napier Rapiers and two rectangular wings with bracing wires. Its undercarriage retracted into the back of the engine nacelles, and a long slim fuselage with the open cockpit just behind the nose gave the pilot a clear view ahead and below. It was a promising configuration like later bomber destroyers, the Bf 110 and the brilliantly successful de Havilland Mosquito.

The Bristol 129 had an unusual single high swept wing. The short fuselage carried a rear-mounted engine driving a pusher propeller. The tailplane was carried between twin fins at the end of long tubular booms and its fixed, spatted undercarriage was supported on a complex network of struts. It strongly echoed the FE2b of the First World War (apart from having one less wing) and would have caused little comment over the Somme in 1916 rather than in a fighter design competition in the 1930s.

Wisely, Bristol also fielded less adventurous designs. Two monoplanes were effectively the same design powered by either an air-cooled Bristol Mercury radial engine in the type 127 or a steam-cooled Rolls-Royce Goshawk in the type 128. Both had a fixed undercarriage in spats carried on struts and the untapered wings were braced by wires.

The Bristol 123 was a Goshawk powered biplane, with landing wheels in trouser type fairings and a short, deep, portly fuselage to allow the pilot a wide field of view. This met horrendous problems with the steam cooling and eventually this very last Bristol biplane was cancelled.

The company's final entrant was the radial-engined Type 133 monoplane. This had a semi-retractable undercarriage partly disappearing into fairings at the lowest points of its cranked wing and had the Armstrong Whitworth Hydra of the Type 132 replaced by a Bristol Mercury radial. This seemed to be the favourite but was eliminated when the prototype underwent final spinning tests. The pilot forgot to raise the undercarriage before entering a spin at 14,000 feet. It refused to recover and when the engine stopped, he bailed out, leaving the sole prototype to be destroyed.

One of the oddest candidates was another biplane, the Blackburn F3. This looked like a mid-wing monoplane with a second wing attached underneath the fuselage rather than on top. The pilot perched in an open cockpit above the upper wing like the front seat of an open-top double-decker bus, with an unrivalled view. However, the cumbersome network of struts and supports, added to the fixed and spatted undercarriage, created stupendous drag. Even with the Goshawk engine, this would have had an unacceptably slow top speed of 190 mph. It completed ground testing in the summer of 1934, but was rejected because of metal fatigue in the fuselage.

Perhaps the most conventional looking entrants of all were the Hawker High Speed Fury private venture and the Gloster SS37. The High Speed Fury was a re-engined Goshawk version of the existing RAF Fury biplane fighter. It failed because by the time it was ready, the contract had been awarded to the Gloster machine, a conventional fixed undercarriage radial engine biplane with a closed cockpit and a surprisingly brisk top speed. However, the experience provided by the High Speed Fury helped its designer Sidney Camm develop the similar Hurricane monoplane to join Mitchell's revised and redesigned Spitfire in Fighter Command before the German onslaught.

And where did Mitchell's original Spitfire, the Supermarine Type 224, fit in all this? While it lacked the grace of its later namesake, it seemed more sensible than its competitors. Like several other entrants, it used Rolls-Royce's latest and most powerful aero engine, the supercharged Goshawk, which delivered a promising 660 hp. However, by this time supercharged engines were delivering more power thanks to higher octane fuel. Tetraethyl additives developed in America eliminated pre-ignition, but the engine produced huge amounts of heat, which had to be removed to prevent overheating.

The Goshawk had an ingenious steam-cooling system, holding water under pressure to prevent it from boiling as it absorbed heat from the engine. Only after it ran through pipes to a wing-mounted condenser did it evaporate into steam. Inside the condenser the cooling airflow turned it back into water, collected in a tank before being pumped back into the engine. This was claimed to produce less drag than conventional cooling radiators, but it caused several severe design headaches.

The first was in the wing. This had to hold a condenser tank large enough to remove the heat generated by the engine delivering full power, with a large enough radiating surface. This meant a thick leading edge, and a corrugated outer skin to increase heat dissipation, reinforcing Mitchell's natural caution over wing stiffness. His Schneider Trophy racers had had ultra-thin, low-drag wings which fluttered badly at high speeds, so the final S5 and S6 designs needed bracing wires to prevent vibration.

A relatively thick wing section would increase stiffness and avoid the extra drag of bracing wires and also provide room for a condenser tank. Unfortunately, it was almost half as wide again as that of the 109, creating additional drag to guarantee a low enough wing loading for a high service ceiling and deliver enough manoeuvrability to match earlier biplanes.

There was one meagre advantage. It was likely the wing could deliver sensible landing and take-off speeds without flaps or anti-stall slots, which suggested a low top speed. In addition, the 224 was one-fifth heavier than the 109 prototype and suffered from high induced drag in creating enough lift to fly. The fixed undercarriage and open cockpit made things worse, but Mitchell wanted a simple design.

To limit drag he fitted a small windscreen to the open cockpit, with a fairing behind the pilot's head. The fixed undercarriage legs were enclosed in trouser fairings and to reduce their length, the wing was cranked in a 'W' configuration to give enough clearance for the large propeller. Two machine guns were mounted on the slab-sided fuselage, firing forward through the propeller arc, and two more were fitted under the wing and along the inner sides of the undercarriage fairings.

The design seemed like a dog's breakfast – it was – but its severest limitations were not of Mitchell's making. The Supermarine 224 was the only monoplane entrant to use the Goshawk. In a biplane the condenser could fit in the upper wing so water could drain back to the engine under gravity. In the 224 the condenser was actually below the engine, so water had to be pumped back to the cylinders under pressure. Unfortunately, sending almost boiling water through a pump causes a change in pressure, turning it back into steam so the pump fails and the engine boils.

The Type 224 boiled on almost every climb, especially at full power, essential for an interceptor fighter. When it reached 15,000 feet, steam would spurt from relief valves in the wings as the condenser filled. The pilot had to level off to reduce the load on the cooling system before resuming the climb. Inverted flight increased water loss and threatened engine failure, and was banned, a crippling restriction for a fighter.

Problems like these meant Mitchell's first attempt at a high-performance monoplane fighter was a failure. Chronic overheating problems blighted its future. Inadequate engine power and the colossal drag of a fixed undercarriage, a thick wing and an open cockpit meant it was too slow.

When they ran up the engine before the first flight on 10 February 1934 at Eastleigh, the cooling system leaked. Test pilot Mutt Summers finally managed to fly the 224 ten days later for just ten minutes. When he made a second flight, the top speed was only 230 mph instead of the predicted 245. The unpleasant truth and a bitter disappointment for Mitchell, was that his Type 224, then officially renamed the Spitfire, was outclassed by one of the biplanes. The Gloster SS37 was a totally conventional design, but its Bristol Mercury radial engine not only delivered 840 hp compared with the Goshawk's 660, but did away with all the cooling system limitations. It was as large and as heavy as the 224, but the more powerful engine gave better performance.

The extra wing area of a biplane let it climb more quickly, and in the nine and a half minutes the Type 224 took to reach 15,000 feet, the SS37 could reach 20,000 feet. It carried four machine guns and the pilot had a closed cockpit. But the most surprising fact was that in spite of the drag of an extra wing, the struts and bracing wires and the fixed undercarriage, it was still faster. The SS37's top speed was 253 mph, some 10 percent faster than the 224. It won the competition, and entered RAF service as the Gloster Gladiator.

To Mitchell's credit, he knew his Type 224 was not nearly good enough. Within weeks he had strengthened the wing spar with additional ribs, though the thick wing section was part of the problem. Within five months, he had revised the design with a conventional dihedral wing, retractable undercarriage and a closed cockpit, but as this retained the Goshawk, it was not a viable fighter. Indeed, flight tests with a more powerful Goshawk showed the wing's corrugated skin was overheating and expanding, causing the wing to buckle and increasing drag still further.

So Mitchell began a completely new design, the Type 300, with a narrower wingspan, a closed cockpit and retractable undercarriage to raise top speed from 235 to 265 mph. The Air Ministry felt that this was still only slightly better than some of the other designs on offer. Mitchell then proposed a still thinner wing, covered with stressed skin rather than fabric for a 280 mph top speed. He considered replacing the Goshawk with the Napier Dagger, already developing 700 hp with 800 predicted for later, but at a meeting on 6 November 1934, Vickers, Supermarine's parent company, turned this down as they felt using the Dagger would actually reduce top speed by 20 mph. Instead, they chose Rolls-Royce's new private venture engine design, the PV12, which would eventually appear as the Merlin. This would have conventional but highly effective cooling, with a power objective of 1000 hp.

On this basis, Mitchell was given the go ahead to design his new Type 300, backed by the company as a private project. Fortunately, the Air Ministry had realized the Gladiator could only be a stopgap, and Mitchell and Supermarine were the right people to provide the long-term answer. They issued a new contract to back the project on 1 December 1934, after only a month as a private venture.

This was just in time. After the disappointment of the 224 Spitfire, Mitchell was determined to produce a world-beater, but his days were already numbered. His doctors diagnosed cancer and ordered him to reduce his heavy workload. Instead, driven by his fear of the Nazi threat, he strove even harder to avoid the mistakes of the earlier machine. He would only have one more chance to develop the supreme fighter and could afford no more mistakes. Everything had to be right first time.

Fortunately, the gap between the previous specification and the one issued for the new fighter contract allowed him the freedom to follow his own inspiration, enough to secure the success he yearned for. When Air Ministry specification F37/34 was issued on 3 January 1935, it was written around his new Supermarine fighter project, just as the Luftwaffe followed the Heinkel and Arado proposals.

So once again inspiration was followed by specification. While Messerschmitt's 108 gave him an ideal test bed for the ideas for his new fighter, the Type 224 Spitfire

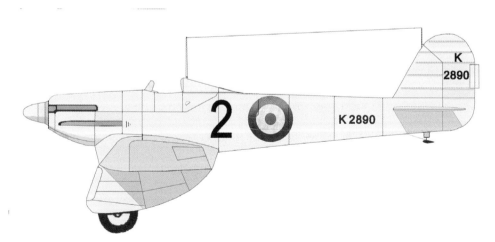

Mitchell's first fighter, the Supermarine 224, was blighted by the limitations of its steam-cooled Rolls-Royce Goshawk. *(Author drawing)*

Supermarine 224 undergoing engine trials at the Woolston factory in early 1934. *(via Alfred Price)*

provided the opposite inspiration. Mitchell turned his back on his earlier design so completely that its part in the story ended after its first appearance, such was the furious pace of development.

His new Type 300 would have the features the 224 lacked. The cranked wing would be replaced by a straightforward dihedral structure, thinner and more efficient. It would have the elliptical plan with the lowest induced drag for its size of any wing shape, with room near the root for guns and retractable landing gear. It would have a closed cockpit with a sliding transparent hood. All it shared with its predecessor would be the Spitfire name, to lend it immortality in aviation history.

Meanwhile, there was the remaining test flying of the 224. Twice the engine cut out, forcing the pilot to glide to an emergency landing, and it was still banned from inverted flying. It was still officially known as the 'F 7/30 Type 224'. Sir Robert McLean of Vickers wanted to call it 'Spitfire', but the Air Ministry insisted on the existing designation until acceptance by the RAF. Nevertheless, it was accepted at RAE Farnborough under its new name.

Its future remained bleak. Work was suspended from 9 January 1935, though test flights continued to generate information for its successor. The tailplane had been clipped after spinning trials, and models had been tested in the wind tunnels at Farnborough and Weybridge. An underslung radiator was tried to ease cooling problems, but this caused so much airflow change that the attempt was abandoned, though tests were carried out to determine the ideal shape for the cockpit windscreen.

Finally on 24 July 1935, the 224 prototype began its tests at Farnborough. On 2 December it was abandoned, as its performance was still unacceptable. It was

The 224 engine installation was revised to incorporate separate exhaust ducts, and the short pipes at the wing roots were carburettor air intakes. *(via Alfred Price)*

finally transferred to Martlesham Heath on 25 May 1937, and a month later its engine was removed and the airframe taken to the Orfordness firing range as a gunnery target. By then its brilliant namesake had been completed, and was on its way to glory.

Chapter Five

Power Struggle

Despite their differing problems, Mitchell and Messerschmitt had one overriding need in common. They desperately needed a genuinely powerful engine, a problem for aircraft designers dating back to the beginning of powered flight. A decade before the Wright Brothers' first flight, another aviation pioneer, Sir Hiram Maxim, had insisted it was 'neither necessary nor practical to imitate the bird, but give us a motor and we will very soon give you a successful flying machine....'[1] Sadly, the crude petrol engines of the time were too heavy for the power they delivered to enable an aircraft to fly.

Instead, Orville and Wilbur Wright decided what they needed and told their assistant Charles Taylor to make it; a four-cylinder water-cooled petrol engine with 4-inch bores and stroke, weighing no more than 200lbs, but producing 12 horsepower. No specialist engine maker had done this and Taylor's only mechanical experience

The engine which set off the revolution in powered flight: the crude but effective power unit of the Wright Brothers' 'Flyer' in Washington's Smithsonian Museum. *(Author photo)*

---

1. R. D. Grant, *Flight*, p.15.

had been trying to repair a broken down car. Nonetheless, he produced an epic of homespun craftsmanship.

He turned four cast iron pistons on a small lathe. The crankcase was cast from aluminium alloy to save weight. The crankshaft was a block of tool steel. Taylor traced the outline of the shaft on to it and drilled holes around the outline before punching it out with hammer and chisel, finally turning it on the lathe. A cast iron block was machined into a flywheel and the entire engine took just six weeks to manufacture, in time to help the Wrights make their historic flight in December 1903.[2, 3]

The First World War spurred progress, and by the 1918 Armistice, two engines had proved particularly successful. One, designed by Professor Ferdinand Porsche for Austro Daimler, was a liquid-cooled, in-line six blessed by Porsche's legendary attention to detail. It had overhead inlet and exhaust valves in inclined rows, operated by push rods and rockers. The long crankshaft ran in seven main bearings, and copper water jackets covered the upper parts of the cylinders where temperatures were highest. Within four years, copper had been replaced by welded sheet steel, and the engine found customers and copiers all over Europe. With bores of 130mm and a 175mm stroke, total capacity was almost 14 litres and initial power output was a conservative 120 bhp at 1200 rpm.[4] But Porsche's design inspiration guaranteed smoothness and reliability, both qualities vital for aircraft.

Success bred imitators and Porsche's design inspired a series of Mercedes in-line sixes. The most powerful, the DIVa, had 160mm bores, a 180mm stroke and a capacity of 21.7 litres, delivering 260 hp at 1400 rpm. A single overhead camshaft drove inclined valves, and cylinders were made from machined steel forgings screwed together, with other components welded on.[5] As with Porsche's design, the power output was conservative and stresses low.

On the Allied side, the Hispano-Suiza HS-8 V8 combined two in-line fours. It was tailored for reliability; instead of machining each cylinder separately and bolting it to the crankcase, designer Mark Birkigt used a one-piece alloy block casting into which thin steel cylinder liners were screwed to make the engine lighter, stiffer and easier to build. Instead of setting cylinder banks at 45 or 60 degrees to one another, Birkigt set them at right angles to stiffen the engine and lighten the crankcase further.

Moving parts were enclosed in an oil bath. Components like magnetos, valve springs and spark plugs were duplicated. Total weight was just over 400lbs for 170 hp but development was just beginning. By 1915 the Hispano Suiza HS8-Aa boosted the performance of French SPAD VII fighters, so the Germans began losing aerial

---

2. Charles Taylor, *My Story*, Airline Pilot, April 2000 edition.
3. Gunston, *The Development of Piston Aero Engines*, pp. 104–5.
4. Ludvigsen, *Ferdinand Porsche: Genesis of Genius*, p. 243.
5. Gunston, *op.cit.*, p. 119.

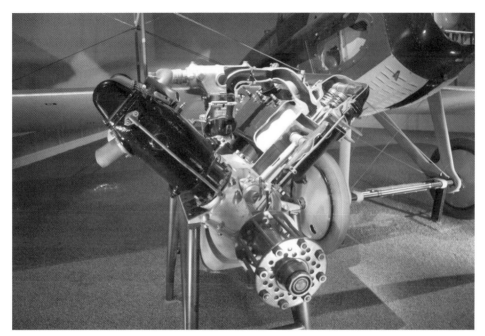

Birkigt's inspired design for the Hispano–Suiza V8 represented a major advance in aero–engine technology. *(Author photo in Smithsonian Museum)*

superiority. The higher compression HS-8Bb produced 235 hp for the SPAD XIII and the British SE5.

Rolls-Royce finally and reluctantly followed Mercedes into making aero–engines. The head of the Admiralty's Air Engine Section, Engineer Commander Wilfred Briggs and his assistant, Lieutenant W.O. Bentley decided to examine the latest Mercedes racing car engine. The 1914 French GP winner had been bought by a wealthy brewer and sat in a dealer's showroom on London's Shaftesbury Avenue. On the first Sunday of the war, Briggs and Bentley drove through the West End to the showroom, requisitioned the car and took it to Rolls-Royce's Derby works to test and dismantle its engine.

Mercedes' combination of individually forged steel cylinders surrounded by a welded steel water jacket dated back to 1903. Sir Henry Royce adapted it for the 1906 Silver Ghost luxury car and now he used it for his first aero engine. He fitted two in-line sixes side-by-side in a 60-degree V12 named the Eagle. With 4.5-inch cylinder bores and a 6.5-inch stroke, the capacity was now 20.32 litres. Cylinder castings had flanged bases with holes drilled for bolts to fix them to the light-alloy crankcase.[6] Each bank had a single overhead camshaft for two rows of valves, and the twin ignition systems had two plugs per cylinder with twin magnetos.

---

6. Gunston, *op.cit*, p. 119.

Farnborough originally stipulated air cooling, but Royce persuaded the RAE to trust his liquid cooled engine experience, and the steady improvement which made his cars so successful worked well in the air too. Compression ratio was 5.22 to 1, and later versions rotated faster to produce more power. To limit propeller speed to 1100 rpm, it was linked to the engine through an epicyclic gear train.

Already a gap was opening between Rolls-Royce and Mercedes. Where Mercedes set conservative targets, Rolls-Royce were pragmatic. Where Mercedes engines were designed to do just what the engineers predicted, Rolls-Royce would set a higher output target, make the changes needed to attain it, and then build and test the engine thoroughly. Whenever a part failed, it would be strengthened and replaced, and testing would continue.

Bench testing the Mercedes engine showed it produced 100 hp at 2000 rpm, but Royce decided the Eagle could do better. At the start of bench testing in February 1915, the prototype weighed just over 900lbs and turned out 225 hp at 1600 rpm. By August the engine produced 300 hp at 2000 rpm for short periods, and the final Eagle IX post-war version turned out 360 hp at 1800 rpm. Its only drawback was relative rarity. The company had difficulty meeting production orders, but refused all requests to license other makers in case reliability suffered. When production ceased in 1928, just 4,681 had been made.[7]

After fifteen years of progress between the Wrights' first flight and the end of the First World War, engines now hit another barrier to progress. As aircraft flew higher, falling air pressure left them short of breath, so an engine running at 30,000 feet only delivered one third of the power produced at sea level! Moreover as they rotated more quickly, the fuel-air mixture had less time to reach the cylinders and the burned gases less time to exhaust.

This meant forcing gases in and out of the engine under pressure. Racing car designers had already solved the problem and Daimler filed a patent for an engine driven supercharger by 1900. By the early 1920s most racing cars used superchargers to increase power. Ironically, two Mercedes supercharged sports cars were the first sold to the public when German companies were still forbidden to make aero engines under Armistice restrictions.

Superchargers in aero engines faced varying power demands over a single flight. Improving performance at operating height risked over-boosting at low altitude where stresses were high. Only as the aircraft climbed and air pressure fell could boost be safely increased, and lower drag allowed the aircraft to fly faster. Only if it climbed still higher and air pressure fell further would power fall again.

Unfortunately increased power produced more heat, so better cooling was needed to avoid pre-ignition of the fuel-air mixture inside the cylinder, which would increase stress and reduce power. In time, falling supercharger efficiency at height would be solved by multi-stage blowers. Pre-ignition would be reduced by high-octane aviation fuels, and rising temperatures would be controlled by pressurized cooling systems.

After the Armistice, Rolls-Royce happily turned its back on aero engines to refocus on luxury cars. Wartime rivalry had vanished with Germany forbidden to make any aero engines at all, having already fallen behind on technical grounds. Yet within twenty years Rolls-Royce and Mercedes would again be close rivals with engines astonishingly similar in configuration and capabilities. Both would make liquid-cooled supercharged V12s with pressurised cooling systems for the Bf109 and the Spitfire to fulfil their potential at last.

In 1925 Air Chief Marshal Sir Hugh Trenchard, Chief of the Air Staff, asked Rolls-Royce to develop a new engine for the RAF's Fairey Fox biplanes. These relied on American Curtiss D12 engines, which had been used for the winners of the 1923 and 1925 Schneider Trophy races. Fifty of these liquid-cooled 18.8 litre V12s had been imported, giving the Fox a top speed of 150 mph, so fast by the standards of the mid-1920s that on exercise they were ordered to fly slowly to give opposing forces a chance! But Trenchard disliked imported engines and wanted a British replacement.

The company originally named their new V12 the Falcon, as a smaller and more modern version of the Eagle. Improvements in casting complex shapes meant each bank of six cylinders could be cast as a single unit, and bolted to a stiff but light crankcase cast in light alloy. With 5-inch bores and 5.5-inch stroke, the engine had a capacity of 21.25 litres. Breathing was improved by two inlet and two exhaust valves per cylinder with a single overhead camshaft for each cylinder bank, and pressurized cooling kept the boiling point at about 150 degrees C.

Running on 87-octane fuel and renamed the Kestrel, the first production version delivered 450 hp in 1930. Rolls-Royce recruited a supercharger specialist, Jimmy Ellor, as they wanted the Kestrel to be supercharged from sea level up to high altitude. The final production version produced more than 700 hp in 1940, when 100-octane fuel allowed boost to be raised further without pre-ignition. To prevent stresses from sharp changes in engine speed when opening or closing the throttle, the supercharger was linked to the engine through a centrifugal clutch and planetary gears. Different gear ratios increased boost pressure, and by 1934, blowers ran at almost ten times engine speed without harming reliability.

This was still not enough power, but the Kestrel would prove vital for both fighters. The prototype 109 could only make its maiden flight with a Kestrel VI sent to Germany as part of a barter deal to supply Rolls-Royce with a Heinkel He 70 monoplane for high speed engine testing. It was also the basis for the Goshawk which ruined Mitchell's first Spitfire. However, it later inspired the engine which powered his revised Spitfire for the greater part of its operational life, the superb Rolls-Royce Merlin.

Since the Kestrel's pressurized cooling could not cope with much greater power in 1933, Rolls-Royce tried an evaporative cooling version in the Goshawk. The theory seemed promising, but coolant leaked in flight or boiled when it did not. Large radiators were heavy and vulnerable, and the need for a condenser tank restricted it to biplanes or to straight and level flight. Only a few prototypes were made, delivering between 600 and 700 hp, and offering no marked improvement over the Kestrel.

Time was running out for Sir Henry Royce. Just as knowledge of his approaching death spurred Reginald Mitchell to complete the Spitfire, Royce was equally eager for an engine to break the power barrier once and for all. With no RAF pressure he began the private venture PV12, an uprated Kestrel with bores widened to 5.4 inches and stroke lengthened to six inches, producing a total capacity of 27 litres.[8]

The Rolls-Royce Kestrel, a smaller but basically similar predecessor to the Merlin, enabled the first flights of Germany's Junkers 87 dive-bomber and the Messerschmitt Bf109 thanks to the lack of a suitable German power unit. *(Author photo)*

Royce himself died in April 1933, weeks after Hitler seized power in Germany, but the initial drawings for the PV12 were already finished. Originally intended as a fighter engine, its designers considered one radical idea. Inverting a conventional engine, with valve gear at the bottom and crankshaft at the top, had two big advantages. The drooping nose contour gave the pilot a better view and kept items needing access for maintenance, like plugs and valve gear, in easier reach. Bizarrely, manufacturers rejected the design as the engine was 'upside down'. It was redrawn the conventional way up, but a visiting German delegation saw a mock-up of the inverted engine and many felt their thoughts were influenced by this. In Bill Gunston's opinion, a great opportunity was missed. 'The fact remains that by the late 1930s the inverted V-12 was universally regarded as a Germanic style of engine. Had this

---

8. Gunston, *Rolls-Royce Aero Engines*, p. 64.

The Daimler-Benz DB605 was a slightly larger version of the DB601 which powered the later versions of the Bf109. *(Author photo in Smithsonian museum)*

not been the case, the author believes there is a chance that later Merlins, and unquestionably the larger Griffon, might have adopted this configuration. As it was, any notion of 'copying the Germans' soon ruled out such a possibility. The PV12 therefore went ahead at the beginning of 1933 as a conventional upright engine....".[9]

At the time, Rolls-Royce and the RAF had a powerful consolation. If the gap between the Kestrel and the 1000-plus horsepower needed for modern fighters seemed daunting, the Germans had an even steeper hill to climb. Only in 1926 did German aero-engine development restart with the amalgamation of Mercedes' parent company and their rivals at Benz to create Daimler-Benz. The new company re-entered the aircraft engine market with a revised version of the wartime D series, now seriously out of date. Next came a larger liquid-cooled V12, which set the pattern for the future. This highly sophisticated design became the Daimler-Benz (DB) 600, but first intended for bombers rather than fighters.

Germany already had an equivalent of the Kestrel made by another company. The Junkers 210 was another inverted V12. Each cylinder bank was cast as a single unit with half the crankcase, so two castings bolted together made a complete engine. Each cylinder had three valves, with two intake and a larger sodium-cooled exhaust valve operated by a single overhead camshaft. With 124mm bores and a 136mm stroke, the capacity was smaller than the Kestrel at 19 litres, but a two-speed supercharger mounted on the starboard side of the engine was driven by a transverse link from the crankshaft.

The 210 flew for the first time on 5 July 1934. It missed its 700 hp target, but was the best engine available for the Bf109 prototype. Delays forced them to use the Kestrel instead, but later 109 prototypes used the 210 with poorer performance, while early 109Bs and 109Ds used the later 210D. The Bf109C was powered by the 210G with a fuel-injection system, but power was still too low for first line combat aircraft and production ceased by 1939.

The DB600 had more long-term potential. With 150mm bores and a 160mm stroke, its capacity was 33.9 litres. Each cylinder bank was cast in light silicon-aluminium alloy, with a duralumin crankcase. Connecting rods ran in roller bearings, and the engine was supercharged. Two prototypes were completed by 1931. In 1933, contracts were issued for four examples of the improved F4B, still earmarked for bombers.

For the first time, take-off power neared the magic 1000 hp mark, but the DB600 was only the first of a series of more successful power units. Its immediate successor, the DB601, was meant for fighters from the start. Bosch direct fuel injection replaced the carburettor to cut fuel consumption and allow negative-g manoeuvres without interrupting the fuel supply. The supercharger was driven through a hydraulic clutch for an infinitely variable ratio between engine speed and supercharger speed. Based on the Daimler Fluid Flywheel car transmission, it ensured the supercharger turned relatively slowly at ground level, where boost requirement was low and clutch slippage high. As the aircraft climbed, a barometric sensor adjusted the slippage through an engine-driven pump fed with oil from the main pressure filter, so at operational height, slippage was almost zero and boost approached its peak.

The early DB601A produced 1050 hp at take-off, and 1100 hp at 10,000 feet. The supercharger was mounted on the port side to cut the length of the engine and allow air to be fed from an intake on the port side of the nose. It was driven by a transverse shaft through bevel gears from the tail end of the crankshaft.[10]

The DB601 would prove ideal for the Bf109 and improved versions would power later variants of the fighter. The production DB601A produced 1175hp, and the engine went into production in 1936 for most Bf109E variants. Some 109Es had the DB601N with flat-crown pistons to raise compression from 6.9 to 8.2 to 1, using C-3 100-octane fuel to deliver 1275 hp. This was also used for earlier variants of the 109F while later 109Fs reverted to 85-octane fuel with the 1350 hp DB601E, a redesigned propeller shaft and reduction gear.

---

10. Fernandez-Sommerau, *Bf109 Recognition Manual*, pp. 93–103.

Meanwhile, work continued on Rolls-Royce's PV12. With bores widened to 5.4 inches, and stroke lengthened to 6.5 inches and a moderate capacity increase to 27 litres, this was still smaller than the DB601. At first, cylinder banks and crankcase were cast as a single unit. The propeller drive used double-helical reduction gears, and the cooling system was a strange hybrid using both water and steam and including both a radiator and a condenser.[11]

These were three steps too far. When the prototype PV12 had its first bench test on 15 October 1933, a year after the DB600, it suffered a series of failures. Accurate double-helical gears were difficult, even for Rolls-Royce. Failures of water jackets and cracks in the complex engine castings forced a return to separate blocks, heads and crankcase. Straight cut gears solved the reduction gear problem and the cooling system switched to ethylene glycol like the racing seaplanes.

For each failure, a solution would be found or a different approach would be used. If components failed, they were strengthened and the engine was reassembled and retested. Repeat failures received similar treatment, and little by little, reliability and power improved. Ramp cylinder heads were tried and rejected, pistons and supercharger bearings were strengthened, and the Merlin II went into production to power the early Spitfires and Hurricanes.

Rolls-Royce Merlin. *(Will Owen photo)*

---

11. Gunston, *Rolls Royce Aero Engines*, p. 65.

The engine was now increasingly reliable and delivering a useful 890 hp at take-off, rising to just over 1000 hp at 3000 rpm and 16,250 feet. Inlet and exhaust valves had double sets of springs with sodium-cooled exhaust valves seated on Stellite coated steel rings, screwed into the cylinder heads. Most superchargers had two-speed drives with different gear-sets connecting engine and blower through hydraulic clutches to let the pilot select one or other as necessary. At low altitudes, low ratio cut manifold temperatures and avoided over-boosting, but shifting to high gear as the aircraft climbed allowed performance to be maintained.

The Kestrel had a two-speed supercharger by 1934, and work began on a two-speed Merlin blower in 1935. In 1938 the Merlin IV cooling system was revised to eliminate persistent leakages caused by pure glycol, which was also highly inflammable. A mixture of pure water with 35 percent glycol, maintained under pressure like the German cooling systems to retard boiling, kept cylinder temperatures some 20 degrees C lower and able to operate up to 135 degrees C with ease.

Other improvements resulted from the Speed Spitfire (see Chapter Eight). This had streamlined bodywork and a strengthened Merlin II burning an exotic fuel mixture to attack the World Landplane Speed Record. Pistons, connecting rods and gudgeon pins were all strengthened, and when the engine ran at 3,200 rpm with 27lbs of boost, burning a mixture of 60 percent benzole, 20 percent straight-run California petrol and 20 percent methanol, it produced a massive 2160 hp.

Yet this was no temperamental and fragile racing engine like that of the Bf109R or even the Rolls-Royce 'R'. It developed 1800 hp on a fifteen hour endurance run with no trouble, and the strengthened components were incorporated in normal production Merlins, making it in Bill Gunston's opinion, 'The toughest engine Rolls-Royce had ever made, and better able to stand up to long periods at full throttle than any previous engine.'[12] This extra strength let Merlins tolerate higher manifold pressures than German engines and Rolls-Royce would later develop still more powerful superchargers, though without the Daimler-Benz automatic coupling.

Pre-war racing showed supercharger efficiency was poor at high boost levels though higher octane fuels coped better. The Air Ministry began stockpiling 100-octane fuel on the eve of war while the Luftwaffe still used 87-octane. Notwithstanding the risk of tankers facing U-boat attacks on Atlantic convoy routes, this vital traffic of 100-octane fuel was never seriously interrupted though none was produced in British refineries.

---

12. *ibid*, p. 70.

When RAF fighters finally switched to 100-octane in early 1940, pilots were simply told that instead of 6lbs of boost they could use a full 12lbs producing 30 percent of extra power to just over 1310 hp. This extra performance only began to fall above 9000 feet, and even this would later be improved. One surprising benefit was because most early Spitfires captured and tested by the Germans still ran on 87-octane fuel, the Luftwaffe underestimated Spitfire performance for most of the Battle of Britain.

Changes to the shape of Merlin exhausts tested in Rolls-Royce's Heinkel 70 added another 150 hp. Revised contours for the underwing radiators turned drag into extra thrust from a basic ramjet effect.[13] This was vital as tests showed the drag of a fully extended radiator consumed some 30 percent of engine power.[14]

So the Merlin held its own against larger German engines for most of the war. It had only one real disadvantage. A negative-g manoeuvre, like pitching the nose down to follow an opponent diving to safety, would make the carburettor float chamber lift, cutting off the fuel and causing power failure. Though pilots could maintain fuel flow by rolling their aircraft inverted in chasing their opponent, this usually gave time for the enemy to escape.

This reflected the German lead in the extremely precise engineering needed for injection systems. However carburettors retained one advantage. The temperature of the mixture entering the cylinders of the Merlin was some 25 degrees C cooler than in the DB601, which meant an extra 60 horsepower and added 6–7 mph to the Spitfire's top speed, potentially decisive in combat. Help was also provided by what pilots called 'Miss Shilling's Orifice'. This was the work of Beatrice Shilling, known as 'Tilly', a Farnborough engineer who designed a carburettor washer with a hole large enough to admit fuel for full engine power even under negative-g. Pilots could not fly inverted for long but it was enough for the fast movements and changes of attitude in combat, and by March 1941 all Fighter Command aircraft had it.

Later Merlin variants delivered more and more power like the Merlin 61 used in the Mark IX Spitfire to outfly the formidable FW190. These are described in later chapters, but the next major changes produced the DB605 and the Rolls-Royce Griffon.

The DB605 appeared in 1941 with cylinder bores widened from 150mm to just 154mm to avoid weakening the block castings. Capacity was 35.7 litres and peak power 1475 hp at 2800 rpm, thanks partly to a larger supercharger. The DB605 powered all versions of the Bf109 from 1942 onwards up to the Bf109K, but later improvements used power boosting systems rather than engine modifications.

---

13. Morgan and Shacklady, *Spitfire: the History*, p. 25, footnote.
14. *ibid*, p. 37, footnote.

Rolls-Royce Griffon. *(Author photo)*

The first time this had been done was in the tenth prototype of the DB601A, fitted to the record breaking Me209V1 (see Chapter 8). Power was raised to 2300 hp at 3500 rpm for the minute needed to cover the course by injecting methanol and water into the supercharger air intake. This gained new appeal in the mid-war years when, according to Bill Gunston, 'The Luftwaffe was so concerned at not being able to win by sheer engine power that it funded two power boosting systems which were fitted to virtually every German piston-engined fighter and fighter/bomber by 1944.'[15]

German writers see this as a positive step rather than a desperate expedient, criticising the implication that 'Germany developed power-boost systems as a makeshift solution to match the level of performance developed by Allied aero engines, which was not the case. Development of power-boost systems kept the engine small and compact, which also provided aerodynamic advantages. The German policy of 'on-demand power' through the use of diverse power-boost systems proved most successful.'[16]

GM1 injected nitrous oxide and compressed air into the supercharger to boost power and aid high altitude cooling. The 109E-7/Z carried four gas bottles behind the pilot and the 109F-4/Z and high-altitude 109G variants had four bottles in

---

15. Gunston, *Rolls-Royce Aero Engines*, p. 158.
16. Fernandez-Sommerau, *op.cit*, p. 93.

each wing between ribs six and eight. Variants with wing-mounted guns had no space for the bottles, but later versions stored nitrous oxide in a fuselage tank with compressed air propellant in the starboard wing and oxygen bottles in the port wing. The liquefied gas was stored under pressure and needed careful handling to avoid explosions in contact with water or oil.

The other system used methyl alcohol (MW) or ethyl alcohol (EW) mixed with water and an anti-corrosion agent. From July 1944, this was used with special spark plugs and 100-octane fuel on all new fighters, producing a 300 hp power boost but with corrosion shortening operating life. Later 100-octane fuel was dropped, but changes in valve timings and improved superchargers still raised 109 take-off power close to 2000 hp.

Rolls-Royce still produced more power from smaller engines. Though lacking the precision engineering and superb finish of German engines, they were consistently stronger, more reliable and efficient, and developed to a much greater extent. Had the Germans produced more compact engines of equivalent power and reliability, the performance of the Bf109 might have been improved still further and been better able to cope with increased armament.

Rolls-Royce then began developing the Griffon. The Royal Navy wanted a larger and more powerful engine than the Merlin, reliable and easy to service, so the company based its layout and measurements on the 'R' engine from the Supermarine racing seaplanes with 6-inch bores and 6.5-inch stroke, and a capacity of 36.7 litres. The first Griffon prototype was produced late in 1934 using many features and components of its predecessor, so development ran more smoothly than the Merlin. It was essential it matched the Merlin's size as closely as possible, so it could be fitted in the same types of aircraft for greater performance.[17]

The Griffon II had magnetos moved from the rear of the engine to the front and camshafts driven from the propeller reduction gear, also at the front of the engine to reduce overall length, torsional stresses and vibration. The water pump was placed beneath the engine, with a smaller single entry blower. Weight was cut by more than 200lbs by late 1941.[18] At the end of this process the Griffon was actually seven inches shorter than the Merlin, half an inch narrower and just six inches taller! The frontal area of the Griffon, was actually 7.9 square feet, compared with the 7.4 square feet of the Merlin.[19]

A hollow crankshaft fed oil to the main and big-end bearings in an idea later used for the Merlin. Because the Griffon rotated the opposite way, pilots had to use port rudder instead of starboard to counteract the take-off swing. The new engine made

17. Morgan and Shacklady, *op.cit.*, p. 133.
18. Gunston, *op.cit*, p. 86.
19. Flight, 20 September 1945, *Rolls-Royce Griffon 65*, pp. 312–4.

its first flight in a Spitfire at a time when the new FW190 made the need for good low-altitude performance especially urgent. Chapter 19 shows how the Mark XII became the first Griffon-engined Spitfire in service, with the Griffon VI delivering 1815 hp at 15lbs of boost. This later powered the naval Seafires XV and XVII as a belated reward to the Navy for initiating Griffon development.

By 1943, Griffons had two-stage superchargers in the 60-series versions used in the Mark XIV Spitfire, and a Merlin-engined Mark VIII modified to take a two-stage blower Griffon 65. Just as the Merlin-engined Mark V had been transformed by the two-stage blower Merlin 61 into the Mark IX which won air superiority back from the Germans in 1942, now the Griffon 65 delivered more than 2000 hp at 7,000 feet and 1820 hp at 21,000 feet, demanding five-bladed propellers to cope with it.

The final wartime Griffons turned out 2340 hp at low altitude and 2145 hp at 15,000 feet for the latest marks of Spitfire and Seafire. Carburettors were replaced by the Bendix-Stromberg fuel metering device or the Rolls-Royce injection pump. Increasing power brought stronger take-off swings so the last examples had twin contra-rotating propellers, and the engine continued in service into the 1980s with the contra-prop Griffon 58s of the Avro Shackleton maritime patrol aircraft.

Though the Griffon was larger and more powerful than the Merlin, the difference was less than their relative capacities might suggest because it ran more slowly. It still produced more power for its weight than the updated DB605 which was only a little smaller, and without the complex boost systems introduced when greater output became vital. Even after the war Griffons and Merlins continued in development, raising performance still higher, until only the gas turbine forced their obsolescence.

Chapter Six

The Final Steps

By the mid-1930s, Mitchell and Messerschmitt had both met with almost insuperable obstacles. Designing a fighter to catch and shoot down a fast modern bomber was appallingly difficult, a point underlined when Mitchell's Type 224 Spitfire was beaten by an obsolete biplane. Messerschmitt had been barred from the Luftwaffe fighter contract, with so bad a reputation that he could expect no concessions and remained on probation. His Bf109 could only win by stealth.

Both men had to unlearn methods which had brought Mitchell success, but left Messerschmitt implicated in fatal crashes. A successful fighter needed high performance, but paring away at weight and drag could erode safety margins. Fortunately, Messerschmitt had one ace up his sleeve. His M37 four-seat tourer for the 1934 Tourism Challenge was effectively built under false pretences. The contract specified a new design, begun after the document was signed, and imposing an impossible deadline. Had he openly failed to meet it, the Ministry would have eliminated a nuisance, but recasting the M37 as the Bf108 evaded their trap and left him in the running.

Given its dramatically different purpose, his opponents failed to realize the 108 was an engineering rehearsal for his new fighter. Believing Messerschmitt could not catch up with the competition, the Ministry completely underestimated him. Though the performance and purpose of the 108 and 109 were worlds apart, a slimmer fuselage with a single-seat cockpit, weapons and a much more powerful engine turned one into the other, though only the first requirement was under Messerschmitt's control. Meeting the rest depended on others.

One German historian was certain of the 108's crucial role in creating the 109. 'The construction of the Bf108 and the experience gained, certainly played a factor

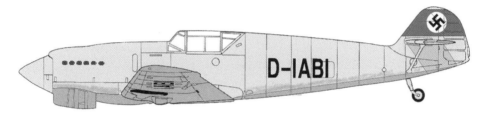

The Bf109 prototype made its first flight with the help of a Rolls-Royce Kestrel engine. *(Author drawing)*

Messerschmitt's four-seat Bf108 was an engineering rehearsal for his 109 fighter. This example in the markings of the Swiss Air Force was photographed at Zurich's Dubendorf Museum by Sandstein.

in accelerating the work on the Bf109 fighter. A considerable number of fundamental questions, such as retracting the undercarriage ahead of the wing's main spar and diverting the airflow around the (necessary) wells in the wings, fitting the control surfaces, and many others, had already been thought through and resolved from a design standpoint. The Me108 [sic] can almost be considered a prototype of the Bf109 or vice versa; the 109 was an adaptation of the 108 concept into a fighter.'[1]

Messerschmitt's ideas were simple. Gliders taught him weight and drag forced them back to earth too soon, but his 109 design added strength to the mixture. Making it small reduced drag, but the airframe needed a compact central core. He described his approach in a post-war letter sent in March 1955 to Professor Dr. Ing. Paul Brenner of the *Vereinigte Leichtmetallwerke GmbH* [United Light Metalworks Limited] Bonn:

'The secret of the [Bf109's] success wasn't so much an extremely fine design, something which would have led to costly manufacturing and reduced tolerance, but rather a true application of light construction principles: minimal dimensions, minimal surface area, lightest materials, equal distribution of stress loads, and above all the combination of multiple functions within the same component. The consequential application of these principles not only resulted in an extremely lightweight design, it also led to simplification of construction. In spite of its extremely light weight, we were able to keep

---

1. Rüdiger Kosin, *Die Entwicklung der Deutschen Jagdflugzeuge, Die Deutsche Luftfahrt (Vol 4)*, p. 108 (quoted in Ebert, Kaiser and Peters, *Willy Messerschmitt: Pioneer of Aviation Design*, p. 116.

construction time down…, with the Me109 [sic] requiring less man-hours per kilogram of weight to build than any contemporary design….'[2]

The 109's core combined the engine firewall with the adjacent fuselage bulkhead and the instrument panel in a tough structure to carry engine bearers, cowling machine guns and wing spars. Detail design cut weight by making every fabricated part combine several existing components. A single bracket united the lower engine mounting and the pivot for the undercarriage leg. Both fixing points for the main spar were connected to a large forging on the engine firewall. Other aircraft used different components on different parts of the structure. Messerschmitt's design concentrated them, strengthened them and simplified them.

German wartime shot of Bf109 fuselage production via Airframe Assemblies

Narrowing the fuselage was simple. In both aircraft, the two fuselage halves were butt-jointed together along the top and bottom centre lines to form a monocoque shell. The job involved reshaping the 108's oval cross-section bulkheads to the narrower and deeper contours of the 109 and adapting skin panels to match. Bulkhead frames were alternating T-section and I-section flanged strips. One type had holes drilled in the flanged edges so the roll-formed longitudinal stringers could pass through them. The remaining frames were shaped to allow skin panels to follow a curved profile, when flush riveted to the frames.[3]

The slab-sided cockpit enclosure was skinned in 2mm thickness duralumin alloy sheet, fixed to the framework with countersunk rivets, and supporting the windscreen and cockpit canopy on top. The rear section of the fuselage was similar, with 1mm panelling. A cast tailplane support was bolted to the rearmost fuselage frame and the fin, tailplane and tail wheel leg were all fixed to this. The fin was made from two almost mirror-image halves, with a slight asymmetry to ensure the fin's vertical cross-section was a modified aerofoil, creating a sideways reaction to engine torque, so the pilot could control the aircraft without rudder trim.

2. Letter quoted in Ebert, Kaiser and Peters, *op.cit.*, p. 116.
3. *Messerschmitt 109 Owners' Workshop Manual*, Haynes Publishing, 2009, p. 66.

The 108 and 109 shared a simple wing with a single box-section main spar, and straight-taper leading and trailing edges, easy to mass produce and efficient enough for high performance. Its only problem was combining a high top speed with a low enough speed for approach and landing, so the 109 inherited its trailing edge flaps and leading-edge Handley Page anti-stall slots from the 108.

The undercarriage was attached to the fuselage box rather than the wing structure, insulating it from landing shocks. Also *Rüstungsflugzeuge IV* required the aircraft to be transportable with wings removed on standard railway wagons. Unfortunately, the narrow fuselage meant a narrower track between the lowered wheels. With distorted steering geometry and a high nose-up ground attitude, this made take-offs and especially landings, difficult for novice pilots.

The all-metal wings had duralumin alloy skin panels flush riveted to metal ribs, the leading and trailing edges, and the main spar. They were bolted to the fuselage frame at three fixing points: at the top and bottom of the main spar, and at the wing leading edge. At the front of the central box, each Y shaped tubular engine bearer had the upper ends of the Y bolted to the wing root bracket and the top of the firewall, and the bottom leg of the Y fastened to a fixing point on the side of the engine at its forward end. Later versions would have cast light alloy bearers with two engine mounting points apiece.

The original 109 had a radiator under the nose. More powerful engines, from the 109E onwards, generated enough heat to need a larger radiator under each wing, inboard of the landing flaps. On these versions, lowering twenty degrees of flap made the ailerons droop as well to improve control. Flaps were controlled by a hand wheel on the left-hand side of the cockpit on the same shaft as the elevator trim control wheel. Moving both wheels together cancelled out the pitch change from lowering the flaps. The port wing flap was marked with flap settings between 10 and 40 degrees so the pilot could see to check.

Coping with a wide speed range also meant changing propeller blade pitch since increasing power rendered fixed pitch propellers less efficient. Variable pitch propellers used electrical or hydraulic power to change the angle of the blades to the airstream. The pilot could select fine pitch for take-off, climbing, approach and landing, and coarse pitch for high-speed flying, so the propeller worked more efficiently.

The first two-speed variable pitch propeller was the American Hamilton-Standard of 1932. This helped the Boeing 247 civil transport take-off in four-fifths the normal distance, climb 22 percent faster and cruise 5.5 percent faster. Ultimately the constant-speed propeller would adjust pitch automatically to manage changing power demands and keep the engine at the right speed for peak performance. Developing a reliable system took time, but in Germany the engineers at VDM produced a constant speed variable pitch propeller which Messerschmitt used for the 109. In Britain, de Havilland produced the Hamilton-Standard propeller under licence.

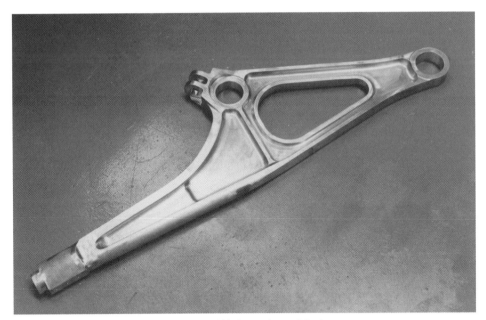

Design simplicity: a Bf109 engine bearer. (Airframe Assemblies)

The 109 pilot could change propeller pitch electrically through a switch on the instrument panel or later on the throttle. With a constant speed propeller he could select automatic operation by another switch whereupon an automatic governor adjusted pitch to match throttle settings. Other electrical services included the fuel pump, the engine and flight monitoring systems, the flight and navigation instruments, and cocking and firing the guns. The undercarriage was raised and lowered hydraulically, together with the radiator flaps and the oil cooler flap.

Each underwing radiator had its own coolant tank. Radiator flaps front and rear were controlled by thermostats but overridden by the pilot when required, and could be closed to limit drag when speed was needed. Coolant was a mixture of one part glycol to one part distilled water. The undercarriage legs were attached to special mounting brackets bolted to the central fuselage box and the cockpit sidewalls. If hydraulic failure or combat damage stopped the system working, a cockpit lever released the mechanism and let the legs drop into the lowered position with enough force to lock them into place. For greater directional control during take-off and landing, the tailwheel could be locked in position and released before taxiing on the ground.

Messerschmitt wanted nothing to spoil the efficiency of the 109 wing, and designed extremely narrow section wheels and tyres to fit the restricted space. Tolerances were so tight that when special parts were not ready for pre-production prototypes, normal section wheels and tyres had to be fitted into bulges in the upper wing surfaces.

The pilot sat on top of his parachute in an alloy bucket seat pan with a seat back-riveted to the rear cockpit wall panel. Above his head was a three-section cockpit canopy. The windscreen had a curved upper glazed panel and two side windows, behind which was a sheet of armoured glass 60mm thick. The main opening canopy was hinged along its starboard side to allow the pilot in and out of the cockpit, and the rear part of the cockpit was fixed in position and glazed above and on both sides. In an emergency the pilot could jettison the centre section of the canopy to escape by parachute or exit the aircraft after a forced landing.

To start the engine, a ground crewman cranked a handle on the starboard side, connected to the flywheel reduction gears. Twenty seconds cranking would spin the flywheel at some 18,000 rpm. When the pilot pulled the start handle, the flywheel would connect with the tail of the crankshaft turning the engine over until it fired. Both magnetos would be checked before taking off. If the drop in engine speed was within limits, the pilot could pull a cockpit lever to retard the ignition and burn away plug deposits. After a flight the same lever was used for five seconds of retarded running at 1800 rpm to clean the plugs.

The 109 was Messerschmitt's master work, a careful and efficient design, easy and quick to produce in large numbers with speed and manoeuvrability for a formidable combat machine. While its design broke new ground in many different ways, it remained practical, with exhilarating performance. However everything depended on its quality persuading the Air Ministry to forget their hostility and place a large order for the Luftwaffe. To do that it must triumph against other competitors at the Luftwaffe test centres of Rechlin or Travemünde. And for that, it still needed a powerful engine.

Time was pressing for all German companies after military contracts. In time the 109 would make as deadly a combination with the DB601 as the Spitfire and the Merlin, but the German engine was still not ready. Even the less powerful Junkers Jumo 210 V12 design, a stopgap for earlier 109s, was not ready. So company test pilot Hans-Dietrich Knoetszch took off in the Rolls-Royce Kestrel-powered first prototype of the 109, from the small airfield at Augsburg-Traunstetten. The undercarriage was fixed in the down position as ground tests had shown that without the vital narrow section high-pressure tyres, the landing wheels could not retract fully. All went well and Knoetszch then flew the prototype to the Luftwaffe *Erprobungstelle* [Test Establishment] at Rechlin north of Berlin, more than 300 miles from Augsburg. He set off on 26 February 1936, but had to refuel at Juterbog-Damm airfield, forty miles southwest of the capital, the home of biplane fighter unit *JG132*, where the future ace Adolf Galland saw his first modern monoplane fighter.

Knoetszch flew on to Rechlin for a short demonstration flight before handing over to the test centre staff. He carried out a series of aerobatics before an unidentified eyewitness [4] described how the 109 'Crashed during a rather hard landing. After

---

4. Quoted in Radinger and Shick, *Messerschmitt Bf109 A-E: Development, Testing, Production*, pp. 27–28.

Messerschmitt's fuselage design greatly simplified the 109's interior framework: modern restoration at Airframe Assemblies.

bouncing rather too high, it came down from a height of one to one-and-a-half metres on to its left landing gear leg and tailwheel. The tailwheel broke off, the fuselage was bent in three places, and the left undercarriage leg was bent inwards. The aircraft then touched down with its left wingtip, tipped over to the right and ended up on its nose' in the kind of error repeated by countless future pupil pilots.

The aircraft was repaired after being sent back to the factory. Knoetsch was sacked and the prototype sent to a different test establishment, Travemünde on Lubeck Bay, where pressure was less at this predominantly seaplane test facility. Here it finished fifty-four test flights to check various design details, but the predicted 'fly off' against other entrants for the fighter contract competition would use the second prototype with the Junkers Jumo 210V12 engine, which first flew on 12 December 1935.

Even when the new BFW chief pilot Dr Ing. Hermann Würster flew the aircraft to Travemünde, it was clear the long run of bad luck was not over. On 1 April 1936 the aircraft took off for an endurance test when the windscreen blew off, forcing the pilot to make an emergency landing. Damage was so severe the plane was scrapped, causing further delay. The third prototype first flew a week later and because the tests included weapons firing, this was the first Bf109 to carry machine guns.

As test succeeded test, the wariness with which Luftwaffe test pilots greeted the 109, due to features like the high wing loading and steep ground angle, was beginning to change. Slowly, as Messerschmitt himself recalled in a *Der Spiegel* interview, 'When the 109 was just about ready to fly for the first time, my friend [Ernst] Udet, who was at that time a Colonel and Inspector of the Luftwaffe, came to me and said, "You're building a fighter, I hear. I'd like to take a look at it, if possible." I showed it to him, and he pulled a funny face. A mechanic opened the canopy and he sat down inside it, the canopy closing over him. Udet looked out, opened the canopy and climbed out again. Slapping me on the back, he said, "Messerschmitt, that will never be a fighter." That was only one or two days before the first flight and we were rather down. "A fighter pilot must feel the speed" he

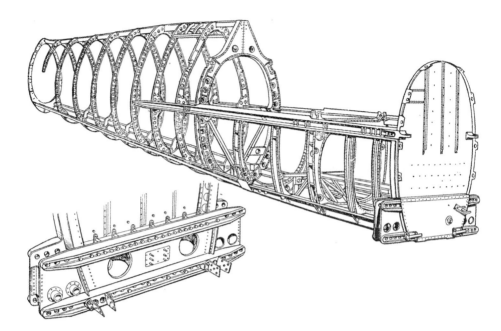

Spitfire fuselage construction – from May 1942 edition of 'Aircraft Production', via Airframe Assemblies.

said, "and the one wing you've got there needs another one above it with struts and wires between – then it will be a fighter."'[5]

Udet was wrong. It would not be a fighter, but a biplane, and biplane fighters were virtually useless. For an expert like Udet to insist on one at this stage was immensely dispiriting, but a decisive opinion shift would follow Dr Herrmann Würster's spectacular demonstration of the 109's qualities at the Travemünde competitive trials in autumn 1936.

Meanwhile, what of Reginald Mitchell and his new Spitfire? In turning from the disastrous Type 224 he faced an even greater challenge than Messerschmitt. First he had to tackle the disastrous Goshawk and the design's excess drag. Rolls-Royce were already persuading their new PV12 to deliver enough power for fighter performance without temperamental cooling arrangements, leaving Mitchell to sort out the aerodynamics.

He began his new design with uncranked wings, retractable undercarriage and a closed cockpit to eliminate the three worst sources of drag. With a shorter span wing, top speed should reach a still unacceptable 265 mph burdened by the Goshawk. Switching to a thinner aerofoil section and stressed skin panelling hinted that 280 mph might be possible, but the Air Ministry remained underwhelmed.

---

5. *Der Spiegel*, 15 January 1964, pp. 34–37.

Only when Mitchell proposed a complete redesign in November 1934, powered by Rolls-Royce's PV12 engine, did the Ministry rise to the bait. After a month of private development work funded by Supermarine, the Ministry paid £10,000 on 1 December 1934 for him to build a prototype by the following October.[6]

Mitchell approached his task carefully. His armament requirement was harder to fulfil than Messerschmitt's. Where German requirements specified twin machine guns, Mitchell had to fit first four, then eight guns to his fighter. Cutting drag meant raising the landing gear once the aircraft was airborne. Finally, the wing must be stiff enough to cope with combat stresses without thickening it and increasing drag, or fitting bracing wires to make things worse.

Fortunately, American research dealt Mitchell three trump cards. He could use the latest American National Advisory Committee for Aeronautics (NACA) aerofoils developed for maximum lift with minimum drag. Rolls-Royce followed American practice with a new glycol-based coolant system for the Merlin, and it would soon have 100-octane fuel.

But the Spitfire's wing would be its greatest advantage and its greatest weakness. To minimize drag, Mitchell chose an elliptical wing rather than the 109's straight edges. Suggestions this was copied from the wing of the He70 were false. Mitchell and his team already knew an elliptical wing created minimal induced drag and provided more internal space for guns and ammunition as well as the retracted undercarriage. The mathematics made perfect sense. A wing's internal space is proportional to the chord and thickness at any point, and on a uniformly tapering wing like that of the 109, it diminishes steadily outwards from root to tip. With an elliptical wing on the other hand, the chord (and therefore the thickness) diminishes relatively slowly at the root, only reducing more quickly as it approaches the tip. Consequently, internal space is greater nearer the root than with a straight-tapered wing of equal span.

Until the Spitfire, elliptical wings were usually symmetrical with leading and trailing edges following the same curve in opposite directions, like the He70. The Spitfire was different. Canadian aerodynamicist Beverley Shenstone joined Mitchell's team in 1931 following an abortive application to Hawker. There he helped create a supremely efficient wing with an inherently beautiful shape which made the most of the monoplane's advantages. It was a deliberately distorted ellipse, swept forward into an almost crescent configuration.[7] It had real advantages, but would be appallingly difficult to build. Even the aerofoil was a compromise, combining two American profiles, NACA 2213 for the inner part of the wing and NACA 2209 for the outer portion.

6. Price, *The Spitfire Story*, pp. 16–17.
7. *Ibid.*, p. 105.

The more complex internal structure of the Spitfire fuselage internals – restoration at Airframe Assemblies.

Shenstone had to plot each of the cardinal points of his wing by careful calculation in 'A sort of elliptical, paraboloid conic entity of rare brilliance.'[8] He created a high-speed wing with enough lift at low speed for take-off and landing without needing leading-edge slots. The narrow wing tips reduced drag further when flying straight and level or in the tightest combat manoeuvres. He also refined the crucial wing to fuselage transition, developing a three dimensional shape accurately tuned to maintain lift and reduce drag over a junction which normally caused turbulence and degraded performance in this area.

This highly advanced wing had another radical feature. A top-class fighter needed high manoeuvrability as well as high top speed. In Shenstone's words, 'A very definite limitation is given by the effective stalling speed of an aeroplane in a turn. One cannot, no matter what 'g' one dares to use, make as tight a turn on a highly loaded type. In fact for every wing loading there is a minimum radius of turn which can be flown.'[9] Mitchell also wanted to ensure young and partly trained pilots could fly his new fighter to its limits in combat. Trying to out-turn an opponent and lacking experience to tell them how close they were to the limit, they might flick into a lethal high-speed stall.

To prevent this, Mitchell provided a subtle warning. He changed the wing's angle of attack at which the airfoil met the airstream progressively from root to tip. Total change was a mere 2.5 degrees, but made the vital difference. With an angle of attack of 2 degrees at the root, this part of the wing would stall first while the tip, at a negative angle of half a degree, would remain unstalled. The pilot would sense a warning vibration while retaining full aileron control at the wingtip. Today this 'washout' feature is common, but in Mitchell's time was relatively unknown. Nevertheless the pilot could balance his plane on the edge of a stall to outfight his opponent.

8. *Ibid.*, p. 106.
9. Article on 'Fighter Fundamentals' by Beverley Shenstone, under the pseudonym B. Worley, in the March 1940 edition of *Aeronautics*, Vol 2, No 2, quoted in Cole, *Secrets of the Spitfire*, p. 121.

Spitfire leading edge D-box – author shot at Airframe Assemblies.

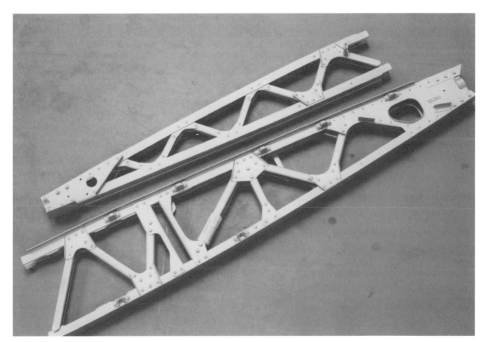

Design complexity: Spitfire wing ribs assembled from a series of small components: Airframe Assemblies.

There was a danger that twisting the wing risked distortion, losing the ellipse's advantage of low induced drag, but Shenstone's revisions prevented this. The slight forward sweep of the wing meant the washout angle could be reduced, speeding up the boundary layer of air over the wing and limiting distortion, producing a wing with unparalleled high-speed performance. In high-speed dives the Spitfire could reach Mach 0.9, or nine tenths of the speed of sound (more than 680 mph at sea level). Even when equipped as an operational fighter, its tactical Mach number was 0.82. As a comparison, the later P51 North American Mustang with a high efficiency laminar-flow wing, had a Mach number of 0.78 while the Bf109, with Messerschmitt's no-holds-barred wing had to be content with 0.75. It was an astonishing achievement, ensuring a fighter designed in the mid-1930s could cope with speeds higher than 500 to 600 mph more than ten years later.

In combat the design worked well. As the war progressed, fighters had to carry more fuel, weapons and ammunition, and increasing weight meant minimum turning radius increased correspondingly, but wing loading was vital in allowing a pilot to turn inside an opponent and bring his guns to bear. The larger area of the Spitfire's elliptical wing gave a loading of 24 lbs/sq.ft compared with 32 lbs/sq.ft for the Bf109, effectively one third greater. Shenstone's own calculations showed that at a reasonable combat speed of 275 mph the Spitfire could turn a full 360 degrees in eight seconds at a radius of 580 feet. The Bf109 took 12 seconds for the same turn at the same speed with a radius of 720 feet imposing a 6g loading just tolerable for the pilots.

Not everyone agreed. In his excellent study of the Battle of Britain, author Len Deighton[10] said the Spitfire could not turn as tightly as the Bf109, claiming the Spitfire flying at 300 mph could turn as tightly as 880 feet while the Bf109 could turn in a radius of 750 feet. However, his figures assumed a maximum g-force of 7.0 for the Spitfire and a staggering 8.1g for the 109. He also used wing loading figures of 25 lbs/sq.ft for the German fighter rather than the true figure of 32 lbs/sq.ft. Another limitation of the Bf109 wing was a combination of higher induced drag and boundary layer airflow turbulence, especially at the wingtip. Much of this airflow flowed along the wing from root to tip, so the 109-based Buchon fighters used by the post-war Spanish Air Force needed boundary layer fences to keep it within limits.

Mitchell's next problem was to guarantee stiffness for combat stresses, especially in the thinnest outer section. Like Messerschmitt he knew a hollow box section was stronger than a solid bar of identical size. He cleverly designed the Spitfire spar as a combination of both. Each spar consisted of an upper and lower boom joined by a vertical web, narrowing from root to wingtip to match the wing's taper. Each boom had five square section extruded aluminium alloy tubes fitting one inside another to

10. Len Deighton, Fighter – the True Story of the Battle of Britain, Triad Panther Paperback, St Albans, 1979, p. 106.

The complex geometry of the joint between the Spitfire wing and fuselage was designed and assembled to be as aerodynamically efficient as possible. *(Author photo)*

tight tolerances. The innermost tubes were the shortest and each succeeding tube was longer, so the booms were strongest at the wing root with five overlapping tubes. At the wingtip where stresses were lowest, only the two longest tubes remained, making a strong but resilient structure similar to a leaf spring

Mitchell enjoyed an extra stroke of luck. At first, expecting to have to use the Goshawk's cumbersome steam cooling, he provided large leading edge boxes to hold its massive cooling radiators. When these proved unnecessary he retained them for additional wing strength. The combination of spar and leading-edge box created great inherent stiffness to meet combat stresses and landing loads.

Sadly, his clever wing had no room for the undercarriage to retract inwards. The retraction gear had to be placed at the wing root since mounting it further out would stress the main spar as it began to taper. This actually meant a narrower track than the 109, but with more room for the retraction gear, he could at least improve steering geometry. Consequently the Spitfire remained more controllable when landing and taking off than the 109. It also had enough internal space for firstly two machine guns and then four to be mounted in each wing, well clear of the propeller arc for maximum fire power, with room for ammunition as well.

The problem was that the Spitfire wing's complex three-dimensional shape contained few straight lines. Had Mitchell opted for a purely elliptical wing, even the main spar would have had to curve with the greatest thickness from root to tip.

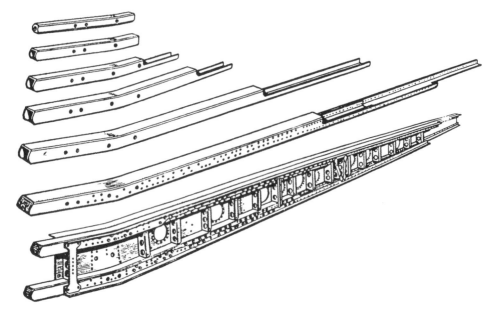

Spitfire spar construction – from April 1942 edition of 'Aircraft Production'. (via Airframe Assemblies)

Every wing rib would meet the spar at a different angle, a nightmare to assemble. Fortunately, this proved impossible. As the Merlin was heavier than the Goshawk, it shifted the centre of gravity forward, and the wing's leading edge taper had to be reduced to move the centre of lift forward to match. A straight main spar could have the ribs fitted at right angles to it, simplifying the structure and producing the classic Spitfire profile. In the words of one member of the design team, a curved spar 'Would have made the wing joints almost impossible to make (bending the booms for dihedral was bad enough). Manufacture would have been most difficult if the spar was not at right angles to the ribs.'[11]

The designs of both fuselages were similar in essentials but different in detail. Where Messerschmitt's fuselage frames were pared to the minimum to reduce weight, the Spitfire's bulkheads, stringers and longerons were thicker and more closely spaced. This added a weight penalty partly redressed by thinner skin panels in areas where stresses were lower. Spitfire wing ribs were more complex than those of the 109. A typical Spitfire rib was made in two halves, one joining the leading edge to the front of the spar and the other joining the rear face of the spar to the aileron hinges. Each was a Warren truss with upper and lower edges connected by fourteen cross members at different angles, riveted to them. In just one rib of the outboard section of the wing more than seventy rivets were involved, but the 109 rib was a simple one-piece pressing.

---

11. (Morgan and Shacklady, *op.cit.*, p. 19).

Each Spitfire undercarriage leg was mounted on the rear face of the wing spar, next to the joint connecting it to the fuselage. If the system failed, the pilot could send carbon dioxide gas to force the gear into the down and locked position. Spitfire flaps had only two positions – up or down. Compressed air-driven jacks lowered and locked the flaps. When raised, they were pushed upwards by air pressure and a reinforcing spring till they locked into position.

Like the 109, the Spitfire fuselage consisted of three sections. Its engine mounting consisted of a U-shaped I-section beam to support engine weight, fastened to the fuselage through steel tube frameworks on either side. Two tubular members joined at the forward end where they were bolted to the engine block, and at the rear were bolted to the top and centre of the main bulkhead at the front of the cockpit section. The strong cockpit section had fifteen frames connected by five longerons, two at the bottom, one at each side at the bottom of the canopy level, and one at the top from the back of the cockpit opening to the tail. Stringers linked them into a dense framework to carry the outer skin. The tail section had its front frame bolted to the rear fuselage frame. Two frames were extended upwards as fin spars. The control surfaces, like the 109, were covered with doped Irish linen, while the fin and tailplane were skinned in alloy panels.

Bearing in mind the differences between the engineers responsible and their lack of experience in fighter design, it remains surprising their efforts matched so closely in concept and execution, differing only in detail design and production techniques. Though aircraft design and performance had leapt forward so spectacularly in the early 1930s, methods in both the British and German aircraft industries had changed little since the previous war. Nevertheless, the tests they faced would demonstrate their exceedingly close achievements in performance and capability.

Chapter Seven

First Flights

With the designs complete, both fighters now had to be accepted by their intended users. Messerschmitt knew that even a narrow win over his competitors might still let his Air Ministry opponents bury his fighter. His 109 needed a decisive victory, but what kind of opposition did he face? The same invitation to tender had gone to Arado, Heinkel and later to Focke-Wulf, producing three designs which proved promising, unsuccessful and truly bizarre.

Closest in performance and potential to the 109 was the Heinkel 112, which appealed to pilots wanting the familiar. It had more in common with Mitchell's 224 than the Spitfire or the 109. It had retractable undercarriage, but a wing 40 percent larger than the 109. It was 18 percent heavier, but its wing loading was 20.3 lb/sq.ft, against the 32 lb/sq.ft of Messerschmitt's design.

This should have made the He112 more manoeuvrable, but slowed its rate of roll, a potentially lethal drawback in combat. Its thick high–lift aerofoil created more drag than the 109's thin wing and performance suffered even with an identical engine. The 109 prototype recorded 290 mph on its demonstration flight, but the original He112 could only manage a disappointing 273 mph.

Its open cockpit added to the overall drag, though pilots were delighted with the splendid view. It had a cranked wing like the Junkers 87 dive-bomber and the Supermarine 224. Good ground handling was due to a sturdy and reliable undercarriage, placed far enough from the fuselage for a wide track and greater stability.

However, all would depend on how the two designs performed in competitive trials and in particular, how they met the strict official requirement for spin recovery. This was included as unpressurized cockpits meant pilots might lose consciousness in high altitude combat, allowing the aircraft to fall into a spin. Recovery was essential once the pilot regained consciousness in the denser air at lower altitude. To ensure this was possible, the Ministry required each fighter to be spun at least ten times in each direction while remaining stable and allowing the pilot to regain control afterwards.

A whole series of changes were made to the Heinkel fighter to improve its speed and agility, but there were signs that spin stability would remain a problem. Prototypes two and three had clipped wings and a closed cockpit, but the second prototype crashed when test pilot Gerhard Nitschke bailed out after losing control during spinning tests. This left a question mark over the He112 and offered the 109 an opportunity as the Ministry began hedging their bets. No longer was the 109 the last place outsider. Its new position as one of two front-runners led to the issue of two development contracts, one for Messerschmitt and the other for Heinkel. These required ten pre-production prototypes from each, and rescheduled competitive trials for autumn 1936.

Heinkel then fitted a smaller, low-drag wing, straightening the trailing edge and curving the leading edge, rather like a Spitfire wing back to front. The tailplane was smaller and the fin taller, while the fuselage was rebuilt around a lighter and simpler frame. By prototype number nine the 112 was almost a new aircraft, the He112B, with yet another reshaped wing making its resemblance to the Spitfire even closer. It now had a single main spar like the Spitfire and the 109 and its wingspan was even less than Messerschmitt's compact fighter. It had a 20mm cannon in each wing to reinforce the two cowling guns, but it was sadly all too late. At the start of the series of competitive trials, most of these changes were either being made or still being planned. For once, events would favour Messerschmitt and his 109 against a competitor preferred by the Air Ministry from the beginning.

When the Bf109 entered the competitive trials, Messerschmitt's new test pilot Dr Herrmann Würster produced a triumph. In the all-important spin recovery trials, he spun the 109 at its most tail-heavy setting through twenty-one revolutions

Luftwaffe pilots were initially suspicious of the 109's high ground angle and poor forward view – this 109E has its tail supported on a cradle to reduce the normal ground angle. *(Will Owen photo)*

to the right and seventeen to the left, followed by a tail slide with the fighter falling backwards into a vertical dive from 25,000 feet to just above the ground. As this was not long after Nitschke's escape from the earlier He112 when spin recovery proved impossible, many began to look more favourably at the 109. Furthermore, when Würster was offered a trial flight in the Heinkel, he was warned not to try such testing manoeuvres.

This left the He112 at the post. Messerschmitt held all the cards and seeing his 109 triumph before his opponents must have been sweet revenge. Already, Ernst Udet had revised his original opinion after flying an early 109 prototype in January 1936. Now others, including future General of Fighters Ritter von Greim, joined the 109's supporters, and after detailed discussions it was decided to award the huge production contracts for Germany's future fighter to the 109.

Meanwhile, what had happened to the other two entrants, the Arado Ar 80 and the FW159? Both were effectively development dead-ends. The Ar80 was another cantilever low-wing monoplane, though its makers knew little of stressed skin construction or retractable undercarriage, so simply avoided them. The single spar wing had light alloy panels on top and fabric underneath. The fuselage had a heavy welded steel-tube framework covered by metal panels and strips rather than a monocoque, and a fixed undercarriage represented a bad case of wishful thinking. Using cowled wheels on legs shortened by a cranked wing like Mitchell's 224, Arado hoped that reduced weight might balance extra drag. They were wrong. The plane weighed more than predicted and the landing gear added yet more drag. The first prototype with the obligatory Rolls-Royce Kestrel flew in spring 1935, but crashed before testing finished. A second prototype used another of Germany's dwindling stock of Kestrels. It was 16 percent overweight and drag cut maximum speed to just 255 mph.

Arado then predicted a top speed of 264 mph with a constant speed propeller. This was not nearly enough for the Air Ministry, who told them not to bother. A third prototype was finished with an uprated Jumo 210, a constant speed propeller and a revised wing with the crank taken out to cut weight, but before testing could begin the project was dropped.

The Focke-Wulf 159 had a radically different configuration. Designed by *Dipl-Ing* Kurt Tank, later responsible for the brilliant FW190 radial engine fighter which became the 109's most formidable rival, the FW159 was a high-wing parasol design. This had been encouraged by the Air Ministry following the successful Focke-Wulf FW56 *Stösser* [Hawk] advanced trainer, and suggested as an avenue worth exploring.

Tank thought it interesting to follow a different route from his opponents, but his hopes were misplaced. He began with the main fighter requirements except the placement of the wing. It had a carefully streamlined stressed-skin monocoque fuselage with a closed cockpit beneath a one-piece sliding hood. It had retractable undercarriage and the metal-skinned wing with its rectangular plan form and rounded wingtips had a single main spar with an auxiliary trailing edge spar for flaps and ailerons. With this set below the fuselage in the normal way it might have given the 109 a run for its money, but the parasol position incurred huge drag from massive struts connecting wing to fuselage. At least it had an unrivalled view ahead and below, with easy access to engine and fuselage for maintenance and repair. The main fuel tank was set over a hatch to be jettisoned in emergency if set ablaze by enemy fire. On its maiden flight all went well until it was time to land. Suddenly test pilot Wolfgang Stein applied full power for an overshoot. The landing gear had failed to lock down and would probably collapse on touchdown.

Sadly, the prototype lacked the jettisonable fuel tank or even a radio to talk to ground control. Stein flew around for ninety minutes to burn fuel, trying violent aerobatics to shake the gear into locking down. Nothing worked and he finally made a belly landing with wheels still partly extended. Without a low wing to cushion the impact, it turned end-over-end twice before crashing on its back. Stein survived but the prototype was wrecked.

The engineers had underestimated the drag imposed on the hydraulics. They were strengthened in the second prototype, but its poor performance showed up more clearly. It climbed slowly, its manoeuvrability was poor due to high wing loading and top speed was a mere 252 mph. After three prototypes the project was dropped, leaving Tank free to press on with the ultimately far more successful FW190.

So the Bf109 was chosen as the Luftwaffe's front line fighter. Now ordinary pilots would have to come to terms with its high ground angle and fragile undercarriage, and its need for careful handling. It was hard to turn while taxiing, since applying full rudder and differential braking made the wheels lock. Pilots had to unlock the tail-wheel and push the stick forward to lift the tail with a burst of power, using both rudder and brake. This difficult technique needed practice. Skilled pilots could taxi quickly, but a mistake meant a ground loop and undercarriage collapse.

Once the 109 was lined up for take-off, the tail wheel had to be locked with the aircraft moving slowly straight ahead to keep it in position for taking off and landing. As the pilot opened the throttle for take-off, his first shock would be the

Artist's impression of the Focke-Wulf 159 parasol fighter which was another rival for the Luftwaffe fighter contract. *(FW159 prototype drawing from Flight magazine, 28 Dec 1939 (via Wikipedia))*

Scale model of FW159 shows complex undercarriage linkages which failed on maiden flight. *(Model of FW159 by Juergen Klueser via Wikipedia)*

huge torque reaction from the powerful engine, causing the 109 to swing to the left and tilt on its tall and narrow undercarriage.

Pilots had to be careful raising the tail. Until they did, the steep ground angle blocked their view and the rudder had no effect. Without lifting the tail the fighter remained semi-stalled and could not take off, but lifting the tail too soon lost the effect of the locked tail wheel, and the marked change in attitude caused a gyroscopic reaction to yaw the aircraft. Only with the tail fully raised could the pilot catch the swing and control it with the rudder. If the aircraft was pulled off the ground too quickly on reaching flying speed, the port wing would drop and using opposite aileron could turn the aircraft over with fatal results. Only after accelerating past flying speed was it safe to pull back the stick to make the plane leave the ground.

Once safely airborne the 109 was pleasant enough to fly, especially with fuel injection. Pilots could dive by simply pushing the stick forward and speed built up quickly. On the other hand, Messerschmitt pilots were not particularly comfortable. The hood could not be opened in flight, so many felt claustrophobic. The RAE Farnborough test report on a captured 109E said, 'The cockpit is certainly too cramped for comfort. The width is too narrow to fit in comfortably or allow movement within the cockpit. The headroom is also insufficient for taller pilots. When wearing a seat-fitted parachute, a pilot of average height's head will touch the hood roof. The seat position is also tiring when flying for extended periods.'[1]

---

1. *Pilot's Notes for Messerschmitt Bf109*, p. 6.

The narrow fuselage left little room to move the control column sideways, limiting the aircraft's ability to roll quickly. The seat could only be adjusted on the ground and there was no rudder bar adjustment for pilots of differing heights. Engine noise added to fatigue, but the most serious omissions were the lack of an artificial horizon on early versions for flying in poor visibility, and the lack of any rudder trim on all 109s. At 215 mph it would fly straight with no rudder applied, but left rudder was needed at higher speeds to avoid sideslip. Above 300 mph, holding the necessary 2 degrees of left rudder tired the pilot and could limit his ability to make high speed left turns in combat. On the other hand, elevator trimming was easy, but a pilot using the 109's high diving speed had to avoid trimming into the dive, as this risked failing to pull out before hitting the ground. Many pilots died by forgetting how the controls stiffened up with speed, until it proved impossible to pull out.

In normal aerobatics the slots smoothed out flight close to the stall, though one often opened before the other number. Royal Navy test pilot Captain Eric 'Winkle' Brown found, when flying a captured 109, that on trying stalls at a safe height they opened at around 20 mph above the stall, but were 'Accompanied by the unpleasant aileron snatching as the slats [*sic*] opened unevenly.' He also tried mock attacks on a Lancaster and a Mustang, but found 'The slipstream of these aircraft caused the intermediate operation of the Bf109's slats so that accurate sighting became an impossibility.'[2] Farnborough test pilots agreed, explaining that 'If the stick is pulled back during a tight turn, producing additional 'g', the slots will open at high airspeeds. As they open, the stick will snatch laterally through several inches in either direction. This snatch is sufficient to spoil the pilot's aim during a dogfight.'[3] Some novice pilots mistook the noise and vibration of the slots opening for enemy gunfire, and carried out unnecessary evasive action. Others disagreed, claiming that 'The gentle stall and good control under 'g' enable the pilot to get the most from the aircraft during dog-fighting by flying very near to the stall.'[4] Nevertheless, the higher wing loading of the 109 for similar overall weight to the Spitfire and the Hurricane meant a wider turning circle. Its high diving speed meant it was better suited to dive and zoom tactics than classic dogfighting.

Landing the 109 needed an even more delicate touch. Pilots said that surviving the take-off meant one was also quite likely to survive the landing and the approach was straightforward. Eric Brown found 'The approach was steeper than that of the Spitfire, but elevator feel was very positive and gave delightfully accurate control at 118 mph.'[5] It was in the final landing phase that problems arose. If speed fell too low on finals the 109 could drop a wing without warning. If the pilot reacted

2. Brown, *Wings of the Luftwaffe*, Crecy Publishing, Manchester (revised edition), p. 212.
3. *Pilot's Notes for Messerschmitt Bf109*, p. 13.
4. *Ibid.*, p. 16.
5. Brown, *op.cit.*, p. 212.

quickly to level the wings, it would often touch down too heavily on one wheel, risking undercarriage collapse. It was also reluctant to slow down. Pilots landing on main wheels risked disaster, as without the locked tailwheel the rudder could not prevent a vicious ground loop. The remedy was to pull back on the stick to hold the aircraft just above the ground until speed dropped to settle into a three-point landing, but this took skill and confidence. For the same reason, overshooting needed great care. Opening the throttle too quickly needed full right aileron and almost full right rudder to keep straight, and too harsh a touch on the throttle could lose control altogether. When landing, Captain Brown found a large change of attitude was needed when flaring for landing, and even after touching down there was enough lift to cause bouncing on rough ground. Only when the tailwheel was firmly on the ground could the brakes be applied hard enough for a short landing run, 'But care had to be taken to prevent any swing, as the combination of narrow-track undercarriage and minimal forward view could easily result in directional problems.'[6]

The RAE Farnborough tests of a captured 109E said, 'After a few practice landings the pilot should become easily accustomed to the landing technique and have no further problems. The centre of gravity is unusually far behind the main wheels and the brakes can be fully applied immediately after touchdown without lifting the tail. The ground run is very short, with no tendency to swing or bucket. As the large ground attitude causes the nose of the aircraft to be very high, visibility is poor for taxiing. Landing at night is very difficult.'[7]

Warnings like these suggest the Bf 109 was a temperamental thoroughbred, lethal to clumsy pilots. However, for those with enough skill and confidence to master her limitations, she proved an outstanding fighting aircraft and the Luftwaffe trials confirmed that little could be faulted on performance grounds.

Compared with the 109, the Spitfire's birth was remarkably painless until it entered production. The only doubt about the first flight was the actual date, recorded as 5 March 1936, though many insist it was on the following day. Dr Alfred Price found a document bearing the handwriting of Supermarine company secretary Henry Duvall confirming the prototype made its maiden flight on 5 March 1936.[8]

However, Supermarine test pilot Jeffrey Quill insisted he had flown from Brooklands airfield to Martlesham Heath on 6 March to pick up Supermarine's chief test pilot, Mutt Summers, and fly him to Eastleigh for the maiden flight. This proved completely routine. The prototype machine, K5054, had a two-bladed fine pitch propeller to reduce the take-off run and the torque reaction swing. The wind was a light breeze and Summers took off with the undercarriage locked down.

---

6. *Ibid.*, p. 212.
7. *Pilot's Notes for Messerschmitt Bf109*, p. 9.
8. Price, *The Spitfire Story*, p. 36.

For some fifteen minutes he checked slow flying and stalling characteristics, and the operation of the flaps, before a final approach and an immaculate three-point landing. At the end of the flight he announced he wanted nothing touched, but contrary to later mythology this simply meant no immediate adjustments were needed.

On 10 March Summers flew again, with undercarriage doors in place and a normal coarse-pitch propeller fitted, and he raised and lowered the landing gear easily. Finally, Jeffrey Quill flew the prototype on 26 March. With a replacement engine to cure starting problems and a normal coarse-pitch propeller, acceleration on take-off was slower than before. The increased torque reaction meant holding full right rudder to keep straight as it rocked on its narrow track undercarriage. Raising the landing gear was by a hydraulic hand lever, which meant changing hands on throttle and stick, without causing the aircraft to porpoise.

Quill also found the approach too flat for good visibility over the nose, due partly to the flaps being set for maximum lift rather than the maximum drag needed to slow the machine down. On his second flight he forgot to raise the flaps, but the fighter lifted off with only slight buffeting. After reaching 200 feet he lifted the flaps and the aircraft sank slightly before climbing away normally. His flight ended with a perfect three-point landing needing so little stick movement that the Spitfire seemed to land itself.

This was very promising, but two outstanding issues remained: name and performance. The Germans gave no formal names to aircraft but the British named them early in development. By 11 May 1936, Quill referred to the F300 prototype as the 'Spitfire' on Air Ministry instructions. Supermarine names usually began with the letter 'S', like the 'Sea Lion', 'Sea Eagle' and 'Southampton' flying boats, but other aggressive suggestions for the new fighter included 'Shrew' and 'Shrike'. Happily, Sir Robert McLean, chairman of the Vickers Group which owned Supermarine, insisted 'Spitfire' embodied the right qualities as he often called his high-spirited young daughter Ann 'a right little spitfire.'

At the end of 1933, his office cabled the Air Ministry to ask the name be kept for the Type 224. The Ministry insisted this remained the 'Supermarine F7/30' until accepted by the RAF. As this never happened, the name remained available and when the new design showed potential the Ministry decided to revive it. Mitchell was unhappy as it recalled an extremely unsuccessful design. By now suffering from aggressive terminal cancer, he expressed his opinion forcefully; 'Just the kind of bloody silly name they would give it.'

The decision was inspired though premature. The new Spitfire still had to win official RAF acceptance with doubts resulting from the performance of its future stablemate, the Hawker Hurricane. In spite of a thicker wing and a hump-backed fuselage, Sidney Camm's rugged and conventional fighter had reached a top speed of 330mph. It would be simple to produce, maintain and repair, so the thoroughbred Spitfire needed a clear performance advantage to ensure its future.

The Spitfire prototype K5054 ready for her maiden flight with unpainted finish, undercarriage fairings removed and gear locked down, and a twin-bladed fixed-pitch propellor. *(Via Alfred Price)*

No speed measurements had yet been made. Two flights by Jeffrey Quill on 27 March made a serious attempt to record speeds around the rated full-throttle height of 17,000 feet, and the results were disappointing, with the measured 335 mph only slightly faster than the portly Hurricane. The first suspect was the coarse-pitch propeller. Running at full speed, the blade tips neared supersonic speed so compressibility effects added to drag. Supermarine engineers made a propeller with modified blade profiles. Quill carried out more tests with this on 5 May, recording an improved speed of 348 mph, closer to Mitchell's objective of 350 mph.

K5054 in flight on 11 May 1936 after application of pale-blue finish, markings and serial numbers, flown by 'Mutt' Summers, with Jeffrey Quill flying the camera plane with R J Mitchell aboard. *(Via Alfred Price)*

On 22 March 1937, the prototype's engine seized from oil starvation on a test flight at Martlesham Heath, forcing the pilot to make a belly landing. Skill and good luck reduced damage to the minimum and the programme was resumed after repairs and modifications, including a camouflage finish and the fitting of heating ducts to prevent the guns freezing at high altitude. *(via Alfred Price)*

Following repairs, K5054 was given camouflage paint and testing resumed – the small projection ahead of the tail was the fitting for the spin recovery parachute. *(via Alfred Price)*

On the second day of the Second World War, 4 September 1939, K5054 stalled on approach during tests at Farnborough, bounced and turned over as she hit the ground. The pilot, Flight Lieutenant White, was trapped in the wreckage and died in hospital from his injuries. *(Via Alfred Price)*

When the Spitfire undercarriage was lowered, the starboard leg blanketed the radiator causing rapid overheating. *(Author photo)*

Some smaller problems were fixed before K5054 went to Martlesham Heath for RAF acceptance trials, including rudder over-sensitivity, cured by fitting a smaller horn balance. Tests showed that the aircraft was marginally stable longitudinally, so care would always be needed over moving the centre of gravity with increasing loads during later development. The first test dive to 465 mph went well, but when repeated, the port undercarriage fairing tore free with a loud bang and damaged the aircraft underside. Quill also reported that aileron control was heavier at high speed, another harbinger of future problems.

The prototype was flown to Martlesham Heath on 26 May by Mutt Summers and handed over to 'A' Flight commander, Flight Lieutenant Humphrey Edwards-Jones. He was ordered to test the Spitfire as soon as it arrived to determine suitability for RAF pilots. He found it easy and pleasant to fly, but his first flight nearly ended in disaster when he followed a Hawker Fury on his landing approach and forgot to lower the undercarriage. With the hood open he could not hear the warning horn and only a vague sense of unease made him spot his mistake in time.

He reported that the Spitfire could indeed be flown by new pilots, and Supermarine were given an initial production contract for 310 Spitfires with Hawker given an order for 600 Hurricanes. In fact, the prototype Spitfire was no operational warplane. Britain was still so short of Browning machine guns that the first had

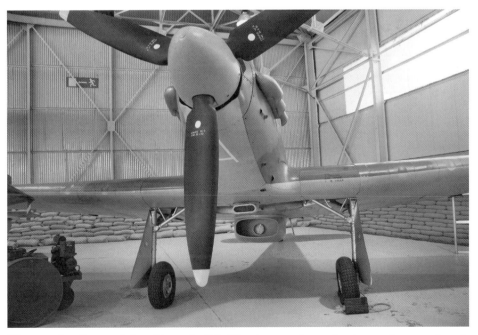

The central radiator and the wide track of the Hurricane made overheating on the ground much less of a problem. *(Author photo)*

been used for the Hurricane prototype. Mitchell's fighter had to carry ballast to represent the weight of guns, ammunition boxes and the radio not yet fitted.

During K5054's first session at Martlesham Heath, RAF testers recorded a top speed of 349 mph and a service ceiling of better than 35,000 feet. It returned from the trials with a request for higher drag flaps to shorten the landing run and a less sensitive elevator control, and in view of Edwards-Jones' experience, a louder warning horn audible with hood open and engine running! It also noted a tendency to instability in the glide with flaps and undercarriage down, but insisted this was satisfactory for a fighting aeroplane and a good compromise between manoeuvrability and steadiness as a gun platform.

The flap setting was increased to 90 degrees. They removed the undercarriage fairings, which often failed to close. As this had no effect at all on speed they were not replaced. An improved oil cooler was fitted and Brownings reclaimed from the Hurricane were fitted in the wings with ammunition boxes and feeds, and access doors in the wing upper surface. Finally the engine was replaced with a brand-new Merlin F, delivering 1050 horsepower.

Following a report from RAE Farnborough that wind tunnel tests suggested spin recovery problems, a small parachute was fitted in a box inside the cockpit. If the Spitfire failed to recover, the pilot could open the hood and throw the two-foot diameter parachute over the side opposite to the spin direction. Anchored to a bolt by the fin, its deployment would pull the tail like a much more powerful rudder to cause immediate recovery.

Jeffrey Quill flew the Spitfire on 11 December to test the parachute, and confirmed all worked well. Two days later he flew with the centre of gravity set to its forward limit, theoretically the best for recovery, and began a spin. The rate of rotation sped up and slowed down in a series of flicks on each revolution, with the nose hunting up and down between a steep upward pitch and a vertical downward attitude. However, this ceased once he began the recovery drill, applying opposite rudder and pushing the stick gently forward, whereupon the spin stopped immediately. As the specified eight turn spins both left and right were completed with the centre of gravity moved progressively aft to the limit of its range, spin recovery was always immediate and the recovery parachute was removed.

Reducing drag was next. Improvements to the radiator's internal shape accelerated the airflow to improve cooling and also generated thrust like a simple ram-jet. The drag from the rows of rivets was checked next. There were two types: complex and costly flush rivets and cheaper and simpler dome-headed rivets. The prototype had flush rivets throughout, but to investigate the effect of domed rivets, halved split peas were glued over the flush rivets and the drop in top speed was measured. Different groups of split peas were then removed and the speed measured again. Results showed domed rivets on the fuselage cut top speed by just 1 mph, but on the wings cut it by more than 20 mph.

The radio mast and aerial were fitted and the tail-skid replaced with a tail wheel to stand up to paved runways. With guns in place the aircraft returned to Martlesham Heath for the next set of trials. On 6 March the fighter was flown to 4000 feet and all eight guns fired with no problems at all. Meanwhile a double tail wheel proved less liable to bounce on the ground, but was quickly blocked by mud and replaced by a free-running single wheel.

On 10 March everything went wrong. An RAF test pilot took the Spitfire up to 32,000 feet to fire all eight guns. One port gun loosed off 171 rounds but the other three failed to fire at all. The starboard guns did even worse: one fired eight rounds, another fired four and the remaining two fired none. The problem was simple to explain but difficult to cure; guns were icing up in the bitterly cold air. There were two more crises. One pilot completed a series of loops, dives and tight turns to check elevator buffeting, but on reducing power to return to the airfield, the engine ran roughly and oil pressure fell away. Doubting his ability to return to base he made a wheels-up landing on a nearby open heath. Thanks to flying skill and good luck, the aircraft stopped short of an eight-foot deep pit, the damage was moderate and the Spitfire was repaired at the factory.

Sadly, it was now clear Reginald Mitchell was dying. After an unsuccessful operation at a specialist clinic in Vienna on 29 April 1937, he declined rapidly. He returned to Southampton and died at home on 11 June 1937 at the tragically early age of forty-two. It seems almost certain his iron resolve to complete the fighter had sapped his strength. It seems equally certain he knew the stakes and was determined to see his design take to the air against Nazi Germany. Fortunately his team was able to take over his role. His assistant, 'Agony' Payn, ran the design team with chief draughtsman Joe Smith taking Mitchell's post as Chief Designer. The smooth takeover was essential as putting Mitchell's design into production as an operational warplane would meet all kinds of difficulties.

The repaired prototype flew on 9 September. Ten days later it flew again with modified exhausts, carefully reshaped by Rolls-Royce to exploit the huge volume of gases ejected from the engine under full power. Calculations showed these emerged from the exhausts at a blistering 1300 mph and deflecting them backwards increased power output by some 70 bhp. The total effect of all these improvements raised the Spitfire's top speed to 360 mph, a figure which would have delighted Mitchell had he lived to see it.

Keeping the guns warm enough to fire proved more difficult. Designing and fitting ducts to channel hot air from the radiator to the weapons took time and effort. In the meantime, tests continued to assess the Spitfire's suitability as a night fighter as required in the specification. On 15 March 1938 one pilot followed two textbook landings with another where he overshot the field and ran on to soft ground, tipping the aircraft forward on to its nose. The cause was a shortage of compressed air to operate flaps and brakes after a series of landings, and a more powerful compressor was fitted.

Repairs took three days, but a week later another heavy night landing caused the port undercarriage to fail, and the aircraft came to rest on its starboard wheel and its port wingtip. More extensive repairs gave time for the new gun heating arrangements to be installed and tested before it flew again. Another major landmark passed on 15 May 1938, with the first flight of the much-delayed first production Spitfire to take over most development testing.

The prototype proved the gun-heating system fit for purpose by mid-October 1938 and she was sent to Farnborough to help develop the record breaking Speed Spitfire (Chapter 8) and then become a flying test bed for engine improvements. Finally returning to Farnborough after a test flight with new spark plugs, her pilot stalled too high on the approach. The Spitfire dropped out of the sky fully stalled, bounced twice and turned end-over-end, killing her pilot.

This was the end of the road for K5054, the only Spitfire her creator ever saw fly. After production headaches which came close to cancelling the whole programme, just twenty machines had emerged from the workshops by the end of the 1938 Munich crisis and time was running short. In an ironic reversal of fortune, the Spitfire's advantage over the Bf109 would melt away under the fierce hostility from the service for which it was designed. Just as Mitchell's fighter seemed on the road to success, the RAF would do its best to bury the Spitfire once and for all. They wanted to switch to more and more lethally obsolescent light bombers or a series of replacement fighters which would never have filled the resulting gap in British defences. At the time when the 109's future was safe at last, that of the Spitfire would become extremely shaky.

Nevertheless its performance was excellent. RAF pilots flying the new fighter were delighted with its handling and the designer's maxim of 'If it looks right, it'll fly right' had been triumphantly vindicated. Where the 224 looked clumsy from the start, the Spitfire was arguably the most beautiful military aircraft of all time. Its impeccable behaviour suggested that Mitchell's artistic combination of subtle curves had been truly inspirational.

It was surprisingly easy to fly. Taking off, like the 109, meant using full right rudder and stick to counter the swing from engine torque. But Spitfire pilots could set the rudder trim (which the 109 did not have) to 3 degrees right to help this process. Once the speed was high enough to raise the tail, directional control was easy and she would lift off by herself at flying speed. Pilots careless of her over-sensitivity in pitch who pushed the stick forward too clumsily, risked hitting the propeller blades on the ground.

On early Spitfires, retracting the undercarriage meant tightening the throttle screw to hold it open as the pilot's left hand took over the stick and his right hand started pumping the undercarriage handle some twenty times. If not screwed down sufficiently the throttle could close, turning off the power with the aircraft low down and the undercarriage half retracted. With undercarriage safely up and locked, a red bulb lit up on the instrument panel, and on early machines a complex linkage raised an indicator in the wing upper surface.

Once cleaned up and at a safe height, pilots could test the extraordinary sensitivity of Mitchell's wing. Time and again they would report that it was not necessary to move the controls to bank or turn. Simply thinking about the manoeuvre and slightly increasing pressure on stick and rudder bar were enough. It was described in official RAF Pilot's Handbooks as essentially stable and it was, but pilots often said it should be flown with just finger and thumb on the control column to avoid over-correcting.

One design flaw was that the undercarriage blocked airflow through the radiator when wholly or partially lowered. This meant finishing all the take-off checks before the engine overheated. Equally, on a landing approach, pilots would lower flaps as late as possible to avoid raising engine temperature. To keep the runway threshold in view, they flew a curved approach as the speed decayed with the wheels lowered. Ideally they crossed the runway threshold at 80 mph before straightening up and pressing gently back on the stick to start the flare, persuading the aircraft to drop on to the ground in a three-point attitude. There was little tendency to ground loop, but they had to finish taxiing before the engine reached its limiting temperature of 115 degrees C.

Some pilots preferred to keep a speed margin in hand, but this risked the aircraft gaining height and stalling as the speed decayed again. The remedy was to come in lower and slower and closer to the point of no return without crossing it. If the pilot found out he was too high he could open up and go around again and again until he synchronized closing the throttle with back pressure on the stick for a smooth landing.

At first, neither the Spitfire nor the 109 had two-seat trainers, so pilots had to fly the aircraft from first acquaintance. In the RAF they could sit in the cockpit and practise operating controls, switches, buttons and levers. Sometimes this could be done with the fighter on trestles so the wheels could be raised and lowered, but often there was no time. Later they could progress from basic trainers like the Miles Magister or the Tiger Moth to more advanced trainers like the Miles Master or the North American Harvard to reduce the gap to be bridged.

The Spitfire cockpit was compact and for taller or bulkier pilots, undeniably cramped, but the height of the seat and the reach of the rudder pedals were adjustable. Climbing in and out of the cockpit was easy, by sliding back the one-piece transparent hood and opening the small, drop-down door on the port side. The standard drill for take-off and landing was to slide the hood back and partly open the two-position catch on the cockpit door to hold it securely in place.

In the end the designers' priorities were clear. Messerschmitt designed the 109 for performance and production. Mitchell also chose performance, but ahead of ease of production, had placed ease of handling and a level of viceless behaviour that put most other fighters to shame. But the other paramount quality shared by both aircraft would be an astonishing ability to remain competitive, even under the furious pace of wartime development.

Chapter Eight

Record Breakers

On 4 April 1939, BFW's chief test pilot *Flugkapitan* Fritz Wendel, was landing a Messerschmitt at the company airfield of Augsburg-Haunstetten. As he lined up for final approach across the large park of the *Siebentischwald*, he lowered the undercarriage. At that moment, and to his unbounded horror, the oil cooling system failed. The engine cut out and his aircraft plunged like a stone to the earth not far below. Without power, there was little he could do. As an experienced test pilot, his immediate reflex was to trade height for speed, and push the stick forward to steepen the dive. This seemingly suicidal action increased flying speed enough to allow him to lift the plane over the line of trees bordering the *Haunstetterstrasse* before gravity took over again. His machine hit the ground with appalling force and was completely wrecked. Incredibly, Wendel escaped with scratches.

This disaster was the outcome of a series of problems for an aircraft built with two overlapping objectives and a double identity to match. It was originally meant to break the existing landplane speed record of 352 mph set by the billionaire American aviator Howard Hughes on 13 September 1935 over Santa Ana in California. His H-1 monoplane had been carefully streamlined using flush rivets and fitted with retractable undercarriage to minimise drag. It was powered by a 14-cylinder Pratt and Whitney twin-row radial engine delivering some 1000 hp. He made a series of faster runs over the set course, but on the final pass his engine ran out of fuel, forcing him to crash land in a bean field. His record was not an absolute speed record as an Italian Macchi seaplane reached 440 mph in 1934, despite the drag of its twin floats.

After the fierce competition between Messerschmitt and Heinkel for the Luftwaffe fighter contract, it was no surprise that the idea of record-breaking occurred to both at almost the same time. The speed needed to break the existing record by at least 5 mph seemed well within reach of their new fighters. Both were aware they needed to keep their plans secret, but realized they would need Air Ministry backing, so they contacted the Ministry for official funding. As chief of the Technical Office, *Generalmajor* Ernst Udet was amazed to receive two approaches from key people in his service's fighter programme, and had to think carefully about what he should do.

On the face of it, allowing such a needless distraction from developing warplanes for Germany's military should be rejected outright. However, Udet's dilemma was that neither applicant knew the other was in the running. Better to let both have their

The Hughes H1 radial-engine speed record aircraft, photographed in the Smithsonian Museum. *(Via Wikipedia)*

way than face allegations of bias. Surely it would be a straightforward process of fitting a more powerful engine to an existing aircraft, followed by reducing weight and drag as far as possible? Udet's inaction seemed sensible, until he realized the determination of both men. Heinkel had been bitterly disappointed when Udet himself had turned down his He112 for a fighter contract. Instead of accepting rejection of a plane which had come close to the Bf109's attributes and even improved on them in some respects, he set out to produce a new machine with performance spectacular enough to eclipse the 109 and reverse the decision. This meant a complete redesign of the He112, exploiting many principles devised by Messerschmitt for the 109. Simplifying the airframe and reducing the size of the aircraft cut weight and drag. In revising detail design to make production simple, Heinkel cut the number of components by more than half, an astonishing achievement. The wing was no longer the complex semi-elliptical shape which would give Supermarine such difficulties in producing Spitfires, but a simple, tapering straight edge plan like the 109, and the new Heinkel was now appreciably lighter.

It was also faster. It first flew on 22 January 1938 with the Daimler-Benz DB601 and the hard work had clearly paid off with a top speed of 416 mph, faster than other fighters of the time. Originally numbered the He113, Heinkel changed it to the He100. Nevertheless, its new design and dramatically improved performance made it an ideal contender for the record.

For his part, Messerschmitt had no such radical improvement available. Instead, he took over Bf109 prototype number thirteen after it completed taking part in a series of international competitions held at Dübendorf in Switzerland in July 1937. In the factory, it was fitted with one of a series of special racing versions of the Daimler-Benz DB601 with higher compression and maximum speed raised to 2800 rpm, delivering peak power of 1520 hp for up to twenty minutes or 1660 hp for up to five minutes at a time. In the speed record world, reliability and endurance were non-essentials.

The fuel was an exotic cocktail of 100-octane aviation spirit and methyl alcohol. The engine used twenty-four spark plugs, two for each cylinder, simply for starting and warming up, then all twenty-four had to be taken out and replaced with Bosch plugs specially developed for the record attempt, before the engine cooled down. Even the pitot head was removed to cut drag, which left the pilot with no idea of the aircraft's speed – for a speed record attempt! All seams between skin panels were taped over and the aircraft's outer surface coated in a putty-like compound, sanded and polished to a high gloss finish before being spray-painted in all-over grey. It was fitted with a large and carefully streamlined spinner and a three blade variable pitch propeller.

The rules specified how the record flights would be carried out. A straight course of three kilometres (1.8 miles) had to be covered at least four times at a maximum height of 75 metres above ground level, or some 250 feet. As Dr Würster knew the engine delivered its greatest power at low altitude, he set himself a height of 35 metres, or just over 100 feet. This was difficult and dangerous at high speed, but at each end of the course he had to make a wide-radius banked turn to reverse course without gaining height, since this would be seen to give him an illegal speed advantage in diving back to the right height for the next run.

Local research suggested the ideal course would be three kilometres of the main railway line running almost due south from Augsburg to Buchloe. Afterwards Dr Würster explained some of the difficulties. 'The high speed at low altitude makes it necessary to always train one's eyes three to four kilometres ahead of the aircraft, which means that on passing over the starting point the pilot does not see the end of the three kilometre course, instead his eyes are already trained one kilometre beyond it. Consequently, during the approach which was ten kilometres long, the pilot must 'aim' the aircraft precisely at the course.'[1]

With the difficult turns needed to line up precisely for the next reverse leg, all at high speed and low altitude, it was clear careful rehearsal would be needed. On the day of the attempt, 11 November 1937, Würster flew up and down the course twice in a Bf108 to check the weather and the visual cues. The cloud level was down to 1600 feet, but below this blanket visibility was acceptable, marred only by low winter sun. There was very little wind. Special high-speed cameras with precision

---

1. Radinger and Schick, *op.cit.*, pp. 66–67.

timers were sited at each end of the course, and three more BF108s took off with official observers to watch the event.

Finally the Bf109 was warmed up, the plugs were switched for the special racing versions and Würster took off for his record flight. In his words, 'It all went like clockwork.' He finished four runs over the measured three kilometres in just under twenty-three minutes, raising Hughes' old record by 44.83 kph, or approximately 28 mph, to 379 mph. The first deliberate deception was entered on the certificate issued by the *Fédération Aéronautique Internationale.* The aircraft was described not as the Bf109 but as the Bf113R, a designation that did not exist, powered by a Daimler Benz DB600, which was not the engine actually used. The discrepancy was for reasons cited as 'security' and 'propaganda', though the truth was probably a mixture of the two.

This news galvanized Heinkel. Two months after Würster's success, the He100 prototype had already flown faster than the new record, so it was clear Heinkel would try to break it as soon as possible. On June 6 1938, Ernst Udet himself asked to fly an apparently standard He112 round two laps of a fifty-kilometre course near Rostock on the Baltic coast at a height of 18,000 feet. He recorded an encouraging 394.4 mph, but this was yet another deception. The aircraft he flew was the second prototype of the He100, but to prevent the truth leaking out, no photographs of the record breaking machine, which was referred to officially as the He112U (for Udet), were released.

Next in Heinkel's sights was Messerschmitt's new landplane world record, using the same over-boosted Daimler Benz DB601 racing engine. With Heinkel chief test pilot Gerhard Nitschke at the controls, the record-breaking aircraft was being tested in September 1938 when the retractable undercarriage failed, leaving the gear half extended. Rather than attempt a landing, Nitschke was ordered to bail out and the aircraft crashed. Nitschke himself was injured hitting the tail on his way out of the aircraft, but survived.

The next entrant was the eighth prototype with clipped wings, streamlined windscreen and a lowered tailplane. Its record flight was switched to Oranienburg, twenty miles north of Berlin, on 30 March 1939, with war now just over five months away. There, test pilot Hans Dieterle pushed the record up to 746.6 kph (463.9 mph), but the deceptions continued. The aircraft was described as the 112U, though Udet was not involved. As part of the cover-up, the less obviously modified fifth prototype starred in a film of the attempt.

This was a huge improvement over the Bf109 record and actually beat the speed of the Italian Macchi seaplane, making it an absolute speed record for the first time. But it availed Heinkel little, since his He100 was plagued by undercarriage problems even worse than those of the Bf109. One He100 even had its landing gear collapse while sitting quietly on the tarmac, and its evaporative cooling system proved as difficult to manage as Mitchell's Type 224. No orders resulted, and Messerschmitt was determined to reply with an even bolder deception and a plane which was almost unflyable, to lift the record again. But first, it was the British turn for a record attempt, using a modified Spitfire.

This was altogether much more modest. Like the Bf109, the target was the 1935 Howard Hughes record, following a German propaganda boast that the Luftwaffe was being equipped with the fastest fighter in the world. The British decided this needed correcting, but hard work was needed to make it happen. At the time, the Spitfire prototype had recorded a 349 mph top speed at 16,800 feet. In denser air at the low altitude used for the record attempts, this would decrease to some 290 mph. For any record attempt, more power and less drag would be essential.

Rolls-Royce began developing a racing engine and by 7 August a sprint version of the Merlin II was delivering 1536 hp running at 2850 rpm with 18lbs boost for up to four minutes at a time, long enough to complete a record attempt. Like the German engines this required a mixture of petrol, benzole and methanol with small amounts of tetraethyl lead, and two complete sets of plugs, one to start and warm the engine, and another for high speed and high power needed for the record attempt.

Unlike the 'He113', the Speed Spitfire was based on a production machine with stronger engine mountings for the uprated Merlin. The cockpit canopy was lower and better streamlined, the wings were clipped and rounded off, and a fixed-pitch four-blade propeller fitted with a tail-skid replacing the normal tail wheel. Cooling was by a pressurized water system, and the aircraft was flush-riveted, with all cracks filled before a special high-speed paint finish was applied in dark blue for the upper surfaces and pale blue underneath.

By June 1938, the Merlin was delivering twice its usual power at even higher boost settings, but speed was not high enough to take the record by a convincing margin. By spring 1939 the complete cooling system had been removed to save weight, with a water tank and condenser instead of the upper fuel tank. Though water would be lost quickly, the cooling system's range should match the reduced fuel capacity and last long enough for the record. However, making all these changes overran the last days of peace.

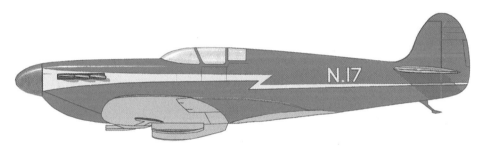

Careful streamlining and a more powerful Rolls-Royce Merlin created the Speed Spitfire for a British record attempt. With a carefully cleaned up surface finish in deep blue and silver, its cooling problems delayed the project until it was overtaken by the outbreak of war. *(Author drawing)*

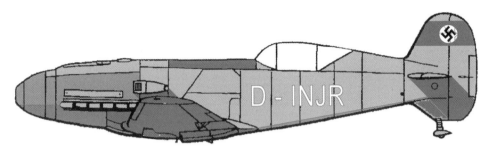

The Messerschmitt Me209 was virtually a flying engine, and almost impossible to fly. *(Author drawing)*

By the time the work was finished and the machine returned to Farnborough, England and Germany had been fighting for almost eight months and record breaking was irrelevant. As test pilot Jeffrey Quill taxied out for take-off at Eastleigh on 30 April 1940 a broken coolant pipe enveloped him in steam. The project was abandoned. The Speed Spitfire was converted back to orthodox configuration and sent to a PR unit. Unfortunately the performance improvements crippled its endurance, and its sixty-gallon fuel tank was insufficient even for a return trip between an English airfield and northern France.

Nevertheless, the potential record breaker had stimulated Rolls-Royce to highly successful efforts to raise the Merlin's power and guarantee Mitchell's fighter a bright future of increasing performance and capability. Yet even this faster and more powerful version of the Spitfire was simple and undemanding to fly. This was at the opposite end of the spectrum from the new and almost lethal 'Me109R',

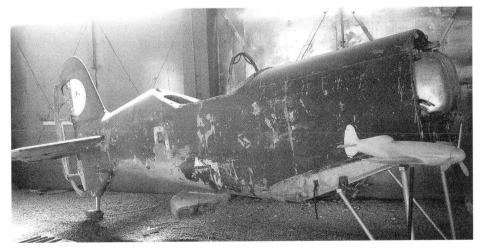

The remains of the record-breaking Me209 with the fragments of the swastika on the tail and the civil registration letters on the side, languishing in the Cracow Museum with a small scale model alongside. *(Wikipedia)*

which Messerschmitt had prepared for his final record attempt and which was far too difficult to handle to make an operational fighter.

Anyone who saw the 109R at rest, rather than flashing past at low altitude would know it was no mere variant of the 109. Though it was also a single-seat monoplane with the massive bulk of a Daimler-Benz engine in the nose, the two designs had almost nothing in common. The Me209, its official designation, was effectively a flying engine with a tiny cockpit for the pilot to try to control it, rudimentary wings and control surfaces, and a wider track version of the 109's spindly landing gear, made to retract inwards for once.

Where the 109 was compact, the 209 was positively petite to cut drag further. Its wingspan was 25 feet 7 inches against the 32 feet 4 inches of the fighter. It was 23 feet 10 inches long, 4 feet 2 inches shorter than the 109, and its wing area was more than a third less. Unfortunately, it still had a full-size cooling problem, with an engine delivering 2300 hp for brief periods before blowing itself apart. Instead of conventional radiators, they used a steam cooling system like that of Mitchell's Type 224. Engine oil was cooled in a separate circuit and ducted through an annular intake hidden behind the massive spinner.

As gross weight was 10 percent more than the 109, the wing loading was 48 lb/sq.ft compared with 29 lb/sq.ft for the 109. This increase of 40 percent did not bode well for flying qualities, and so it proved. Test pilots of the time were used to aircraft with the aerodynamic efficiency of a house brick and the stability of a hippo on roller skates, but this was something special. When the experienced Dr Hermann Würster tried to fly the original prototype on 1 August 1938 from the company airfield, he carried only enough fuel and water for a fifteen-minute sortie. Even then the instrument panel warning lights lit up like a Christmas tree almost as soon as the little single-seater left the ground, and he landed again immediately with the undercarriage still locked down.

Later test flights by Würster and his successor as BFW chief test pilot, *Flugkapitan* Fritz Wendel, saw all kinds of vices emerge, threatening the life expectancy of anyone taking the controls. Where 109 pilots had to deal with appalling visibility over the nose while on the ground, those tackling the 209 had to fly almost completely blind until they could raise the tail on the take-off run. When they did, the aircraft liked nothing better than to treat a change in pitch as an invitation to a vicious ground loop. If the pilot made it leave the ground after its long take-off run while still under partial control, this was purely temporary. If he banked the aircraft to alter course, the nose would pitch downwards, and if the angle of bank increased too far at full throttle, the 209 would flick over on to its back. The cooling system was not up to the job, and the engine would run roughly, while airframe weaknesses were worrying. The landing gear could only be lowered below 155 mph, but any faster manoeuvres usually caused the main wheels to drop out of the undercarriage bays anyway.

The filler caps would flick open at high speed, leaking oil from the hydraulics would cover the windscreen, and exhaust gas seeping into the cockpit forced the pilot to breathe oxygen for the whole flight. At low speeds the controls were sloppy, and at higher speeds they were heavy and tiring. Finally, for the slightest chance of a successful landing, the air had to be still. Even the gentlest touchdown normally triggered a vicious swing, and touching the brakes a fierce ground loop. The normally understated Wendel described it as a 'vicious little brute' and 'a monstrosity.' Nevertheless, strenuous attempts were made to tame these vices, and Würster managed to fly the second prototype on 8 February 1939 and survive the experience.

On 4 April, Wendel tried to repeat Würster's success, but engine failure led to his near death experience on his airfield approach and the machine was written off. Time was pressing and so the Daimler-Benz engine prepared for the record attempt was put into the first prototype instead. Then the weather broke and days passed waiting for any improvement. Hubert Bauer, manager of prototype construction at BFW, described what happened on the record attempt. 'In the early morning hours of April 26 1939, the Me209 yet again took off from the company field of Messerschmitt AG in Augsburg. It was another attempt by Fritz Wendel to break the world's absolute speed record.... The firm's aerodynamic engineers had figured that with its stubby wings and its Daimler-Benz DB601 engine, the Me209 would be able to attain a speed which would break the current record. The fact that the machine was quite difficult to fly, especially when landing on the small airfield, came with the territory. The machine demanded everything a pilot could give it.

But this attempt failed as well – the weather was too uncooperative. When Professor Messerschmitt arrived there was a situational analysis meeting in his office. General Udet, who had the final say for the RLM, agreed with Professor Messerschmitt to call off the attempt. No more flights, the course was to be dismantled and the specialists sent home. It was over for the record flight. With heads hung low and many a curse on their tongues, the people involved slowly filed out of the conference room. The mood of the players sank to zero. All the efforts had been for naught, the risks which had been taken were inconsequential, and the ground crew had laboured night and day on the plane and its engine for nothing.

'Then [Saint] Peter cast a sympathetic eye on the scene. Around noontime he parted the clouds and the sky turned a spotless blue. I couldn't stay in my office any longer; by 3 o'clock that afternoon, I knew that such ideal weather had to be exploited. But none of the senior managers were in their offices, and I wasn't able to get ahold [sic] of anyone by telephone to reverse the decision of that morning. I was able to get the director of test flying on the line, but that wasn't enough. I made a judgment call and, taking responsibility for the consequences, ordered another flight be made ready – so long as Fritz Wendel would volunteer to fly later on that day. Fritz told me that he was willing, and landed around 1800 hrs – everything had gone well. We waited a long time at the *Hotel Drei Mohren* [Three Moors] until around midnight when the results had been tallied and recognized by the gentlemen from the FAI. The absolute speed record of 755 km/h was ours.'[2]

When this faster speed was confirmed by the FAI, the documents showed the designation 'Me109R'. This was the greatest deception of all, to stretch credibility by insisting this lethal racing projectile was a modified version of the Luftwaffe's standard fighter. Only experienced test pilots could handle it and the idea of wartime operations beggared belief. Had it entered service, it would have caused so catastrophic a casualty list that the Battle of Britain would have been won without the RAF having to fire a shot.

The intention was to make potential opponents quake in their flying boots at its massive speed advantage, but this would dissipate once it failed to show up in combat. Nevertheless, a half-hearted attempt was made to tame the 109R/209 by fitting an ordinary DB601 engine and lengthening the wings to improve stability, but this gave no real advantage. During 1940 a final twist to the deception was promoted by photographing the fourth prototype in camouflage and military markings, but this simply delayed the inevitable. When the 209 did appear, it was a genuinely new design intended as a possible Bf109 replacement (see Chapter 23).

By then, Professor Heinkel was incandescent with fury. He realized the lethally specialized 209 had only beaten the best speed of his He100 by a trifling margin of 5.3 mph. He also found the FAI rules specified record flights had to be flown at less than seventy-five metres above the ground (roughly 246 feet), but no mention was made of the height above sea level. The He100 had reached its record speed at Oranienburg, some fifty metres above sea level, while the Me209 had done so at Augsburg, around 500 metres above sea level. The differences in air density at the two sites gave Messerschmitt a speed advantage of some 15 mph. Heinkel wanted to make another record attempt over the same railway line as the Bf109 of 11 November 1937, redressing the balance and restoring the He100's pre-eminence.

Ernst Udet stepped in and put a stop to the whole business. At the time it seemed possible the aviation world had accepted the Bf109R/Me209 deception, but seeing the record broken so quickly by a plane from a different maker with no in-service Luftwaffe equivalent risked exposing the whole setup. Furthermore, Heinkel's hopes of having his fighter join the Bf109 on the wartime front-line were soon dashed. In Nazi Germany, high-level decisions allowed no appeal, and maintaining pressure would have jeopardized Heinkel's orders for aircraft like the He111 medium bomber, equally vital to German plans for the coming air war.

---

2. (Ebert/Kaiser/Peters, *op.cit.*, pp. 144–45).

Chapter Nine

Weapons

During the summer of 1934, two officials from the Air Ministry's Operational Requirements Branch were busy at the Army firing range at Shoeburyness, near Southend on the Essex coast. Squadron Leader Ralph Sorley, head of the Branch, and Major H.S.V. Thompson had set up an old metal airframe as a target. They then assembled a battery of eight Colt Browning machine guns, each able to fire twenty rounds a second, 400 yards away. They then fired a series of two-second bursts with all eight guns, ripping the hulk apart.

For Sorley, the fury of the display confirmed his predictions. Serving as a pilot in the First World War, first with the Royal Naval Air Service on anti-shipping strikes and later in the Royal Air Force in Iraq, earning both a DSC and a DFC, he realized the challenge represented by the new all-metal bombers was beyond the RAF's existing fighters. They relied on one or two machine guns mounted on the

The Bf109G still depended on cowling guns to back up the nose cannon, but heavier weapons needed bulges in the panelling to provide room for the gun breeches. *(Author photo at Smithsonian)*

engine cowling, synchronized to fire through the propeller arc, which reduced their rate of fire.

His wartime experience convinced him they would not inflict enough damage to down enemy bombers. Assuming a fighter could catch one of the new machines, still a forlorn hope in 1934, even the very best pilots could only keep their sights on target for a couple of seconds. What kind of armament could deliver a lethal punch in that brief moment?

His calculations factored in range, rate and spread of fire, bullet weight and calibre, and predicted an average of 266 hits with rifle calibre rounds might allow a moderately skilled pilot to shoot down a metal skinned bomber. Delivering this punch within two seconds meant that two guns, or four or even six, would not be enough. Only eight unsynchronized guns could potentially deliver 320 hits between them in two seconds, enough to do the job. Now his practical test had shown this ideal was possible.

Of course this was a long way from demonstrating this in the air. First World War fighters used cowling mounted guns for a reason. Machine guns tended to jam under combat stress. Many pilots carried hammers to free the gun mechanism, but only when these were within easy reach. Mounting guns on the engine cowling made sound sense, as this was the strongest part of the aircraft structure. However there was only space for two at the most.

For Sorley's eight-gun armament to work, two conditions had to be met. The guns had to be mounted in the wings, and be reliable enough not to jam. They also needed space for the ammunition they fired. A minute's burst in combat would consume almost 10,000 rounds, which meant extra weight as well. Which weapons would give him the qualities needed?

He discussed the problem with Captain F.W. Hill, Senior Ballistics Officer at the Aeroplane and Armaments Establishment at Martlesham Heath in Suffolk. Using the establishment's firing range they compared the Vickers guns used by wartime British aircraft with the Japanese Kirileji gun, the Danish Madsen and French Darne guns, the American Colt Browning weapons and the 20 mm Hispano cannon firing explosive rounds favoured by the French. Tests showed the American Colt Browning .303-inch machine gun was the most reliable, with an average of one jam per 15,000 rounds, and this also had the highest rate of fire. As an additional benefit, the guns could be made to fire British ammunition from the huge stocks of .303 bullets from the First World War.

To ensure future supplies, Major Thompson visited the USA in September 1934 to see the Colt Weapons Corporation HQ at Hartford, Connecticut. He received a licence for the Colt Browning gun to be produced in Britain, and for detail changes to enable them to fire the British rimmed .303 rounds. On his return, tender documents were drawn up and sent to the main British weapons makers, with the exception of the largest and most experienced, Birmingham Small Arms or BSA.

Once this omission was discovered, BSA were sent the documents and their representative, James Leek, was asked to quote for producing 1050 guns at a rate of fifty per week. With eight guns per fighter, this suggested a maximum total production of thirty-two fighters a year! Fortunately Leek had visited Germany and noticed the level of weapons production in progress there. He suggested at an Air Ministry meeting that quoting for 2000 guns per week, equivalent to 13,000 fighters a year, would be a more sensible target. Of course, his suggestion was ignored, but by June 1940 that figure would indeed be ordered.

In addition to numbers, reliability and rate of fire, other problems surfaced. The guns would have to work under very harsh conditions. With combat occurring at higher altitudes, they were likely to freeze. Supermarine cured this problem by adding ducts to divert warm air from the wing radiators to internal boxes around the guns. In the meantime, an extra problem was the wet cold encountered when climbing through cloud.

This was made worse by cordite propellant used in British ammunition rather than the nitro–cellulose used by most other countries. It was less susceptible to tropical heat, but tended to detonate in the breech of a hot gun during a prolonged burst – fatal for a fighter. So the gun mechanism was made to leave the breech open and the chamber empty at the end of each burst for better cooling. The gun lubricant was mixed with paraffin to lower its freezing point, but this made the mechanism vulnerable to corrosion. Moisture was kept away from the guns in the first place by sealing the muzzles with fabric patches, blown away when the guns were fired, but replaced as the fighter was rearmed on the ground.

Two other changes increased the damage done by each round when it hit the target and help the pilot ensure more hits were scored. Fighters sent to attack the large and highly inflammable Zeppelin raiders of the First World War had used special Mark IV Buckingham incendiary rounds. These ignited when fired and left a clearly visible smoke trail all the way to the target, which the pilot could follow and use to correct his aim while firing a burst.

Modern incendiary bullets had a much more difficult job to do. Instead of simply piercing a fabric gas-bag and setting light to massive quantities of hydrogen inside, they would have to penetrate the aircraft's metal skin panels and then the casing of a fuel tank before provoking a really lethal blaze. Two European inventors, a Belgian ammunition manufacturer living in Switzerland named Paul René de Wilde and his Swiss collaborator, Anton Casimir Kaufman, a clerk and amateur watchmaker, showed their incendiary rounds to Air Ministry representatives in December 1938. The Ministry commissioned practical tests, carried out on 11 January 1939. Guns loaded with sample rounds were fired over a range of 200 yards at a target of a sheet of duralumin alloy the thickness of an aircraft skin panel, with a can of petrol six inches behind it. Nine rounds were fired. Three burst through the alloy skin panel and the petrol can and set the fuel on fire. The delighted Air Ministry paid for a licence to produce up to 35 million rounds.

Ultimately, the truth proved disappointing. Another Army officer, Captain C. Aubrey Dixon of the Bedfordshire and Hertfordshire Regiment, had been posted to Woolwich Arsenal in London with responsibility for small arms ammunition for all three services. As an observer at the trials, he was ordered to take sample rounds produced by the inventors and pass them as suitable for mass production. This was doomed from the start. Dixon asked Kaufman what production tolerances were acceptable and was appalled to be told 'none'. Possibly practical for a watchmaker, this meant every production round would have to be handmade to be absolutely perfect, a complete nonsense for a nation building up ammunition stocks for a world war.

Fortunately Dixon made progress by disobeying orders. After talking to his brother, a wartime RFC pilot, he tried to make a working incendiary bullet during his time at Woolwich. The obstacles were daunting. His research budget allowed just £5 on any single project, and official opinion said the task was impossible in any case. When he tried to repeat the tests with ten hand-assembled rounds, two exploded in the barrel of the gun, wrecking the weapon. Dixon made a series of detailed modifications to eliminate these faults, but the improved bullets were not sensitive enough to ignite when they hit the target.

By now everyone was queuing up for a shot at poor Dixon. 'De Wilde was furious and blamed Dixon, accusing him of deliberately tampering with the design with the aim of impairing performance; he had learned from Kaufman of Dixon's prior experiments and saw him not as an ally but a rival.'[1] The Ministry were also suspicious of his changes and the effect they had on the bullets' performance. When he insisted the original rounds had been completely unsuitable for mass production 'With no suitable alternative in sight the Air Ministry placed a firm order for a hundred thousand rounds and a list of acceptance trials was issued.' Dixon began these tests on 23 June 1939, but barrel after barrel was wrecked from premature detonation.

Ironically, this hopeless assignment gave Dixon a smoke screen to hide his own experiments from superiors, but he was running an appalling risk. With a ban on further tests imposed by Brigadier A.E. Macrae, Chief Superintendent of Design at Woolwich, he had to work at night without official authorization. His team included an industrial chemist, J.S. Dick, who searched for a slow-burning compound which would ignite on striking an enemy aircraft and remain ablaze long enough to penetrate a fuel tank and set its contents ablaze.

The search went on up to the outbreak of war with no success, until in a moment of frustration, Dixon kicked an apparently empty tin under a laboratory bench, but found it was full of an anonymous grey powder. Dick revealed this was a compound tested some years before, but rejected as too sensitive. To Dixon this seemed highly promising, but first results were disappointing. Then they tried a final engineering solution. They fitted a metal insert into the nose of the bullet and a ball bearing at the

---

1. Ralph Barker, *Unknown Stories from the Battle of Britain*, pp. 6–7.

rear to press the pyrotechnic mixture against it. As the bullet hit its target, its nose would break up and the ball bearing would detonate the incendiary chemical. After experimenting with different amounts, they arrived at a promising compromise. They built six targets from sheets of aluminium alloy and cans of petrol, and began proper firing tests, after working hours on the empty range. On firing each round, the bullet ignited every time, smashed through the metal and set the petrol ablaze.

Dixon and his team had triumphed, but final victory went to De Wilde and Kaufman. The Air Ministry continued to call the bullets 'De Wilde rounds', to convince the Germans they faced the original ineffective ammunition. De Wilde and Kaufman were paid a total of £20,000. Dixon was paid just a pound for licensing the Americans to produce billions of rounds. After the war, he and Dick were rewarded for their success at turning the completely ineffective Swiss ammunition into an effective weapon, with a payment of £3000 – between them.

With eight guns on each aircraft, the new 'De Wilde' rounds were normally loaded in one gun to attack the fuel tanks of enemy aircraft and provide a flash as each round struck the target to confirm the pilot's aim was accurate. Two guns were loaded with the old Buckingham rounds, providing a smoke-trail 'tracer' effect, and the remaining five fired ball and armour piercing rounds. More Dixon incendiaries were loaded towards the end of each belt to tell the pilot he was running out of ammunition.

All this work was being done by enlightened amateurs in the British tradition. However, the Germans seemed to have frozen their fighter armament around 1917, with no sign of a thaw thereafter. The problem, as with so much else affecting fighter development, was timing. The official *Rüstungsflugzeuge IV* specification appeared late in 1933, months after the Nazis had finally gained power, but well before Sorley and his colleagues had developed weapons to tackle modern bombers. As a result, Messerschmitt had to include a pair of machine guns, mounted on the engine cowling and firing through the propeller arc. As an aerodynamicist rather than an armaments expert, Messerschmitt was happy to have a simple and efficient wing, but the additional delay suffered by the British allowed them to design multiple wing guns into their fighters from the start.

Not that this was straightforward. Sidney Camm's Hurricane had a thicker and more conventional wing, so the guns could be fitted in a closely packed four-gun battery in each wing, which simplified reloading between sorties. The thinner wing of the Spitfire meant guns had to be spaced irregularly from root to tip, to make full use of limited internal space. When the Germans finally realized the British were fitting fighters with eight unsynchronized machine guns apiece, compared with their own twin MG17 7.92 mm synchronized guns, adapted from the German Army's standard machine gun and mounted on the engine cowling, they were appalled. In the case of the Bf109, already the *Reich*'s most promising fighter, there was no real chance of squeezing extra guns into the wing without compromising its performance. Nevertheless, they tried hard. They tried fitting a gun on a special engine mounting to fire through the hollow spinner without synchronization. Indeed, the hole in the 109 spinner misled many into believing a cannon was there from the start. The actual solution tried on the Bf109B was mounting a third MG17 on the engine with its barrel aligned with a tube passing through the hollow propeller shaft, so the airstream passing back through this tube could cool the gun. This let the gun fire at its maximum rate, increasing firepower by more than half. Unfortunately, the weapon tended to jam, and inaccessibility made it worse. Most production 109Bs had the gun removed to save weight, though they kept the hole through the spinner.

Another option was to fit a single MG17 into each wing. They had to be sited with the breeches level with the fighter's centre of gravity, to avoid having to retrim the aircraft every time they were fired. That proved a major challenge, but finding room for the ammunition proved almost impossible. The solution used, up to the Bf109E, was to avoid the bulk of conventional ammunition boxes by running the belts out to the wingtips, over rollers and then all the way back to the wing roots and then back to the guns. The belts held 500 rounds per gun, while the cowling guns with proper ammunition boxes had 1000 rounds apiece. All four guns were harmonized so bullets converged at 400 metres ahead of the fighter.

The installation was tested at Travemünde on the eleventh Bf109 prototype from the middle of May 1937, so that production aircraft fitted with the wing machine guns could be delivered from the beginning of August.[2] A series of long term test firings was carried out on the ground and in the air, amounting to more than 78,000 rounds, up to 7 October 1937, when the aircraft was returned to the factory.

These wing guns were outside the propeller arc, so they more than doubled the 109's firepower, but the long-term aim was to fit 20 mm cannon firing explosive shells to inflict heavier damage. To avoid shaking the wings to pieces, a specially modified version of the MG FF, based on the Swiss Oerlikon anti-aircraft gun produced in Germany under licence, was developed to lessen the stresses imposed when fired. 'The gun was very light and very compact, but the design of the breechblock did not incorporate anything to lock it when the shell fired. The explosion in the breech that fired the shell also opened the breech for the next one. All of this resulted in a very poor muzzle velocity.'[3] In addition, it had other limitations. Its sixty-round drum only held enough for seven seconds firing and the drums had to be fitted into the slender Messerschmitt wing from below, with a drag-inducing bulge to provide room.

To test the guns, a Bf109B was fitted with a starboard wing carrying the MG FF and a port wing allowing the choice of an MG17 or another MG FF fitted in its place. It was flown to Travemünde where ground-based firing tests began on

2. Test data from Radinger and Shick, *op.cit.*, pp. 46–7.
3. Len Deighton, *Fighter*, p. 99.

14 October 1937, but the rows of rivets connecting the forward cannon mount to the wing ribs were shaken loose by intense vibration. Poor riveting by BFW was blamed, repairs were carried out at Travemünde and testing resumed.

During the course of the month, a series of air test firings was also carried out and after twelve of these, rivets at the attachment points of the forward cannon mounts were shaken loose again. Clearly a redesigned mounting would eliminate the problem, but the rivets were replaced with screws so tests could continue. Finally, after nine more air firings and 1999 rounds expended, torsion wrinkles and other signs of damage were found in both wings. The tests were stopped, the aircraft was sent back to BFW by rail, and the C and D series machines were fitted with an MG17 in each wing instead. Not until May 1938 was a modified wing fitted with cannon tested and cleared for the Bf109E.

The lower muzzle velocity MG FF/M version fired a heavier shell with a thinner casing to cripple the target. The M stood for the FFM *Minengeschosspatrone*, [mine shell] and warnings were stencilled on the ammunition drum access panel, as the rounds were not compatible with normal MG FF ammunition. Each shell had thinner walls and a heavier explosive charge with a self-destroying fuze. The end result was a heavier punch against aircraft and a gun which weighed less, with lighter breech slides and a weaker recoil spring.

The cannon were positioned on two mounts, each 2.28 metres from the centre line of the fuselage, with the ammunition feed on the wing root side of the weapon. Springs forced each shell into the breech. 'The empty shell cases were ejected from the weapon opposite the ammunition feed and were collected in a shell casing area, whence they could be removed through an access panel. The cannon were aligned so their trajectories crossed those of the cowling machine guns 200 metres ahead of the aircraft.'[4] During the Battle of Britain, many German cannon shells exploded on hitting the skin panels of British aircraft, perforating the skin rather than blowing large holes in it or penetrating deeply enough to cause lethal damage. All the same, the 20mm cannon shell weighed fourteen times as much as a Browning bullet, but the supply was much more limited.

Later trials were conducted for a belt feed for the MG FF cannon during the development of the Bf109T carrier fighter variant, but persistent reliability problems caused it to be dropped. This was the last attempt to fit guns into the impossibly cramped inside of Messerschmitt's 109 wing. After this, weapons were limited to cowling machine guns, under-wing gondola mountings and a nose cannon from the 109F onwards, when the vibrations associated with engine mounting had been solved.

The MG FF/M was fitted in the nose of the first 109Fs, where stress relief measures to protect the wing structure made its use straightforward. Later,

---

4. Radinger and Shick, *op. cit.*, p. 88.

From the Bf109F onwards, cannon were mounted between the cylinder banks of the engine to fire through the spinner. *(Author photo)*

Mounting cannon inside the Spitfire wing proved unexpectedly difficult as in the case of this early Mark 1. *(via Alfred Price)*

Mauser developed the MG 151/15 and MG 151/20 cannon, which were genuine improvements. These weapons were essentially the same gun firing different calibre ammunition through the appropriate 15mm or 20mm barrel. They were cocked and fired electrically, with electric cartridge detonation, and had a 30 percent greater muzzle velocity than the MG FF with an equivalent boost in its rate of fire. They were much more reliable and used as nose cannon on most 109s after the 109G-6.

Some cannon were slung in underwing gondolas. This allowed muzzle velocity, rate of fire and ammunition stocks to be increased at the expense of performance and controllability. In the end, they were confined to attacks on US daylight bomber formations, but once these had fighter escorts, the heavily armed 109s had themselves to be escorted to and from their attacks by conventionally armed 109s to protect them from Allied fighters.

By this time, the cowling guns had been replaced by the heavier 13 mm MG 131, more reliable than earlier weapons. The heaviest gun of all fitted to the 109 was the advanced MK 108 cannon, which fired a 30mm shell with massive destructive power, in spite of a lower muzzle velocity and rate of fire than earlier cannon. This replaced the Mauser nose cannon and those in the underwing gondolas of the bomber-destroyers, though pilots found it unreliable and prone to jamming. Finally, as the need for even heavier weapons against US daylight bombers became more urgent, some 109s carried the massive 21cm (8.25inch) *Bord Rakete Sondergerät* [Special Rocket Apparatus] which fired a mortar shell electrically from a launching tube. A direct hit on an enemy aircraft was devastating, but the launchers so crippled the 109's performance in fighter to fighter combat, that pilots needed a control button to jettison the system if forced to defend themselves against enemy fighters.

Clearly armament was the biggest weak spot in Messerschmitt's fighter design. Even then, pilots with sufficient experience to aim properly and the courage to approach close enough to their targets, could still use the single nose cannon and twin cowling machine gun combination to bring down opponents with chilling ease. It might have been some consolation had the Germans been aware of the equally frustrating, though ultimately more successful struggle of their RAF opposite numbers to fit cannon to the Spitfire.

By the end of the Battle of Britain it was clear that Sorley's original estimate of 266 rounds to cripple an enemy bomber was far too optimistic. Fitting self-sealing fuel tanks and increasing amounts of armour plate meant that in some cases it had taken as many as 4500 hits to bring down a bomber, and the need for a more powerful gun firing explosive shells was now as much a British objective as it was a German.

Ironically the weapon they chose had been developed from the Oerlikon which gave Messerschmitt so much trouble. The Hispano cannon was based on the same Oerlikon FF S Swiss anti-aircraft weapon as the German MG FF, and shared with it the limitation of a sixty-round ammunition drum. Only in the late 1930s did Hispano engineer Mark Birkigt develop a revised gas-operated action, so the passing of a shell past a port cut in the side of the barrel allowed gas to enter the chamber and press a piston to unlock the bolt. Residual pressure in the barrel drove the bolt backwards and locked it for the next round to be fired, so the gun could be made much lighter.

It also raised the muzzle velocity and the rate of fire by some 200 rounds per minute over its German equivalent, and it was intended to fire through the engine drive shaft and spinner of a single-engine fighter like the French Morane-Saulnier MS406 or the Dewoitine D520. Other improvements were under way like larger calibres, heavier shells and belt-fed ammunition, but all of these fell victim to the German victory in France in summer 1940.

In the meantime, the British had already negotiated a licence to make the gun themselves, and they developed a belt feed for the shells so the weapons could be mounted in the nose of twin-engine fighters like the Whirlwind and Beaufighter, and in the wings of the Spitfire and Hurricane. Nevertheless, mounting the guns in the Spitfire wing proved difficult. The only way they would fit inside the wing without large drag-inducing blisters was to mount them on their sides, but ejected shell-cases tended to bounce back inside the wing and jam the breech mechanism. Deflector plates and rubber pads failed to cure the problem, and stoppages were common, particularly under combat stresses.

Though the much greater hitting power of cannon was a definite incentive, Dowding was wary of meeting the inevitable teething troubles at the height of the Battle of Britain. To avoid this, he sent the first cannon-armed Spitfires to an experienced unit, 19 Squadron, in the quieter surroundings of 12 Group. He explained to the Secretary of State for Air, Sir Archibald Sinclair, 'The present situation is that the guns of about six Spitfires in No 19 Squadron are working satisfactorily, and the defects in the others will probably be rectified in about a week or ten days.

'I quite realise that information concerning the fighting qualities of the cannon Spitfire is required as early as possible and I will take the first opportunity of getting it into action; but I am not at all keen on sending it up against German fighters since it will be extremely badly equipped for that task…, because it has only two guns and even the Me109 has two cannon and two machine guns. Furthermore, it has fired off all its ammunition in five seconds.

'So you will see that the existing cannon Spitfire is not an attractive type, but it has been necessary to produce it as an insurance against the Germans armouring the backs of their engines. They have not done this yet, their engines are still vulnerable to rifle-calibre machine-gun fire, and therefore the eight-gun fighter is a better general fighting machine than one equipped with two cannon only.'[5]

Dowding soon had the information he needed from the August combat reports. 'During the combat on 16 August, both cannon functioned properly on only one of

---

5. Letter of 24 July 1940, quoted in Price, *The Spitfire Story*, p. 77.

Late model Spitfires like this Mark 24 were fitted with four 20mm cannon to deliver a devastatingly effective punch. *(Author)*

the seven Spitfires which engaged; on the 19 it was none out of three, on the 24 it was two out of eight and on the 31 it was three out of six.'[6] The squadron commander suggested it was time to rethink the decision and begged for a trade-off between his cannon-armed Spitfires and some Browning variants from an Operational Training Unit until the problems had been sorted out. Dowding agreed and took action immediately. On 5 September the 19 Squadron machines were flown to Number 7 Operational Training Unit at RAF Hawarden in North Wales, and the elderly and careworn OTU aircraft sent to replace them.

The Squadron pilots were suspicious of the 'wrecks', though at least the guns worked properly. The instructors at the OTU had a completely new lesson to learn, and learn quickly. Just two days after the transfer, on 7 September, Sergeant L.S. Pilkington DFM was carrying out formation practice with a pupil pilot in another Spitfire when he spotted a Ju88 on a photoreconnaissance sortie over Liverpool. He chased it all the way to Machynlleth in mid-Wales where he forced it down to a crash landing despite his port cannon jamming after firing just three rounds and his starboard one after loosing off fifty-seven.[7]

6. *Ibid,* p. 78.
7. David J Smith, *Hawarden – a Welsh airfield, 1939–1979.*

Sopwith Camel. First World War RFC fighter with cowling-mounted machine guns. *(USAF)*

Eventually the problems plaguing the Hispano cannon were traced to its original pedigree. The French had designed it for an engine mounting to absorb the stress and vibration of firing, so its original users had not met the problems of the cannon Spitfires. Wing mountings were the only option, but did not provide a rigid enough support. When cannon were fired in combat, the wing flexed enough to interfere with the ejection of shell cases and the loading of the next round.

Detailed changes produced the Hispano Mk II with a belt-fed mechanism which avoided the 60-round drum and allowed 120 rounds per gun instead. Slightly reducing the rate of fire greatly improved reliability and the gun could now be mounted upright. Only one limitation remained. The tendency to freeze up under icing conditions meant the existing gun-heating ducts could not cope with a pair of cannon in each wing for most wartime Spitfires in Europe. Instead each wing carried a single Hispano cannon in the inner gun bay and either two Browning .303 machine guns or one .5-inch Browning in the outer bay. The heavier machine gun reduced the tendency for the outermost weapons to spread their fire when the wing flexed when turning. Where icing was not a problem, as in tropical Spitfires for the Mediterranean or Far East theatres, four cannon were fitted from the start and this option was extended to European fighters later. When the conflict ended, the single serious limitation of Messerschmitt's design widened into an unbridgeable gap.

Chapter Ten

Spitfire on the Brink

Both fighters seemed to face a promising future. As war threatened they would emerge in the nick of time. Yet their numbers would be cut by crass misjudgements. In Germany these resulted from misplaced over-confidence about the cost of the air war and the resources needed to fight it. On the British side, mistakes bordered on insanity. With world war approaching, senior officers seemed obsessed with cancelling their most promising fighter and replacing it with types later in development with relatively mediocre performance and dismal prospects. Furthermore, even this lethally flawed plan failed to satisfy the bomber lobby fanatics still insisting defensive fighters be replaced by obsolete and inadequate bombers.

The Spitfire's future suddenly darkened on 25 January 1937 when Air Ministry officials visited Supermarine's plant at Woolston in Southampton to check progress with the first production contract for 310 Spitfires signed seven months earlier. Supermarine's chief executive Sir Robert McLean had promised the first production machine in three months with five more aircraft each week thereafter. The Ministry therefore expected to see sixteen weeks production or eighty finished aircraft.

They found none. The factory did indeed contain several production aircraft, but all were biplanes, Walrus amphibians and Stranraer flying boats. Not one was a Spitfire. Of the promised fighters there were half-a-dozen part-completed fuselages, but no single wing. Where were the missing machines and how soon would they appear? This was the moment when initial support for the fighter in official RAF circles began melting away. Where Messerschmitt's Bf109 faced almost fatal official hostility, it was now the Spitfire's turn and Mitchell's design would have to fight for its life against the RAF.

Its most determined opponents included the bomber lobby whose plans would have guaranteed defeat in the Battle of Britain, but few outside the service realized that even those with less extreme views would prefer to sell the Spitfire to overseas customers than meet Britain's desperate need for fighters. Senior officers longed for unproven types late in development, crippled by problems and entirely the wrong aircraft for the imminent conflict.

How had this terrible situation arisen? McLean's over-optimistic forecasts were misguided but logical. The Spitfire's production delays were caused partly by his company, partly by the Government, partly by British voters and partly by Reginald Mitchell, though for the moment McLean was the clearest target. He first clashed with the Ministry as the contract was drafted when he insisted Supermarine could build all the aircraft without sub-contractors. This was nonsense. Supermarine was a small firm with only 500 workers at its two Southampton factories. With no mass production experience, even this first contract would need massive outside help. Yet McLean faced a vicious dilemma. The Spitfire contract was a potential lifesaver, but if he warned of potential problems he risked the Air Ministry cancelling the project. Instead, he produced encouraging but unattainable forecasts.

The Air Ministry made things worse. Delighted with the performance and handling of the prototype Spitfire, they issued a revised specification (F16/36) in June 1936 detailing changes needed for the production version. There had been a reappearance of the high-speed wing flutter which plagued the Schneider Trophy racers. Bracing wires were unacceptable, so drastic action was needed. The engineers moved the vertical web joining the upper and lower main spar booms from the front of the spar to the rear, and thickened the skinning of the leading edge box and the wingtip by less than half a millimetre. These apparently trifling details increased wing stiffness by 40 percent for just 20lbs extra weight, and calculations showed this would eliminate flutter up to an air speed of 480 mph.

Incredibly, this inspired solution nearly killed the Spitfire off. Mass production realities meant the assembly process for the Spitfire wing would have to be revised before production could begin. Other changes – increasing fuel tank capacity from seventy-five to eighty-four gallons for greater range, and increasing flap travel from 57 to 84 degrees for slower approach and landing speeds – caused further delays.

The company made its own detail changes, ironically to make production simpler. Parts assembled from different components would instead be cast, forged, pressed or extruded as single items, which meant more revisions. Updating the drawings in a design office with fifteen technicians and seventy draughtsmen already busy on two other large contracts, meant losing, in the words of one draughtsman, 'About a year. One couldn't conveniently use prototype drawings for the production aircraft, there were so many changes. Though some of the production drawings might have looked the same as those for the prototype, it was much better to redraw and renumber the whole lot. I don't think a single one of the prototype drawings was used on the production aircraft.'[1]

Mitchell's design achieved its brilliant flying qualities without trying to simplify production, so making Spitfires would always be complicated. The fuselage was a simple monocoque structure of three major sections: the engine mountings, the main section from the engine firewall to the stern tail unit attachment and the stern section including the fin. The firewall was an aluminium-asbestos sandwich carrying stub spars for the wings. The cockpit was built separately and connected to

---

1. Jack Davis, quoted in *The Spitfire Story*, Alfred Price (p. 61).

the engine firewall in the fuselage jig before the rest of the skeleton was added with skin panels riveted on top.

The fuselage was placed in a second jig for holes to be bored for wing attachment bolts and for the carefully shaped wing fillets to be fitted before the assembly was painted. Separate jigs were used to assemble the rear fuselage with the integral tail fin and the tubes for the engine mounting framework.

The wing was much more complicated, built around the main spar and the massive D-shaped leading edge torsion box. The two square section spar booms were made by Reynolds Tubes, specialists in racing bicycle frames. They were slightly cranked for wing dihedral, and joined by the vertical web plate at their rear surfaces, with angles for attaching the leading edge skin panels riveted along their front faces.

The strong leading edge skin was made in two sections, each formed on a stretching press from a single alloy blank and supported on twenty-one nose ribs. It was riveted to the spar booms with each rivet hammered into a countersunk hole and the head shaped with a compressed-air chisel. Wing components were assembled in the main wing jig, with each pair of wings built together with leading edges downward. The rib trailing edges were riveted to the rear spar in sections connected by a duralumin trailing edge strip. After the radiator bay and wheel wells were fixed to the framework, the upper surface skins were attached by drilling through the ribs and riveting them into position. The under surface plating was harder to apply and fitted in several sections.

Complexity was inevitable for modern fighters, but the Spitfire's problems were compounded by its elegant curves. Bending aluminium alloy sheet in one plane is easy, but forming double curves needs a powerful hydraulic press and Supermarine lacked tools as well as experience. The rearmost top panel of the fuselage where its convex curve met the concave curve between fuselage and fin was especially intractable, and the only machines to shape these relatively thick panels accurately enough were huge presses. When Supermarine worker Bert Axell saw them in action at Pressed Steel, 'I remember asking the boss man if it would be possible to produce the leading edge in one piece. "No problem at all" was his answer, except for the basic width of the steel which at the time was beyond rolling mill capacity.'[2]

For final assembly the fuselage was lowered on to two jacking points with the tail held down by a canvas loop. The engine mounting framework and the cockpit were bolted into place, the controls connected and the pilot's seat fitted. The oil tank was placed in position, oil lines connected and the Merlin engine bolted to the mountings. After fitting and connecting the fuel tanks, the wings were placed on special trolleys while the wing fixing bolts were tightened, the heating and control

2. (Axell to Jack Davis, Davis papers AC93/14/19, RAF Museum Hendon, quoted in McKinstry, *Spitfire: Portrait of a Legend*, p. 72).

systems connected and the oil, air, hydraulic and electrical systems connected and tested. With the aircraft still on the jacks, the undercarriage was raised and lowered. Finally the propeller was fitted and the tanks filled, ready for final inspection and the first test flight.

The complex process was a huge challenge even for a skilled workforce in a single factory, but sharing work over several sub-contractors made it still more difficult, and the Air Ministry was losing patience. On 2 February 1937 Air Minister Lord Swinton held a meeting where the Director of Aeronautical Production complained the abortive visit to Supermarine eight days earlier showed 'The trouble *which had always been anticipated* [author's italics] over production of the Spitfire has now come to a head....'[3] It was suggested General Aircraft of Feltham in Middlesex, contracted to make Spitfire tail units since November 1936, should produce wings as well. A week later the Air Member for Supply and Organisation [AMSO], Air Chief Marshal Sir Cyril Newall, himself a dedicated bomber enthusiast, insisted the situation was extremely serious and 'The firm frankly admitted they had entirely underestimated the magnitude of the task.'

On 16 February McLean refused the Air Ministry's suggestion to pass work to subcontractors like General Aircraft, but as no completed Spitfires appeared, he had to agree that 80 percent of production work should go to other companies. Samuel White of Cowes would make fuselage frames, Singer Motors of Birmingham would make engine mountings and Folland Aircraft of Hamble produce rudders and tailplanes. The complex wings would be made by General Aircraft and Pobjoy Motors, with Westland Aircraft and the Pressed Steel Company building leading edge boxes, and General Electric the wing tips. Wing ribs would be made by Beaton and Son in disused Thames-side boathouses at Bourne End. Output from all these different companies would be sent to Supermarine and assembled into finished Spitfires in two new hangars at Eastleigh Airport where the prototype had first flown. They would allow Supermarine to produce four Spitfires in December 1937 with six more fighters completed during January and February 1938, eight in March and ten apiece in April and May.

Even this failed to clear the bottleneck. Supermarine were moving from traditional methods to mass producing modern military aircraft. This was harder for smaller companies, many with no direct experience of making any aircraft at all and the Air Ministry's objectives remained elusive. At least McLean now had someone else to blame. For delays in wing production he blamed General Aircraft. They in turn complained Supermarine were late delivering components and drawings. Supermarine countered by blaming the need to work with a subcontractor not in direct contact with the drawing office and claimed no delays with parts they supplied themselves. They insisted the decision that Spitfire components should

---

3. Price, *op.cit.*, p. 61.

be subcontracted was made by the Air Ministry and therefore the faults lay with them. This was too much for both General Aircraft and the Air Ministry, who insisted they had investigated General Aircraft's production methods and found delays were caused by drawings and parts from Supermarine which were late and inaccurate. They suggested General Aircraft make one wing of each aircraft with Pobjoy Motors producing the other. Supermarine would build wings at Woolston, ,but this delayed fuselage production. As a last desperate measure, the delivery date for the first production Spitfire Mark 1 was postponed to early 1938 and later to August 1939, within weeks of war.

Newall blamed Supermarine for failing 'to exercise sufficient control' after previously criticizing them for trying to exercise too much! One problem was the varying experience of sub-contractors. Even aircraft makers had never tackled anything as complex as a Spitfire, and this made a huge difference. Bill Newton of Supermarine noted that an experienced contractor sent in fifty-seven queries during an eighteen-month period, compared with 15,000 from an inexperienced company.[4]

Subcontractors claimed Supermarine parts did not fit and General Aircraft claimed £40,000 compensation for delays caused by Supermarine's inaccurate drawings. The Air Ministry was appalled to find each Spitfire would now cost £7,000 rather than the £4,500 specified in the contract, with the more conventional Hurricane costing £4,000 apiece. After tough negotiations, the price remained 50 percent higher than the Hurricane at £6000 each, but no Spitfires had emerged by June 1938, though three 'might be ready' by the end of July. So desperate was the need for faster wing production that Supermarine engineers flagged up clause (xiv) of the specification (F16/36) which stated, 'At those parts of the mainplane or tailplane where riveting of the covering would be particularly difficult to do, the ribs may be made with their lower booms of wood and the covering attached by stainless steel wood screws.'[5] This implied wing production could be simplified by replacing duralumin skins from the main spar to the trailing edge by 2mm plywood, cutting weight by 7 percent, but tests showed torsional resistance would fall by an unacceptable 8 percent. Instead, wing ribs might be made from metal reinforced plywood, but by then other improvements made them unnecessary.

A second factory visit in summer 1938 found seventy-eight completed fuselages, but only three sets of wings. The first two production machines finally appeared in July 1938, more than two years after the prototype. Two more were delivered in August and two more in September. Production problems seemed to be easing as thirteen were handed over to the RAF in October with forty-nine Spitfires in RAF

---

4. Quoted in McKinstry, *op.cit.*, p. 74.
5. *Spitfire*, Morgan and Shacklady, p. 46.

hands by early 1939. However, they lacked heated gun bays so weapons could not be fired at the altitudes where they would have to fight!

Time had now run out. Something had to give and the shock swept away the top men at both company and Ministry. Sir Robert McLean was promoted to run the Vickers Group before moving to the entertainments industry. Air Secretary Lord Swinton had backed the Spitfire programme from the start. Delays were now a corrosive political issue and Swinton was attacked, though he threatened to summon the Vickers board for Air Council questioning to explain their failures.

In February 1938 he had to admit serious problems to the Imperial Defence Committee, and criticism grew after the Labour Opposition's U-turn, calling for faster rearmament after years of fighting every such move. On 12 May the Opposition opened the Air Estimates debate in Parliament with the unlikely support of Winston Churchill, long a supporter of increased weapons production. Finally, Prime Minister Chamberlain called for the resignation of Lord Winterton, Commons aviation spokesman, and of Swinton himself.

It was a miracle the whole Spitfire project did not vanish with him. In the words of Wellington about Waterloo, it would be 'The nearest run thing you ever saw in your life.' At the very time when Spitfire production was increasing, the Air Ministry decided to abandon it. They would grudgingly allow Supermarine to finish the 310 Spitfires already ordered before switching to Bristol Beaufighters. These would be followed by the next generation of RAF fighters, the Hawker Tornado and Typhoon, and the twin-engined Westland Whirlwind.

While the Air Ministry had every right to tackle the Spitfire's appalling production problems, they were wrong to think any modern design would be easier to produce.

The Bristol Beaufighter, powerful but cumbersome and no substitute for the Spitfire as a dogfighter. *(USAF)*

Their suggestions for Spitfire replacements might seem attractive to politicians, but would have been military disasters. The Luftwaffe would almost certainly win the coming air battles, with incalculable implications on the war.

Swinton resigned on the day the first production Spitfire flew – eight months late. Newall, still yearning for more bombers, was now more powerful and the fighter's future was grim. Only a series of surprises involving the chief architect of appeasement, a lawyer politician seen as lazy, another seen as incompetent, a millionaire motor manufacturer who found aircraft production an impossible challenge and a ruthless Canadian press baron would see it reach front line squadrons.

The first of those helping to save the Spitfire was ironically, Neville Chamberlain. The Munich crisis of September 1938 had called time on appeasement and with no pause in Nazi expansion it was clear massive rearmament was urgently needed. In addition to a sharp change of course, Chamberlain went further. He realized defensive fighters were needed much more than Newall's longed for bombers.

Next of the fighter's saviours was another politician, Sir Thomas Inskip, widely seen as lazy. *Time* magazine in the US mentioned Chamberlain moving 'His friend, slow-moving Sir Thomas Inskip, from the post of Minister for the Coordination of Defence, where everyone agreed he had been a first-class failure.'[6] Fortunately Inskip was sharper and more energetic than that. He demolished RAF claims that the country needed fewer fighters and more bombers; given the lamentable performance of current bombers this was monumentally misguided. The RAF Official History referred to the Ministry view that 'Fighter defence must…, be kept to the smallest possible number. It was, in the view of the Chief of the Air Staff …, only a concession to the weakness of the civilians who would demand protection and cause the Cabinet and even the Secretary of State for Air to do likewise. These demands, he insisted, must be resisted as far as possible.'[7]

Instead, when the Government spotted German rearmament and realized they lacked the means and the public support to reply in kind, they tried to bury the Luftwaffe threat under paperwork. A series of plans from A to J inclusive specified blueprints for an RAF to fight the coming war. With the bomber lobby still in control, all put bombers first. Scheme J for example specified sixty-four squadrons of slow and vulnerable bombers, but only twenty-six of fighters to defend the country.

Given its grip on the RAF's higher ranks, the bomber lobby could only be reined in by its political masters, and so it proved. Inskip refused to give bombers top priority. He assumed the RAF could not deliver a knockout blow against Germany, so fighters would have to defend the country against German attempts to do the same. He ordered the Ministry to produce a new plan K with greater priority for

---

6. *Time*, edition of 6 February 1939.
7. *The Official History, The War in the Air*, Webster and Frankland, i, pp. 54–5.

fighters. Newall insisted fewer bombers would impose an unreasonable burden on the fighters, but political opinion was now moving away. Scheme L shifted the balance further, by boosting fighter numbers, and scheme M in October 1938 insisted 3,700 first class fighters were vital for national survival.

Swinton's successor as Secretary of State for Air had been another lawyer, Sir Kingsley Wood, with as poor a reputation among public and Air Force as Inskip. He astonished the bomber lobby by insisting RAF raids on Germany should avoid bombing civilian targets. When a raid on the Black Forest was planned, he was alleged to have said, 'Oh you can't do that, that's private property. You'll be asking me to bomb the Ruhr next.' Whatever the truth of the remark, Wood stood up to the bomber lobby and refused to amend plan M, discussed by the Cabinet on 7 November 1938. He went further, calling for fighter numbers to rise by 30 percent, issuing production orders for half of these. This was anathema to Newall and those determined on more bombers at any price. He wrote the word 'Christ!' heavily underlined on his copy of the minutes as evidence of his distress.[8]

Regardless of the delays plaguing Supermarine, Wood ordered another 200 Spitfires before production switched to Beaufighters. He contacted Fairey Aviation to issue a provisional contract for another 300 Spitfires. This was a bitter pill for the bomber lobby as Fairey was already busy producing the Battle light bomber. While this proved lethally useless in combat, it still had top priority for the priceless Merlin engines, even ahead of the new fighters!

This was still far short of the production needed, but the government had begun planning 'shadow factories', clones of existing plants run by motor industry experts more used to mass production than painfully expanding aircraft companies. Rolls-Royce built a shadow of its Derby engine plant at Crewe to turn out Merlins. Austin Motors built a plant to turn out still more Fairey Battles, delighting the bomber lobby until Wood dashed their hopes that his policy might mean additional bombing aircraft.

On 19 May 1938 he set up a meeting with Lord Nuffield of Morris Motors, one of Britain's largest car makers, suggesting he build a huge factory to produce 1000 Spitfires for a total of £7 million! The news triggered intense criticism but Wood and Nuffield pressed on. Land at Castle Bromwich near Birmingham was bought from Dunlop for a plant to make at least sixty Spitfires a week, especially welcome from a company with mass production experience.

Construction began in July 1938 just as Supermarine were ordered to keep Spitfire output at a single shift before switching to Beaufighters. So determined was the Ministry to eliminate the Spitfire that the final 200 machines for the RAF between November 1939 and March 1940 would be turned out without engines to release still more Merlins for bombers. A thousand more Spitfires would make no

---

8. McKinstry, *op.cit.*, p. 97.

sense, so Nuffield was ordered to build 600 Westland Whirlwinds instead. His mass production experience told him switching to a new and still untried design was madness, and he refused.

Inevitably, Castle Bromwich met similar problems to Supermarine. Of Nuffield's planned output of sixty Spitfires a week, not one had been completed by May 1940 after millions of pounds of public money had been spent. With German forces in France approaching the Channel, Nuffield followed the usual tactic and blamed Supermarine, but Winston Churchill had replaced Chamberlain as Prime Minister. He appointed the abrasive Canadian newspaper tycoon Lord Beaverbrook as Minister for Aircraft Production, to blast a way through this kind of obstacle. Beaverbrook rang Nuffield on 17 May, heard his complaints and deftly manoeuvred him into resigning. He handed the factory to Vickers, Supermarine's parent company, and sent rival aircraft builder Sir Richard Fairey to the plant to assess the delays.

Fairey's report in the Vickers Archive, and quoted in detail by McKinstry,[9] seems scarcely credible. Although national survival was in real peril, industrial disputes were common. In Fairey's opinion, 'The greatest obstacle to an immediate increase in output is the fact that labour is in a very bad state. Discipline is lacking. Men are leaving before time and coming in late, taking evenings off when they think fit....' Supermarine engineer Cyril Russell said colleagues had found stoppages and wage demands which, although Castle Bromwich workers were paid substantially more than those at Supermarine, had stopped Spitfire production altogether on occasion, which in his opinion 'bordered on treason.'[10]

Fortunately, there was a Draconian remedy. Also in the Vickers Archive is a letter from accountant Alex Dunbar, appointed Managing Director at Castle Bromwich in May 1940 and sent to Vickers director Sir Frederick Yapp on 20 July. It includes the line, 'Incidentally we are sacking at least 60 Jig and Tool draughtsmen next week; we have tried to find out what they are doing but the answer's not a lemon....' Those expelled from the factory faced conscription into the army, though the problem proved persistent.

Parts of the plant were still not finished due to hold-ups from an implacable foreman who reacted to the national emergency by trying to eliminate weekend working to finish the factory. A small minority of more motivated workers caused headaches, convinced they knew better than Supermarine specialists. One fitter wanted to use iron rather than aluminium to make the aircraft stronger and a manager said the elliptical wing was unnecessary so they would redesign it. Air Ministry intervention kept them in line, but a lack of proper record keeping left parts missing on a huge scale and workers paid for overtime not actually worked.

9. *Ibid*, pp. 151–7.
10. Russell, *Spitfire Odyssey* and *Spitfire Postscript*, quoted in McKinstry, *op.cit*, p. 153.

This unproductive shambles finally responded to drastic action and wholesale dismissals, but even when Castle Bromwich output increased, the Spitfire remained vulnerable. In the final weeks of peace the Air Ministry's obsession grew more determined. Sir Wilfred Freeman, a member of the Air Council, had written to the Chief of the Air Staff in the summer of 1939, not to increase Spitfire production but to suggest how the unwelcome situation of having too many Spitfires could be rectified. He thought the increased output rate would enable Supermarine to finish initial orders by spring 1940, leaving a gap before the longed-for date to begin building Beaufighters.

Freeman suggested cutting back Spitfire production to thirty a month, not to supply the RAF with more fighters, but to meet overseas orders. He suggested four Spitfires a month could be sold overseas from October 1939 to March 1940. From then on, he predicted with relief, foreign orders could absorb all of Supermarine's Spitfire production. Given the shortage of fighters in the Battle of Britain this was an astonishing misjudgement, even by Air Ministry standards.[11] In the last days before German troops crossed the Polish border to begin the Second World War, Supermarine received a grudging order for another 450 Spitfires on the understanding that almost half would be sold 'to Dominion and Foreign Governments.'

This astonishing indifference to the RAF's needs could have been fatal. At a meeting on 11 July 1939, less than eight weeks from war, the Air Ministry's Supply Committee agreed Supermarine should complete 450 more Spitfires before turning to Beaufighters. Even at this stage they were set on replacing a fighter whose production problems were almost overcome with designs where problems and delays were still unpredictable. Fortunately the outcome of their thinking was soon revealed by events.

The Beaufighter prototype would not make its first flight for another six days and the first production machine would take another year to emerge. The Tornado fighter prototype would not fly for another two months, would suffer problems with its Rolls-Royce Vulture engine and be cancelled in early September 1941 after just three prototypes had been built. The Whirlwind was furthest advanced, having flown on 11 October 1938 powered by two Rolls- Royce Peregrines, but this too was beset by problems.

The Air Ministry ordered two prototypes for delivery in August 1938 and February 1939. Westland suffered problems with undercarriage, oil coolers, wing castings, fuel tanks and component installation. The company begged for a firm production order, but were warned the project would be cancelled unless the prototype was completed by December 1938. By working twenty-four hour days they met the deadline. Westland planned to produce Whirlwinds at their Yeovil

---

11. McKinstry, *op.cit*, pp. 141–2.

plant from June 1940, but the Air Ministry wanted to switch to another company because 'Production at [Westland's] Yeovil was reckoned impossible in the ill-equipped workshop and by a firm which could not increase its labour force prior to firm contract....'[12]

Eventually, in January 1939, Westland received an order for 200 aircraft, but demanded more factory space. They wanted the engines made to rotate in opposite directions to eliminate swing on take-off, but tests were needed to prove the complexity and delay were worthwhile. Until the results were known the Air Ministry refused to order Peregrines from Rolls-Royce. They also insisted on modifications like tropical equipment even though Whirlwinds would never serve overseas. They criticized Westland for spending too much effort on the Lysander, while Rolls-Royce was spending too much development time on the Merlin to tackle the Peregrine's niggling induction problems with carburation and high temperature vapour lock. Meanwhile, on 4 August 1939, Bristol received production outlines for the Beaufighter, but there was even more delay. The first flight of the second Beaufighter prototype had to be postponed from September 1939 to 22 November.[13]

Delays worsened until 8 May 1940, two days before the German breakthrough, when Rolls-Royce warned Westland their commitment to the Merlin and the Vulture (needed for the Tornado fighter and the Manchester bomber) meant they could only make enough Peregrines for 114 Whirlwinds and the run would end in December 1940. If ordered to make more, they would have to cut Merlin production by two for every additional Peregrine, or postpone production of the Griffon, already scheduled to begin in July 1941.

Air Chief Marshal Sir Hugh Dowding, Commander-in-Chief of Fighter Command, looked on in disbelief. He had condemned the Whirlwind when he wrote to Beaverbrook in June 1940, warning that he expected endless problems with details until Westland gained experience. This also meant that Whirlwind squadrons were not reliable enough to play a full part in the coming battle. Even by December 1940, only 263 Squadron was operational with sixteen machines.

The need for more than 200 modifications and additional engine delays meant 263 had to use Hurricanes until their Whirlwinds finally appeared. Its range was too short for an escort fighter and it only performed well at low altitude. It was finally declared obsolete in early 1944, though pilots admired its flying qualities and its toughness. Had these eagerly-sought replacements for the Spitfire not suffered from similar production problems to Mitchell's ground-breaking design, the RAF might have faced a harder problem. Furthermore, the Whirlwind and the Beaufighter would have suffered the fatal vulnerability of large, twin-engined

12. Michael Bowyer, *Interceptor Fighters for the Royal Air Force*, Patrick Stephens Ltd., 1984, p. 155.
13. *Ibid*, p.165.

A Westland Whirlwind of 263 Squadron, beset with technical and production problems. *(Official photo via Wikipedia)*

Completed Spitfires ready for test flights from the camouflaged Chattis Hill airfield near Stockbridge in Hampshire. *(via Alfred Price)*

Sewards Garage Southampton was chosen to produce Spitfire fuselages and production jigs. *(Via Alfred Price)*

Fuselage assembly in progress at Sewards Garage. *(Via Alfred Price)*

machines fighting agile single-engined fighters. While the Luftwaffe's previously successful Bf110s suffered such severe losses that they had to be escorted by 109s for their own survival, the RAF's new machines would have had no Spitfires to beat off attackers in turn.

So the Spitfire had been saved – just. Had production problems been tackled more effectively the safety margin in the Battle of Britain might not have been so narrow, but wilful blindness is hard to overcome and bomber enthusiasts continued to criticize. Even after the war, Air Chief Marshal Sir Philip Joubert de la Ferté blamed Beaverbrook for providing the means of winning the Battle of Britain at the price of postponing final victory by many months through disrupting the bomber programme. For a senior RAF officer with the advantage of hindsight this was astonishingly misguided. Had fighters not been available, the Battle of Britain would have been lost and the Luftwaffe caused far more disruption by flattening bomber factories unhindered by defending fighters. Moreover, delay at the time would only reduce production of obsolescent types like the Fairey Battle, the Blenheim, the Hampden and the Whitley. The heavy bombers needed to deliver punishing blows against Germany would not be ready for years with the technology they needed to find, and bomb their targets accurately enough to inflict real damage.

Chapter Eleven

Tables Twice Turned: 109 Production

As the Spitfire teetered on the edge of cancellation, two factors finally transformed the 109's prospects. Because its margin of victory over the other candidates for the Luftwaffe's contract was so wide, the Milch faction was rendered powerless. Had it faced similar production problems to the Spitfire it might still have been dropped. The irony of its brighter future was that Willy Messerschmitt's obsession with lightness and simplicity helped it avoid the delays which nearly eliminated its main opponent, and fitted perfectly into Nazi rearmament plans.

Both designers climbed a technical mountain to build their prototypes, replacing wooden frameworks and doped linen fabric with complex stressed-skin metal airframes. Now the workers making those planes would face an equal challenge. Where their First World War predecessors built aircraft by hand like the Wright Brothers, they now had to manage the complexities of high-performance stressed-skin airframes.

This presented a huge challenge to any aircraft industry; but under the Armistice, Germany was originally barred from making military aircraft at all. The Nazi takeover eliminated this barrier, but finding vast numbers of skilled workers for large-scale production, recruiting them and training them in new production methods would take time, and once war began they would be liable for military service. Sophisticated methods would have to square the circle, producing large numbers of warplanes quickly with mainly unskilled manpower.

Messerschmitt's design fitted these requirements well. While he would have been delighted with a large production contract to reward his efforts, he never had the chance originally given to Supermarine. To meet ambitious output targets, Bf109 production would be shared between different firms, many of them Messerschmitt's fiercest rivals, while arguing with Nazi instructions was bad for the commercial and personal health of those who tried.

Establishing mass production in a single company with experienced workers was possible. Repeating it all over the *Reich* at the same time was not. Finance however, would be no problem, as public opinion no longer mattered. 'In the autumn of 1936, Göring was appointed head of a Four-Year Plan organisation to create the economic foundation for war. Two-thirds of industrial investment between 1936 and 1939 was devoted to military production. By the spring of 1939 one-quarter of the German labour force was working on military contracts.'[1] It was fortunate that Messerschmitt's design would be simpler to produce than the Spitfire. Though the basic assembly of both aircraft was similar, details made the vital difference.

Mitchell engineered every component of the Spitfire using principles developed in the solid and reliable flying boats and the racing seaplanes on which he built his career. Messerschmitt pared everything to the bone as he had been forced to do when building gliders. The simplicity of its fuselage made 109 construction easier and quicker. Bulkheads were replaced by frame shells, metal hoops stretch formed on special tools that forced them into shape while the metal was relatively soft. To harden them against flight stresses they needed heat treating and quenching in a water spray. To avoid distortion they developed a 360-degree spray system for even cooling.

As with the Spitfire, the heart of the 109's structure was a strong box section formed by the firewall behind the engine and the cockpit instrument panel. Messerschmitt made this box carry the wing spar and engine bearer mountings as in the Spitfire. One of the greatest challenges in building the 109, again like the Spitfire, was forming the complex double curves of skin panels to cover the framework and increase overall strength. The solution was also the same, massive hydraulic presses combined with semi-automatic riveting machines to fasten the shaped panels to the frame and to one another. At first, 109 production was laid out in what German engineers called the traditional *Takten* (timed or measured) system. Different small teams turned out sub-assemblies for the finished aircraft in a synchronized process, brought together for final assembly. This was soon replaced by an assembly line, like car production. Each fighter was carried on a conveyor belt through a series of stages where parts and pre-assembled units were fitted in sequence. This meant parts could be fitted with little or no delay, and was faster but noisier than traditional aircraft building. One worker on the outside of the fuselage pressed countersunk rivets into pre-drilled holes. The metal skin panels were forced against the rivets while another worker on the inside used a compressed air hammer to squash the opposite end of each rivet to complete the fixing. Later, semi-automatic riveting machines made things quicker and quieter, but as demand rose, even major manufacturers had difficulty keeping up. By 1940, as with the Spitfire, each individual aircraft builder had a network of sub-contractors to supply raw materials and finished parts, but the simplicity of Messerschmitt's design meant production time fell steadily in a masterpiece of efficiency.

One of the flagship assembly plants for the 109 would be Messerschmitt's equivalent of Castle Bromwich, the purpose-built Regensburg factory, opened in 1938 and planned from the beginning for large scale mass production. Five years later, it would be turning out the more powerful and complex 109G in an assembly line system. Sheet metal was cut in a piecework system to feed the large and specialized presses where components for wings and fuselages were formed.

---

1. Professor Richard Overy, *Russia's War*, p. 35.

Messerschmitt's clever detail design employed cast and pressed components to reduce the need for time consuming milling and turning procedures. Wing spars were bored and riveted by sophisticated machines while instrument panels and cable clusters were assembled prior to fitting. Because of the high degree of automation, factory hands called up for military service were replaced by male and female forced labourers, until four-fifths of the workforce used pressed workers in a highly organized system.

'For space reasons, the conveyor belts in the pre-assembly area were U-shaped. The separate manufacturing groups for the wings were pre-assembled and thereafter, in moving and swivelling structures, were screwed and riveted together. All the parts needed for this process were held in readiness directly beside the conveyor belt, and fuselage assembly was likewise conducted on the conveyor belt method. In special working operations, individual half-fuselage sections were joined together by longerons, and subsequently clamped and riveted. These two conveyor belts ended in a construction unit for fuselage final assembly, where up to twenty fuselages were located. Engines and propellers were likewise equipped on conveyor belts for the final assembly of the aircraft. The fuselages arriving from the metal working shop were lifted on to rolling conveyances in the final assembly area and turned through a right angle to the conveyor belt, which moved at a speed of some thirteen inches per second. Between the fuselages were the working platforms from which the engine, electrics, control rods etc, could be mounted. Each worker had his fixed workplace and set tasks to perform.

'Having ready all the aircraft parts that were needed, ensured a continuous work process. Since the rolling conveyances raised the level of the fuselages, engine installation was greatly simplified. After final assembly the fuselage was turned through another right angle and placed on special transport stands which enabled the aircraft to be raised horizontally and suspended on one of the three conveyor belts in the final assembly hall. The wings, also coming from their own assembly lines to left and right, were attached to the fuselages, and all leads and conduits attached. In the final phase, all necessary equipment in the shape of fuel tanks, instruments, armament and radio were installed. The fuselage and wings had already been lacquered with RLM *Hellblau* (light blue) and the engine received its upper and lower covering panels, and the oil cooler in the ventral part of the engine casing was attached to the engine circuit. Undercarriage tests followed before the aircraft was lowered on to its wheels and rolled out of the final assembly hall. It was then led to the firing stand to test its weapons and also for centring the compass on the rotatable compass stand. After fuelling and final checks, the engine was given a test run, after which the Bf109 stood ready for a works flight.'[2]

---

2. *'Nest of Eagles'* [originally published in Germany as *'Die Messerschmitt-Werke in Zweiten Weltkrieg'* in 2004] by Peter Schmoll pp. 55–57.

Though similar to that used for the Spitfire, the level of automation and the simplicity of the design made it quicker and more efficient. The historian Corelli Barnett claimed a Mark VC Spitfire took more than three times as much work to complete as a Bf109G (13,000 man-hours compared with 4,000 for the 109) though official figures tell a rather different story. German records show the first Bf109Es took some 9,000 man-hours each to produce a finished fighter, though this was cut to 6,000 man-hours from the 109F onwards.[3] Other analysts suggest Barnett's figures might have come from small and atypical factories, while the largest British assembly plants like Castle Bromwich achieved productivity levels equivalent to American (or the best German) figures over longer runs.

However, mass production works best while turning out identical products, and time and effort are needed to make changes. This meant the German system took longer to change to a new aircraft type and they underestimated this extra time. Though the Spitfire was harder to produce at first, developing new versions of the aircraft essential to keep it a viable front line fighter proved quicker and more straightforward. Professor Richard Overy described how 'The British system was more wasteful in terms of quantity, but maintained a higher average quality. Major modifications were introduced as and when the military situation justified them or in response to changes like increases in engine power. These were screened by an MAP [Ministry of Aircraft Production] Aircraft Modifications Committee to make sure only the most urgent changes were incorporated, and with the minimum interruption to output. Such a situation was much less wasteful of effort than the complete design and introduction of new aircraft types. [The]Vickers Supermarine company spent 330,000 man-hours on the design of the Spitfire Mark 1, but only 620,000 man-hours on all the subsequent fifteen Spitfire marks. For jigging and tooling-up for the Spitfire Mark 1 some 800,000 man-hours were expended. The average time for all subsequent marks was 69,000.'[4]

But the weakness which limited total 109 output was a toxic blend of propaganda and over-confidence. Dr Goebbels' Ministry of Public Enlightenment successfully portrayed the Nazi regime as limitless in power and resources, and opponents were all too ready to believe his message. Triumphal parades during the build-up to war suggested the German Army was a mighty armoured and mechanized force, delivering irresistible punches and advancing at speed. Endless flypasts of modern bombers and fighters suggested the Luftwaffe was modern enough, large enough and powerful enough to crush opponents quickly and effectively.

The truth was rather different. When the Nazis came to power, Germany's strongest suit was her manpower. In almost every other respect the economy was smaller, with fewer natural resources than Britain and quite incapable of meeting

3. Figures from Schmoll, *op.cit.*, p. 35.
4. Richard Overy, *The Air War, 1939–45*, p. 178.

Hitler's ambitious rearmament plans. The Army had the highest priority, but was not the armoured juggernaut propagandists described and enemies feared. Wartime events would re-emphasize the skill and effectiveness of the German combat soldier in outfighting every other army in the world, man for man, but the real problem was not its fighting manpower. It was the material limitations of what German industry could achieve during the prelude to conflict.

Outside the panzer divisions and motorized infantry formations, the German Army marched the hundreds, or even thousands of miles to its objectives on its own feet. Most German soldiers carried rifles like their First World War predecessors. Most supplies, equipment and ammunition were moved in horse-drawn wagons and even lorries were scarce. German officers inspecting equipment abandoned by the British in the sand hills of Dunkirk were astonished at the resources of their defeated opponents.

Even as Germany geared up for war, weapons had to be prioritized to enable victory against opponents with larger industries and more resources. For the *Kriegsmarine*, prestige meant huge quantities of steel and other metals for obsolescent capital ships rather than far more effective U-boats. In the air the situation was worse as Luftwaffe priorities were shaped by Army needs. For example, there was a shortage of artillery which could be moved quickly enough to keep up with armoured units. With most German field artillery towed by horses, they needed weapons to be in action in minutes. The solution was 'flying artillery' or precision dive-bombers, called in by Army commanders against a specific target.

Pre-war Curtiss Helldiver biplane dive-bomber. *(USN photo)*

The idea began with Ernst Udet six months after the Nazi takeover. He earned a precarious peacetime living as a stunt pilot and air racer in countries from Greenland to South America. While visiting the USA in September 1933 he saw precision attack demonstrations by Curtiss F8C naval fighter-bombers. These were the Helldiver variants of conventional biplanes introduced in the mid-1920s but adapted for the relatively new technique of dive bombing, and he was astonished at their accuracy.

If an aircraft was slow enough or slowed by dive brakes, it could dive nearly vertically. This let the pilot aim directly at the target, rather than dropping his bomb load in level flight, which required a lot of practice. Skilled dive-bomber pilots could place bombs within thirty yards of their targets time and again. Udet managed to raise funds, possibly with a secret loan from Göring, to buy and bring back two planes to Germany for trials, and their promise persuaded the fledgling Air Ministry to sponsor a design competition for a German dive-bomber.

The result was the Junkers 87 *Stuka*, a single-engined monoplane carrying a pilot who aimed the aircraft at the target, and a rear gunner who kept enemy aircraft off his tail while he was doing it. Fitted with a thick, heavily cranked wing to reduce the length of the drag-inducing fixed undercarriage, it had a slight resemblance to Mitchell's Type 224 but with even more limited performance. With guaranteed German air superiority, it was a devastating weapon against ground troops and targets, and large numbers were produced. However against high performance fighters where escorts failed to protect it properly, it suffered heavy casualties and finally had to be withdrawn from combat.

The result was that the *Stuka* crippled the Luftwaffe. It left insufficient resources to build a force of strategic bombers like RAF Bomber Command. The Luftwaffe's own bomber enthusiasts used bombers to support Army ground operations rather than major campaigns against enemy heartlands. When Junkers followed the *Stuka* with the twin-engined and much faster Junkers 88, a potentially outstanding multi-role aircraft, too much time and effort was wasted turning this into another dive-bomber. As a high-speed medium bomber it was excellent, but adding dive brakes to limit speed in a steep dive increased wing stresses and ruined performance.

Finally the Luftwaffe became a prisoner of its own contradictions. Intended primarily as a ground-attack force for army support, its offensive capability was limited to medium bombers with reasonable performance but restricted bomb-loads. To deliver knockout punches against enemy targets, bombers needed a higher priority. Though fighters would have to protect them and shoot down enemy fighters and bombers in turn, there was no information to decide how many would be needed.

The Germans believed sophisticated production methods could produce far more fighters than cumbersome British industry. They were absolutely right, but mistaken in their conclusions and misled by their own propaganda. Where the British over-estimated German potential, the Germans underestimated their

Cowling panels and tailplanes in Bf109 wartime production. (Via Airframe Assemblies)

opponents. Fear of the results of German rearmament proved useful to politicians and civil servants wanting to provoke Britain into increasing defence expenditure.

Another drawback of being ahead in the race was that early 109s had less powerful engines and were deficient in speed and armament. The first production version, the 109B or '*Bertha*', appeared from February 1937, first from Haunstetten and later from the large new plant at Regensburg. It was powered by the Jumo 210Da producing 720 bhp and driving a two-bladed fixed-pitch propeller later replaced by a variable pitch equivalent and armed with twin cowling machine guns, with 500 rounds apiece. These were split between the needs of home defence and the *Kondor Legion* operating in Spain (see Chapter 12).

During 1938 the Bf109C '*Clara*' was given the Jumo 210Ga with fuel injection and peak power of 960 hp. The radiator below the engine cowling, a conspicuous feature of all early 109s was larger and two machine guns were mounted in the wings to double the firepower, later replaced by two MG FF cannon. To complicate matters, the 109D '*Dora*' was produced alongside its stablemate, but combined the four-gun mixed armament with a temporary return to the carburettor engine. All these changes reflected steady improvements, but the definitive variant of the early wartime years would be the Bf109E '*Emil*' with the more powerful DB601 engine. With higher speed, fuel injection and the cowling radiator replaced by one under each wing, this appeared early in 1939.

Before the war, the British guessed Luftwaffe strength fairly accurately but assumed that once war began warplane production would soar. This never happened as resources had to be shared between fighters and bombers, but the Air Ministry never realized these limitations. German aircraft production hid beneath the thickest security blanket, so opponents had to accept Nazi boasts of fighter strength at face value. Even when Frank Foley, SIS chief in the Berlin Embassy, took on an agent who supplied him with photocopies of Luftwaffe documents at fortnightly meetings, the appeasement-minded British Ambassador, Nevile Henderson found out and ordered Foley to break off contact.

Sadly, this information would have revealed that the Luftwaffe was still operating at broadly peacetime levels, so their advantage in fighter numbers would not last. Two factors were responsible and inevitably one was Hitler. An astonishingly successful opportunist, he exploited Allied uncertainty to absorb the Rhineland, the Saar, the Sudetenland, Austria and Czechoslovakia into the Reich. He made the fatal mistake of believing his own infallibility, and assumed there would be no obstacles to his march of conquest. Even with French and British pledges to support Poland, he remained convinced success would continue.

He also decided his war preparations would be complete by 1942, with a large surface fleet, hundreds of U-boats to throttle Britain's Atlantic lifeline and German air strength fit for a world war, but he wanted to avoid a proper war economy with restrictions and shortages. In 1918 the German Army claimed it had not been beaten by the Allies, but 'stabbed in the back' by civilian pressure for an armistice, so Hitler wanted civilian support. Given Nazi oppression, his fear was misplaced and the German people would enjoy a better standard of living than their opponents until well into the conflict.

Finally, the Allies called Hitler's bluff three years early. Even then the Anglo-French declarations of war could not affect the Nazi-Soviet envelopment of Poland from both West and East, his first resounding military success, but longer-term implications were worrying. Once war began, German strength would have to increase while fighting was in progress.

As casualties mounted, Army reinforcements became essential and industry was raided for replacements. Aircraft factories had to widen sources of manpower and press workers from the defeated nations, reaching between 45 and 50 percent of the entire labour force by the final months of the war. While foreign forced labourers might have kept German civilians out of the factories, the men needed retraining and there was still little sense of urgency. Many German aircraft factories worked single shifts in 1944, while Russian and British factories worked around the clock from 1940 and 1941.

Even able professionals in the German military found it difficult to contradict Hitler. Yet he was unable to assess the weapons placed in his hands as Germany's supreme warlord. Churchill, conscious of his own limitations, sought advice from experts, but Hitler relied on charlatans and advisers pursuing their own interests. Blinded by easy victories, and his ignorance of British determination to exceed his air strength, Hitler and his commanders underestimated the need for any increase

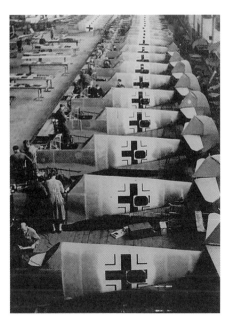

Bf109 engines and fuselages in wartime production. Via Airframe Assemblies.

in aircraft production. Once the Spitfire's production problems had been overcome, British output climbed steadily and continuously, overtaking the 109's lead.

Now it was the Germans' turn to misunderstand their opponents. Luftwaffe intelligence decided in May 1939 that Britain could not build more than 3,000 military aircraft over the next year. They were wrong by a factor of three to one. British aircraft production overtook German output by the end of 1939, and during 1940 the British and French between them produced double that of the Germans. Their second and more damaging assumption was that, as one spectacular victory was followed by another, Luftwaffe strength could cope with steady losses of pilots and machines. By the end of May 1940 it had overcome the air forces of Poland, Scandinavia, the Low Countries and France, not to mention that part of the RAF operating on French territory. Yet even these seemingly one-sided victories exerted a huge cost in men and machines that would never be repaid.

The Luftwaffe lost a fifth of its strength in an apparently easy victory in Poland. The Norwegian campaign cost almost as much in losses and the Battle of France saw almost 1500 aircraft destroyed. The RAF was only involved in the last of these campaigns where its own losses were 959 aircraft, of which only around half were fighters.[5]

When the opponents faced one another for the first time in the Battle of Britain, matters became even worse. During the whole campaign from early June 1940 to

5. Bungay, *The Most Dangerous Enemy*, p. 93 *passim*.

Later Bf109 production adopted the assembly line principle in the purpose-built Regensburg plant. *(Bundesarchiv image Via Wikipedia)*

the end of October, the Luftwaffe received 916 replacement fighters while the RAF received almost three times as many. As the battle opened, Luftwaffe fighter strength numbered 1107 Bf109s, facing 754 Hurricanes and Spitfires, a margin of almost 50 percent. In addition only some 40 percent of RAF fighters were Spitfires, capable of meeting 109s on fairly equal terms, but this fell far short of what the Luftwaffe needed to enable an invasion of England.

In contrast to German complacency, Beaverbrook's role as Minister for Aircraft Production was to boost RAF fighter numbers at any price. He set deliberately unattainable targets at 15 percent above the maximum to force aircraft factories to work harder and faster. He combed every potential source for more fighters to reinforce squadrons. Chasing up RAF storage units helped push production for the first four months of 1940 to 638 fighters. Between then and August 1940, another 1875 fighters were issued to Fighter Command, together with some 1900 damaged aircraft returned to service by repair depots.

German losses during the battle (from 10 July to 31 October) totalled 1733 of all types against 915 RAF planes shot down, predominantly fighters, a ratio of nearly two to one.[6] Because the Germans held the initiative, they could choose

---

6. Overy, *The Air War 1939–1945*, pp. 33–4.

targets and order the strength for each attack, while the RAF could only defend itself as best it could from each incoming threat, but numbers would finally tell. The Germans had limited production at the very time British fighter production was still climbing, and even the rate of increase outmatched German output. In May 1940 German air strength reached a peak of 3692 aircraft of all types. After that production never kept up, so strength continued to fall until the truth could no longer be ignored.

Chapter Twelve

First Blood for the 109 –
Spain and the *Legion Kondor*

On 22 July 1936, Adolf Hitler was enjoying a night at the opera. His love for the sprawling epics of Richard Wagner dated back to teenage years and brought him back each July for the opening week of the Bayreuth Festival, devoted entirely to Wagner's works. As he left his box at the end of the performance, two men were waiting for him. One was the local party boss, the other a German businessman with interests in Spanish Morocco. They brought a letter from a Spanish Army officer, General Francisco Franco Bahamonde, currently an exile in the Canary Islands, and desperately short of weapons.

Five days before, rising tension between the Spanish Republican Government and Nationalist army officers had burst into open revolt. After five years of weak administration, the Spanish authorities had realized the danger and moved suspect officers like General Franco to overseas postings or premature retirement. This triggered the very outcome they sought to avoid. On 17 July 1936 Army garrisons seized power in Cadiz, Cordoba, Seville, Oviedo, Pamplona, Jerez and Zaragoza.

The original Nationalist leader died in a plane crash on 20 July and command had passed to General Franco, but both sides were desperately short of modern aircraft. The Republicans appealed to France and the Soviet Union. Two days later the French promised modern fighters and bombers to keep the Nationalists at bay. On 24 July the Italians would agree to back the Nationalists. For Hitler there were sound reasons to help, for what he would pretend was pure altruism.

Though Franco's letter arrived at Bayreuth rather than Berlin, Hitler responded immediately. Senior Nazis knew a liking for Wagner was a sound career move, so both the Luftwaffe commander Hermann Göring and General Blomberg, head of the Army, were also in Bayreuth where they discussed the letter and agreed to support the uprising. Hitler knew a Nationalist Spain would threaten France with a third hostile power on her borders. It would also offer a chance to test men, weapons and tactics in a real war. It would deter Mussolini from Anglo-French peace initiatives and it would help ensure vital war materials like iron ore. Later, Hitler admitted he preferred a long-drawn out civil war rather than a Franco victory to maintain Germany's bargaining position.

On 26 July, the Communists promised massive aid to the Republican side, but the Germans had already taken the lead. On 19 July Franco had flown to Tetuan in Spanish Morocco, to take command of 25,000 garrison troops comprising Moorish regiments and the Spanish Foreign Legion. They were tough and well trained, but useless unless transported to the homeland. His letter begged Hitler for ten passenger planes so his troops could evade Republican warships patrolling the Straits of Gibraltar.

Hitler needed a respectable façade for his assistance. He suggested Germany was simply helping Franco's forces face their opponents on fairly equal terms. With Republican backing from France and Russia, he was simply restoring the balance of power. When German servicemen appeared in Spain, they were training and advising their Spanish colleagues on using the weapons provided.

On 6 August 1936, the 22,000 ton German steamship *Usarama* of the Woerman Line, trading between the Fatherland and East Africa, tied up in the Spanish port of Cadiz at the end of a five-day voyage from Hamburg. Officially, she carried a group of German engineers, photographers and salesmen on a '*Kraft durch Freude*' ('Strength through Joy') official cruise organized by the Nazi holiday and leisure administration, but her real purpose lay hidden within her hold; eleven crated Junkers Ju52 three-engined transport planes from *Luft Hansa* and six Heinkel He51 biplane fighters from the new Luftwaffe.

Her cargo also contained 88mm anti-aircraft guns, shells, bombs, spare parts and ammunition, and her passengers included eighty-six Germans of military age. They carried papers showing they were tourists, travelling with the Union Travel Company, but they were the pilots, ground crew, support staff and gunners to man German weapons and aircraft for General Franco.

Dictatorships move quickly when needed and within hours of the Bayreuth decision, the aircraft were sent to Hamburg Docks for loading. Another nine Ju52s flew direct from the Junkers factory at Dessau to Tetuan. Forty Spanish soldiers were crammed aboard each plane instead of the normal seventeen passengers. Each aircraft made five flights on that first day, to the airfield at Tablada near Seville, base

Junkers Ju52 in *Luft Hansa* civil livery. *(Author photo)*

for the German operations. During the next two weeks, they shifted almost 2,500 troops to the mainland. When they were joined by the other Ju52s unloaded from the ship and assembled and tested at Tablada, the pace quickened. By 11 October, 13,523 troops with 36 guns, 127 heavy machine guns and more than 250 tons of military supplies, had crossed to Spain to take part in ground fighting.

The six He51 biplane fighters had been assembled and handed over to Spanish pilots. Sadly, they proved a daunting challenge and half crashed within days. With only three left, German pilots took over and in the last week of August, their three senior fliers shot down nine enemy aircraft. *Oberleutnant* Trautloft was shot down by a Republican fighter and parachuted to safety while his comrades chased the enemy away. Even so, German pilots

Interior of Ju52 cabin. *(USAF)*

shot down twenty-one enemy aircraft for the loss of only one of their number by the end of September.

On 13 October, a Soviet freighter, the *Stari Bolshevik* ('Old Bolshevik') docked in Cartagena harbour on the Republican side. She carried eighteen crated Polikarpov I-15 biplane fighters, and pilots and ground staff to fly and maintain them. A second Soviet freighter rendezvoused at sea with a Spanish ship, to transfer a dozen more I-15s, and fifteen more were landed later in the month. The balance was tilting too far against the Germans and their allies, but the situation soon worsened. Further shipments included thirty-one Russian medium bombers and the same number of Polikarpov I-16 monoplane fighters, the most formidable warplanes yet to appear in Spain.

The I-16 was one of the few warplanes[1] designed inside a prison. Stalin's lunatic logic held the best way to cure a man of unpunctuality was to have him shot. When the completion of Nikolai Polikarpov's I-5 and I-6 biplane fighters for Stalin's air force was delayed, punishment followed. In October 1929, Polikarpov was among some 450 aircraft engineers and designers arrested as saboteurs and counter revolutionaries. These trumped-up charges killed off untold numbers of

---

1. The Tupolev SB2 (p147) was another!

Polikarpov I-16 in Spanish Republican markings of the 4th International Brigade Squadron, in Madrid Museum de Cuatrovientos. *(Andrea and Petronas via Wikipedia)*

innocent Russians, victims of a dictator's paranoia, and Polikarpov was sentenced to death. He spent two months in prison, waiting for his executioners. Only when he was clearly more valuable to the Soviet Union alive than dead, was his sentence commuted to ten years forced labour at the special aircraft design bureau number 39, run by the OGPU secret police inside Moscow's notorious Butyrka prison.

There he drew up the design of the I-16. After the prototype flew successfully he was released on parole. He set up his own design bureau in 1938, but then fell back into Stalin's disfavour and was sent on a mission to Nazi Germany. When he returned from that, he found his bureau closed, and his staff transferred to a new firm headed by his chief engineer Mikhail Gurevich and by Artem Mikoyan, conveniently the son of Anastas Mikoyan, a senior member of Stalin's Politburo. Polikarpov was finally cleared of all charges in 1956, three years after Stalin died and twelve years after his own death.

In essentials the I-16 was a crude but effective fighter, as technology was slowly developing. Its low cantilever wing had a main spar of chrome-molybdenum alloy steel, duralumin ribs and a combination of metal and fabric skinning. Ailerons covered almost the entire span, and could be extended downwards as flaps for landing approaches. The stubby fuselage was a wooden monocoque to reduce weight.

Like the Bf109 it began life with a British engine, in this case the licensed version of the Bristol Jupiter air-cooled radial developed by Gnome-Rhone in France and simply copied by the Russians. The pilot had a lever to pump the retractable landing gear up and down. An Aldis tubular gunsight was held in place with elastic bands, and because the cockpit hood flew open during aerobatics, most pilots fixed it in that position. Like the early 109s it had a pair of 7mm machine guns, but fitted in the wings outside the propeller arc for a higher rate of fire. Unfortunately, they were fitted upside down and often jammed.

The unorthodox little I-16 came in for as much official suspicion as the 109, but like its German rival, won over opponents by its good behaviour. On test flights, it recovered automatically when forced into a stall, and simply centring the controls brought it out of a spin. But climbing in and out of the cockpit was difficult, poor quality cockpit glazing hampered visibility, and the undercarriage tended to jam. Its instability and lack of trim tabs also made it tiring to fly.

Yet it proved a formidable combat machine, reasonably fast with a useful rate of climb, and able to complete a full barrel roll in less than a second and a half and once pilots realized its agility their confidence grew.[2] The Russians called it the *Ishak* or 'Little Donkey', and the Spanish the '*Mosca*' or 'Fly', while its opponents called it the '*Rata*' or 'Rat' for its agility and elusiveness. No less than 475 were shipped to Spain as fighters and bomber escorts, and they changed the face of air combat.

To meet this threat, Germany sent six more He51s, with more pilots and ground staff. Aware of the growing odds, they got their retaliation in first by striking Republican formations, shooting down three fighters, a reconnaissance plane and a transport on 19 October. As Hitler dreaded being associated with defeat, the fiction of German aid and advice was ditched. German soldiers, sailors and airmen would operate under German command, using German tactics in a new organization. On 7 November, the *Legion Kondor* (Condor Legion) was formed under the command of the portly but able *Generalleutnant* Hugo Sperrle who arrived in Spain eight days later, with *Oberst* Wolfram *Freiherr* von Richthofen, a cousin of the Red Baron, as his Chief of Staff. Six thousand German troops had already disembarked with more Ju52s, modified into stopgap bombers.

The next clash came on 13 November. Five Ju52s and three He46 two-seat reconnaissance machines escorted by nine He51s, raided Madrid. Near the target, they were attacked by twenty-four Republican fighters, sixteen I-15 biplanes, and eight I-16s making their first appearance. The German pilots beat them off, shooting down three biplanes and a pair of I-16s, but the *Legion* lost two fighters with their pilots in addition to one Ju52. The German pilots faced a much tougher future.

2. Jackson, Robert. *Aircraft of World War II – Development – Weaponry – Specifications*. London, Amber Books, 2003.

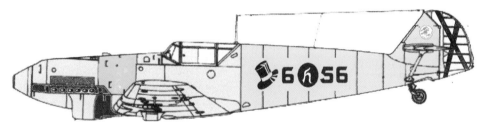

Bf109B of *Legion Kondor* bearing the personal markings of *Hauptmann* Gotthard Handrick, *Gruppenkommandeur* of *J/88*. *(Author drawing)*

Sadly, their commanders drew the wrong conclusions. German biplane fighters could indeed shoot down I-16s but only because of the skill and experience of their pilots. As the opposition caught up, the wide gap in performance would prove fatal. When another big reinforcement arrived at the Tablada airfield on 18 November, with parts for another sixty fighters, they were all He-51s.

For the present, the *Legion* punched above its weight. Reconnaissance flyer Wilhelm Balthasar had been turned down for Luftwaffe pilot training with poor eyesight, but learned privately instead. He had been accepted by the *Legion Kondor* as an observer/gunner, but after taking over as a pilot in emergencies and proving his ability, he was moved to *Jasta 88* in September 1937. He would later become one of the most successful Bf109 pilots in the opening campaigns of World War II with forty victories to his name in addition to the seven he scored in Spain.

On one reconnaissance mission on 23 September 1936, Balthasar reported ships and cargoes in the Republican supply port of Cartagena. Sperrle led thirty-six Ju52s to hammer this tempting target as heavily as accuracy would allow. However, Sperrle knew all too well the *Legion's* future in Spain would be grim without more modern aircraft and began lobbying senior commanders in Berlin. It worked. On 19 March 1937, twelve early production Bf109B 'Berthas', (out of a total production of just thirty) were sent to Tablada to re-equip *2/J88*. Pilots drafted from *II/ JG132* in Germany were rushed through a rapid conversion course to teach them to land, take off and fight in these radical new machines. The *Berthas* were powered by uprated Jumo210Da engines of 720 hp driving Schwarz fixed-pitch propellers. Armament was limited to two synchronized cowling-mounted MG17s with 500 rounds apiece. In spite of the huge performance leap, conversion went smoothly and the re-equipped *2/J88* was operational by the end of April. In two years of furious aerial combat, the 109s would show for the first time exactly how good they were, facing real enemies in a genuine war.

Their first chance came on 6 July 1937 when Republican defenders of Madrid attacked Brunete to the west of the capital to raise the siege and take the pressure off Republicans in Northern Spain. *2/J88* was ordered off at dawn from a dusty strip at Herrera near Santander to fly more than 200 miles to a new airfield at Avila, closer to the fighting. Their comrades from the biplane *staffeln* attacked advancing Republican columns, but the 109s were soon put to work protecting the Ju52s from large numbers of I-16s. Because Republicans usually flew at around 20,000 feet, the I-16s had little chance of fighting 109s on equal terms. In spite of agility and firepower, they were vulnerable to the fast diving attacks of the German machines.

Typically, the German pilots would remain several thousand feet above, following their adversaries and choosing the moment to attack. A simple radio command told each pilot to pick a target before entering a dive. So quickly did the 109s accelerate that by the time they appeared out of the blind spots behind and below their targets they were too fast for any retaliation. Provided their aim was accurate enough in the few seconds allowed, the combat had but one ending. The 109s used their speed to pull out of their dives and climb back to their starting height in what were later called 'dive and zoom' tactics. At worst, they might miss. At best, another Republican fighter would be plunging earthwards, and Republican fighter groups often returned to base with gaps in their ranks.

These tactics meant *2/J88* suffered no losses for several weeks flying several missions each day, provided they ignored their opponents' attempts to lure them into a fight at lower altitudes where the I-16s' manoeuvrability would tip the scales the other way. As it was the Republicans claimed to have shot down a 109 on 8 July, prior to their move south. An expatriate former US Marine pilot, Frank Tinker, claimed another on 17 July, when he spotted a 109 which lost too much speed and was climbing back to safety. Tinker had time for a simple zero–deflection shot. His burst hit the target, and the 109 fell out of the sky and crashed, killing its pilot.[3]

By now the Republican air force were trying to cripple the lethal 109s on the ground. With no early-warning system, the 109Bs had to fly standing patrols over the approaches to Avila to break up incoming raids. After almost four weeks of brisk fighting, the Brunete campaign was abandoned and the raids ceased, leaving the 109s free to return to Herrera. Astonishingly, they suffered only the single casualty. Not one priceless 109 had even been damaged on the ground. Nevertheless this was not the crushing blow the *Legion* wanted to deliver with their new machines. However their time would come. Soon.

Meanwhile I-16 pilots had severe reliability problems. The average service life of each aircraft was less than ninety days, though some lasted longer. To keep flying in harsh Spanish conditions, a sixth of their time had to be spent on repairs and maintenance. The saving grace was their abundance. The Republicans received 276 I-16s from their Soviet allies over two years of fighting, but by the end of the war, only eighty-nine remained, all in poor condition.[4]

---

3. *Messerschmitt Bf109 at war*, Armand van Ishoven, (Ian Allan 1977) says this happened four days earlier.
4. Maslov, Mikhail A. *Polikarpov I-15, I-16 and I-153*. Oxford, UK: Osprey Publishing, 2010, p. 32.

The remaining months of 1937 involved fighter pilots on both sides having to escort their own bombers. On 4 December *Feldwebel* Polenz's 109 ran out of fuel, forcing him down on a road behind Republican lines. The French offered to release much needed supplies if the Republicans made the 109 available for testing. *Capitaine* Rozanoff, chief test pilot of the French *Centre d'Essais en vol* (Flight Testing Centre) at Orleans-Bricy, test flew the machine at Sabadell airfield, near Barcelona.

The opportunity was wasted. The aircraft had a fixed pitch propeller, so its performance was poorer than newer machines with variable pitch airscrews, giving the French false comfort that the 109 was less formidable than it really was, had they been told, that is. Rozanoff's detailed test reports were judged too secret to be shown to the French Air Force or the French aircraft industry. Both could have profited from the information, though their ignorance made little difference to events two years later, when French and German fighters met in combat.

Meanwhile, the Republicans switched attacks to the Nationalist base of Teruel in the mountains between Valencia and Zarragoza. The 109s and their crews moved to the small airstrip of Calamocha, hidden in the hills north of the town. On bitter winter nights, they had to run up each engine every few hours to prevent freezing. But on Sunday, 7 February 1938, they had their chance to strike back.

The 109s were escorting a formation of new Heinkel He111 bombers against Republican positions. As they neared the target, they saw aircraft approaching from the south. When closer, they saw twenty-two Tupolev SB2 twin-engined medium bombers, supplied by the Soviets. These aircraft could outrun Heinkel 51 biplanes, but the 109s could catch them easily. This was the largest group ever seen over the front, and the unusual lack of a fighter escort made them an ideal target.

Tupolev SB2 bomber in Russian Central Air Force Museum, Monino. *(Photo by Gonzolito via Wikipedia)*

The 109s plunged on their opponents. The Berthas could manage 289 mph, but this was enough. As the German pilots saw the targets expanding in their gunsights, they began firing and within minutes bombers were dropping out of formation. The Republican fighters appeared at last, desperate to join the fight, but that brief delay cost them dearly. Within five minutes, ten bombers and two I-16s were destroyed, and the Republicans turned and fled. Not a single German fighter had even been damaged. To underline the importance of the 109s, more Republican bombers with a fighter escort attacked *J88's* own base. It made no difference at all. In six minutes *Leutnant* Balthasar brought down three bombers and one fighter, raising his Spanish score to seven victories. Nor was this a mere flash in the pan. By the time Teruel fell back into Nationalist hands, *J88* pilots had shot down thirty aircraft for no losses to themselves.

However, Republican reinforcements outnumbered the few replacement 109s spared from Germany's own rearmament, as Hitler worried his European conquests might trigger international action to rein him in. In April 1938 both sides received new aircraft. The *Legion* had five 109C 'Claras' for *3/J88*. Performance was improved with the fuel-injected G version of the Jumo 210, and armament was more than doubled with an additional pair of machine guns mounted in the wings. The new Republican I-16s were type 10s, also carrying four machine guns. But the most significant German reinforcement was a pilot, *Oberleutnant* Werner Mölders. His insight and ability would transform German combat tactics. On 15 July he shot down his first opponent, an I-15. On the seventeenth, he shot down another. Flying one of the four-gun 109Cs, his skill made these combats one-sided, but fighting quickened with the second Ebro campaign in early August. Heavy Republican reinforcements would make it difficult for the 109 pilots to cling on to their lead, though they also received five new 109D-1s to bring *3/J88* to full strength.

The decisive factor was Mölders' clear thinking, producing tactics to stop their opponents seizing the advantage. At the time, most fighter forces flew a basic three-plane V-shaped formation with the leader followed by a wingman on each side. The RAF called this

Werner Mölders in November 1940.
(Bundesarchiv photo via Wikipedia)

a 'vic' and the Luftwaffe a '*Kette*' or 'chain', but both suffered from wingmen concentrating more on staying close to their leader than watching for attacks.

Mölders' new tactics used the '*Rotte*' or 'pair' in a twin-fighter line abreast formation with two aircraft spaced some 600 feet (or twenty wingspans) apart. Each pilot watched his comrade's tail to warn of an enemy approach from astern or below. If either pilot was attacked, he would break towards the attacker, and his wingman would turn in the same direction. Should the enemy continue the attack on the leader he would be attacked by the wingman closing in behind. If it was the *Rotte* leader who attacked the enemy, he could concentrate on his aiming, knowing his wingman was watching his blind spot. Later, with more aircraft, two *Rotten* made a four-fighter tactical formation called the *Schwarm*.

Improvements were made. The leader of the first *Rotte* flew slightly ahead of his wingman, so he gave him a better view of the leader's blind spots. The second *Rotte* flew slightly further back, echeloned in the opposite direction, in the '*Vierfingerschwarm*' or 'Finger four' formation, which likened the positions of the four fighters to the fingertips of an extended hand. To complete the analogy, the hand was rotated slightly at the wrist with the thumb side lowered, reflecting the fact that the second *Rotte* flew slightly higher on the down-sun side for a better view of an enemy attack out of the sun.

This arrangement had one drawback. With four aircraft in staggered line abreast, any turn to launch an attack or deflect an enemy move meant all four turning together. With an overall breadth of some 2000 feet, this needed too high a speed difference between the aircraft on the outside of the turn and that on the inside. The solution was the cross-over turn, where the aircraft on the outside of the formation would turn first and fly over the others. Each succeeding aircraft would do the same, until the formation rolled out on the new heading, with the relative positions reversed, but no marked loss of speed.

This idea had already been used in the previous war. The RAF experimented with multi-plane formations with narrower spacings, but found too high a collision danger. Now the advent of radio and the wider gaps between aircraft made it more practical. Using a full *Staffel* of twelve aircraft in three *Schwärme* stepped up in line astern or in echelon away from the sun applied greater force. Once the attack was launched, each formation would use dive and zoom tactics first used by Russian pilots over Spain, and copied by the Germans to counter the manoeuvrability of their opponents.

The idea was promising, but needed practice to make it work. On April 4, 1938, *1/J88* and *2/J88*, both equipped with Bf109s, were transferring to a new base airfield. *Leutnant* Fritz Awe was leading a *Schwarm* from *1/J88*, when he ordered a 90-degree turn to port. His Number Two, *Unteroffizier* Adolf Borchers, left Awe insufficient clearance. The two aircraft collided, with Borchers' propeller smashing into Awe's cockpit, decapitating the pilot and tearing the fuselage in half. Borchers managed to survive a crash landing when his damaged 109 turned over, and suffered only minor injuries. However, by grouping fighters further apart and deliberately widening the formation on sighting hostile aircraft, the risk was reduced. Mölders also trained his pilots to fly on instruments in cloud or in darkness, a major challenge when early Bf109s lacked an artificial horizon.

Meanwhile, Mölders brought down an enemy fighter in each of the fourteen weeks of his service in Spain up to his recall to Germany on 3 September. His score of fourteen victories was the highest of any German pilot in Spain, though seven of his comrades scored another fifty-seven victories. The combination of the 109s and the Mölders tactics helped bring down twenty-two Republican fighters in just five days while suffering no losses of their own at all.[5]

The second Ebro campaign was the fiercest in the whole Civil War, but their opponents' efficiency forced the Republicans back. Besieging and capturing Teruel from the Nationalists (and ultimately losing it again) took too much of their strength. Now the *Legion* increased the pressure by bombing the bridges over the Ebro, and the Republicans' desperate attempts to repair the damage each night would bring more attacks next morning.

By this time, the *Legion Kondor* was also under strain and the pace slackened. Both sides knew the outcome would be decided by one last campaign, and reinforcements were needed. For the *Legion*, now all three *Staffeln* of *J88* were equipped with 109s, the only sensible solution was to introduce the latest version of Messerschmitt's fighter now emerging from German factories. The 109E or *Emil*, powered by the fuel-injected Daimler-Benz DB601 driving a three-bladed variable-pitch propeller, had performance far beyond the I-16.

Concerns over international reaction over the Sudetenland meant fighters were still needed over Germany. This was solved by a deal whereby the *Legion's* existing 109s went to the Spanish air arm, replaced by 109Es. Germany would invest in Spanish iron ore mining, sell Spain five million *Reichsmarks*' worth of iron ore mining equipment, and be paid in iron ore supplies.

The final act of the Civil War came with the Catalan campaign on 23 December 1938, when six Nationalist armies attacked flimsy Republican defences. The Republicans began retreating towards Barcelona, under fierce attacks from *Legion* aircraft. At the same time *2/J88* were busy converting to the first of the 109Es to be unloaded and assembled from shipping crates. Forty had been sent, of which half were ready, but potential opponents were vanishing too quickly to be dealt with. Barcelona fell to the Nationalists on 26 January 1939, and *Legion* fighters were tasked with preventing their opponents escaping the final collapse. The last loss of a Bf109 occurred on 6 February, but on 5 March another Bf109 shot down an I-15. The *Legion* flew its last combat mission on 27 March 1939 and the Republicans surrendered the following day, a victory confirmed by General Franco on 1 April 1939.

---

5. *Air Power* by Stephen Budiansky, Viking Penguin Books 2004, Page 213–214.

German fighter pilots claimed 409 enemy aircraft, losing nineteen of their pilots including the victims of anti-aircraft fire. Five were killed in flying accidents, three from mid-air collisions and five from illness. Other sources claim the Republicans lost as many as 372 of their aircraft or as few as 277. Whichever figure is closest to the truth, the loss ratio was overwhelmingly in favour of the Nationalists, due to Bf109 performance and the tactics to exploit it. However, the figures also show that more than two-thirds of the *Legion's* losses were accidental, due to the harsh operating conditions in Spain and the difficulties of handling the 109.

With the fighting over, how did the participants profit? The Germans spread the gospel of their tactics throughout the Luftwaffe by sending successful pilots back to Germany to pass on their experience. Those on the receiving end did not learn the lessons so well. The Italians tried to exploit the manoeuvrability of their biplanes and their aerobatic expertise, which proved a blind alley. The Russians learned the lessons earlier, but few spread the message on returning home. Serving overseas qualified them in Stalin's demented paranoia as ideal targets for liquidation.

Meanwhile the RAF, still within the grip of the bomber lobby, saw the Spanish war as irrelevant. The service's all-important Bombing Committee referred to it in a June 1937 meeting to consider the use of fighters to escort bombers by both sides in Spain. After discussing the matter it was decided it would be 'unwise to base conclusions on these reports as the conditions were dissimilar in many ways to those expected in a major air war.'[6] This preference for unsupported predictions rather than learning lessons from events persuaded the RAF to ignore its own intelligence experts.

Senior RAF officers assumed the Luftwaffe could not support a long war, and insisted its only option was to deliver a killer punch with bombers. Since this was the bomber lobby's own plan for dealing with the Germans, everything matched convincingly. However, in spite of the terror of the Guernica bombing, Spain actually underlined the limits of bombing power. Even when Air Marshal Victor Goddard, the RAF's own intelligence expert, insisted the evidence showed the Luftwaffe was primarily intended to support the German Army, the RAF ignored him.

Kenneth Strong, British Military Attaché in Berlin and later Chief of Intelligence to General Eisenhower in Normandy and North-West Europe, clearly knew what he was talking about. When he said the same thing, he too was ignored, and warned to keep out of Air Force business. All this meant the Luftwaffe became the most objective and best prepared of the major air forces. Had RAF senior commanders been left to their assumptions, it would have remained so. As it was, facing up to better enemy tactics and greater experience would challenge Fighter Command pilots to the limit in the Battle of Britain, and overcoming those advantages would impose a terrible toll in machines and lives.

---

6. Budiansky, *ibid.*, p. 212.

Chapter Thirteen

The Contenders Meet at Last

By late summer 1939, the Bf109 and the Spitfire had followed separate but similar paths. The designs, the prototypes, the test flights, the proving of power units and weapons, and the transition into mass production, had made both fighters ready for first-line service. They were now approaching the final and most pitiless verdict; how they would cope against one another in the lethal and unforgiving world of aerial combat?

This was their intended purpose from the start. They were sophisticated modern machines with exhilarating performance, exciting responses and staggering potential, but their ultimate purpose was to shoot down enemy aircraft and kill the crews flying them. Far more important than the beauty of their lines, the quality of their engineering, the efficiency of their production and their speed, agility and armament, was how effectively they carried out this sinister but vital purpose.

As the world slid inexorably into war, the time was fast approaching when they would meet. Only then could they show their worth as fighters. The time for claim and counter-claim, propaganda and prediction was over. From the moment that fighting began, they would operate against a permanently changed background. That first confrontation was still a long time coming. The secure peacetime world vanished on the first day of September 1939, when 1.5 million German troops in five armies poured into Poland in a vast pincer movement. They were supported by massed squadrons of the Luftwaffe, bombing and strafing Polish soldiers and civilians alike. With British and French forces hundreds of miles away on the *Reich's* western frontiers, there was little they could do to affect the situation. Poland would pay the price of her isolation in a crushing campaign to demonstrate the power of the Nazi war machine. Five weeks after the *Wehrmacht* entered Poland, the fighting was over, and German and Russian troops met as triumphant victors. This campaign therefore produced no confrontation between the Luftwaffe and the air forces of France and Britain. The Germans were too busy and their enemies unable and unwilling to interfere. German bombers found little opposition, but fighters had difficulty finding targets at all. Carefully, the Polish air force fell back under the German assault, trying to husband its strength. In the cauldron of fighter combat, and despite the courage and dedication of the Polish pilots, their fighters were no match for the 109s.

Yet if Germany had been teaching the world how good its army and air force were at crushing smaller nations, they learned a lesson of a different kind themselves, but failed to realize its significance. Even against a technically outclassed adversary, the Luftwaffe suffered heavy casualties that would take much replacing. Worse, the failure to understand the implications would bring the Hitler regime to ruin.

The relatively underpowered and poorly armed 109Bs and Cs had done well in Spain, against indifferent opposition. In Poland the much more formidable 109E, with the more powerful fuel-injected Daimler-Benz DB601 and wing-mounted cannon, shattered the courageous but poorly equipped Polish fighter squadrons whenever they met. After a fortnight of combat, the remaining Polish fighters had to retreat into Rumania. However, those two weeks of high intensity fighting had two radical effects. Polish pilots who survived the onslaught had a legacy of experience and a deep hatred for the enemy, which made an almost irresistible combination once they had their hands on first class fighters. In addition, the Polish war accounted for more than 500 German aircraft destroyed or beyond repair, more than 14 percent of the Luftwaffe's total strength, and 60 percent more than the losses of their beaten opponents. As the campaigns in Norway, Denmark, France and the Low Countries added to this cost, the Luftwaffe's numerical advantage would be less than Hitler, Göring or their opponents would believe.

Even after 10 May 1940, when the RAF was drawn into the bloody campaign over German attacks in France, the Spitfire and the Bf109 would not meet for almost two weeks. RAF fighter squadrons sent to France as part of the defences were Hurricane units. As modern fighters, the rugged Hurricanes held their own against German attacks better than most Luftwaffe opponents, but mounting losses soon began to threaten Britain's own defences.

Bf109E in Battle of Britain colours. *(Author photo)*

As the French pleaded for more RAF fighters to help stem German attacks, Air Chief Marshal Sir Hugh Dowding warned the Air Ministry that sending more units to France risked the 'final, complete and irremediable defeat of this country.' His warning was heeded, but the situation on the ground worsened each day. As the Germans broke through to the Channel coast, the British Expeditionary Force began retreating to the beaches of Dunkirk, to evacuation and to safety.

However, as the British perimeter shrank, one consolation resulted. RAF stations in southern England were finally within range of French beaches, and fighters could intervene directly from their home airfields. Safe at last from the moving front and the need to retreat from airfield to airfield, they could fight on equal terms. For the first time, Spitfires confronted their opponents in a truly daring rescue.

This first real clash, on 23 May 1940, involved fighters and pilots from the RAF airfield at Hornchurch in Essex. It began badly for the RAF in the morning, with Squadron Leader Francis White, CO of 74 Squadron, attacking a two-seat Henschel 126 reconnaissance machine. A lucky shot from the rear gunner hit his Spitfire's cooling system. With glycol pouring from the fractured pipes, he was forced to crash land on French soil. The airfield of Calais/Marck was in sight and within range, though the German Army was driving past the main gate. So when his Spitfire came to a stop, he hid in a perimeter ditch and waited for his opponents to take him prisoner.

Instead it was decided to mount a rescue operation with 54 Squadron, another Spitfire unit based at Hornchurch. They had one possibility in the form of a Miles Master advanced trainer with two great advantages and one disadvantage. It was a two-seater powered by a Rolls-Royce Kestrel with relatively good performance. Unfortunately its flanks, undersides and tail were painted in RAF Trainer Yellow to make it stand out clearly against any background. However, Flight Lieutenant James Leatheart devised a plan which involved him flying the Master to Calais airfield to pick White up, escorted by two 54 Squadron Spitfires flown by Al Deere and Johnny Allen. All three aircraft would cross the Channel at low level to reach the airfield, fortunately sited right on the coast. With sufficient speed and surprise it might succeed. Leatheart would land on the airfield and wait with engine running for White to emerge from hiding and climb aboard, assuming he was still at liberty. Meanwhile the two Spitfires would orbit the airfield to warn of approaching aircraft. But wartime plans rarely survive the first contact with the enemy, and so it proved this time.

Deere thought 'It sounded like a piece of cake. There was broken cloud over the area, which meant there was a likelihood of being surprised from above. I decided therefore to send Johnny above cloud, at about 8,000 feet, while I remained below circling the airfield.' As he continued his turn, he watched the Master taxiing towards a hangar when a frantic bellow over his headphones from Johnny Allen warned of a dozen German fighters making straight for him. Since the Master had no radio, there was no way of relaying the message. Deere flew past it on the tarmac, waggling his wings to warn of an emergency. By the time he lined up, the Master was starting its take-off run and a Bf109 was diving down to attack. Deere fired a burst to deter the German pilot from the Master and turned inside him to bring the 109 within his sights. 'In a last desperate attempt to avoid my fire, the Hun pilot straightened from his turn and pulled vertically upwards, thus writing his own death warrant; he presented me with a perfect no-deflection shot from dead below and I made no mistake. Smoke began to pour from his engine as the aircraft, now at the top of its climb, heeled slowly over in an uncontrolled stall and plunged vertically into the water's edge from about 3,000 feet.'

Climbing up through the cloud to help Allen, Deere narrowly missed another pair of 109s which both turned to attack. 'Again I found no difficulty in keeping inside the turn and was soon in range to fire.' A long burst at the number two caused bits to fly off his aircraft, which rolled on its back and careered earthwards…, but the leader was still there and must be dealt with. Reversing his turn very skilfully he too dived towards the ground. Momentarily I lost distance, but I had got in range again before he flattened out above the tree tops and headed homewards.' Deere was out of ammunition, but made contact with Allen who said his Spitfire had been damaged by another 109, and both pilots joined up in mid-Channel and flew back to Hornchurch.

After landing they were delighted by the approach of the Master, with Leatheart and White. When Deere and the first 109 roared past, Leatheart had an appalling shock. His intelligence report told what happened next. 'The moment I left the ground, I saw from the activities of Red One [Deere's call sign] that something was amiss. Almost at once a Me109 [sic] appeared ahead of me and commenced firing. I pulled around in a tight turn, observing as I did so the Messerschmitt shoot past me. I literally banged the aircraft on to the ground and evacuated the cockpit with all possible speed, diving into the safety of a ditch which ran around the airfield perimeter. Just as I did so, I saw an Me109 come hurtling out of the clouds to crash with a tremendous explosion a few hundred yards away. Almost simultaneously another Me109 exploded as it hit the sea to my left. It was all over in a matter of ten minutes, but not before we observed a third enemy aircraft crash in flames. We waited about ten minutes after the fight ended and when it seemed safe we made a hasty take-off and a rather frightened trip back to England and safety.' Despite the noise and mayhem, German troops passing the airfield along the main road failed to notice anything unusual.

This was extremely encouraging for Spitfire pilots facing 109s for the first time in the Luftwaffe's otherwise successful campaign, but combat reports soon showed it was no flash in the pan. Later that same day, 54 Squadron pilots claimed to have shot down eight Bf109s when they attacked the Spitfires en route to defend British troops making their way to Dunkirk. Though not all claims might have been confirmed in the chaos of a dogfight, thanks to a timely warning over the radio of the approaching German fighters, all RAF fighters returned safely afterwards.

From the German point of view this was most unwelcome. They saw enemy air forces as victims for their superior machines and tactics, not opponents able to fight back to such damaging effect. The day after the dramatic rescue at Calais/ Marck, the head of the German General Staff, General Franz Halder, wrote in his diary that 'Enemy air *superiority* [author's emphasis] has been reported by Kleist' [General Ewald von, commander of two Panzer corps, anxious to maintain the headlong advance to the sea which forced the Dunkirk evacuation]. Nor was he alone in his reaction to RAF attacks. On the same day the German XIX Korps [then advancing on Calais and Dunkirk as part of *Panzergruppe Kleist*] entered in its War Diary that 'Enemy fighter resistance is so strong that our own air reconnaissance was practically impossible.' The following day they were ordered to halt their advance, which gave the British Expeditionary Force more time to reach safety.

Statistics confirmed this impression. German official figures showed that over the course of the evacuation, the Luftwaffe lost 189 aircraft. The RAF lost 131, though this included light and medium bombers, which suffered particularly badly from German flak and fighters. Only 99 were fighters. Actions like these boosted the reputation of the Spitfire and had a powerful effect on morale, but several factors were involved. The RAF pilots were more experienced than average squadron pilots who would have to replace casualties once battle was joined in earnest. Secondly these were skirmishes. Groups of fighters blundered into one another by chance before proper control systems could vector squadrons on to escorted German

Spitfires of 19 Squadron, first to be issued with the new fighter, 31 October 1938. *(Via Alfred Price)*

bombers. And they usually involved fighter free-for-alls with opponents trying to out-turn one another at relatively low altitude, which favoured the Spitfire's strong suits of manoeuvrability and armament.

On other occasions, when the 109s had the advantages of height, sun and surprise when diving on opponents, the flying boot could end up firmly on the other foot. When the formal and inflexible tactics laid down by RAF training were added to the mix, the future for British fighter pilots was grim. They had to stick to tight formations and a set routine whereby their leader ordered a particular type of attack and the fighters under his command would line up to fire at the target in succession.

It remains difficult to imagine a system to place pilots in greater danger of being ambushed by their German opponents. Formation flying involved everyone concentrating on the leader, or those closest to them in the formation. This left them unaware of stealthy approaches by German fighters, particularly from above or out of the sun. Adding the need to wait one's turn to carry out an attack, focussed each pilot's attention everywhere but the directions from which lethal threats would appear. Yet these lamentably unsuitable tactics were logical enough to begin with. They were drawn up when German bombers could only reach targets in England by flying from bases within the *Reich*, far beyond the range of Bf109 escorts. This created two problems. Defending fighters delivering uncoordinated attacks risked collisions. Secondly, each attack might be so fleeting that more fighters might be needed to bring a bomber down. The need was to inflict lethal damage as quickly

K9789 was 19 Squadron's first Spitfire to arrive at Duxford on 4 August 1938. *(Via Alfred Price)*

as possible, and with no real possibility of enemy fighters, these tactics could have proved effective.

By the time the Luftwaffe and the RAF clashed in earnest, the Germans had already rewritten the script. By knocking out the French in a few weeks of *blitzkrieg*, they gained new bases on the other side of the Channel. Suddenly, London and southeast England were within range of 109s and the bombers they escorted to attack targets in the Channel, along the South Coast and across the south-east.

Had RAF tactics been changed to deal with different conditions, all might have been well, but large and complex organisations, even fighting services, develop a momentum of their own. Changing course is like turning a laden supertanker, with a long delay between altering the helm and waiting for the ship to respond. In that interval, many pilots would pay a deadly price for following orders that had suddenly become outdated and dangerous, but remained part of official doctrine.

Just how severe a disadvantage was imposed on pilots by RAF tactics was shown on 18 August 1940, when *Oberleutnant* Gerhard Schoepfel of III *Gruppe* of *JG26 'Schlageter'* used German 'dive and zoom' tactics against Hurricanes of 501 Squadron with devastatingly one-sided effect. His tactics suited the 109's advantages of high climbing speed and a fast dive, rather than a turning dogfight where British fighters would have an advantage.

The Germans called rigid British formations, with bleak sympathy, *Idiotenreihen*, or 'ranks of idiots' and it was easy to see what they meant. At lunchtime on 18 August 1940, 501 Squadron was ordered from Hawkinge to patrol at 20,000 feet over Canterbury. They flew an immaculate formation of four sections of three aircraft apiece with a thirteenth fighter weaving backwards and forwards as a look-out. High above and behind them was *III/ JG26 'Schlageter'* led by Schoepfel. Ordering his pilots to maintain course and height he dived down to attack the rearmost Hurricane. His aim was true and his stricken target plunged earthwards, but no other British pilot saw what had happened. Schoepfel used his speed to soar back upwards to safety. Astonishingly, he repeated his attack three more times while the rest of 501's pilots, still concentrating on keeping formation, failed to see him. He shot down three more Hurricanes and only broke off when his windscreen was coated in oil from the engine of his final victim. He had attacked a formation of thirteen patrolling fighters single-handed and shot down four of them in a matter of minutes without the others noticing he was there.

In this case his victims had been Hurricanes, though Spitfires using the same tactics would be equally vulnerable. Schoepfel's tactics exploited the 109's virtues and his unit's height advantage. Had the British fighters flown a looser formation, more pilots would have spotted the approaching threat and spoiled his attack. They would have forced Schoepfel into a turning dogfight, where the British fighters' agility might have imposed a fatal disadvantage.

Fortunately for the RAF, events were not always so one-sided. Increasing bomber losses during the Battle would force 109s to fly as close escorts to maintain bomber morale. In doing so, they lost the advantages of height, sun and speed, and almost guaranteed they had to dogfight with more manoeuvrable RAF fighters, so conditions were much more even. As more RAF units copied German formations, at first in the teeth of service disapproval and later under a more enlightened policy, their chances of survival improved.

Events had already shown the Spitfire and the 109 were closely matched. Both sides captured enemy aircraft in good enough condition to be flight tested, so comparisons could be made. Unsurprisingly pilots from both sides preferred their own aircraft to those of the enemy, but similarities emerged in top speed, roll rate and aileron stiffness. British pilots found the 109 agile but tiring to fly due to no rudder trim, claustrophobic due to a cramped cockpit and prone to ground looping if not handled carefully. German pilots described the Spitfire as 'childishly easy' to fly and thought the Hurricane beneath contempt, yet brave and skilful pilots like the Poles of 303 Squadron shot down many 109s with well-handled Hurricanes.

The Spitfire could climb faster than the 109, but the German fighter could escape by diving faster than its opponents. The Spitfire hood was bulged to avoid pilots hitting their heads when operating from rough grass airfields. It had fewer blind spots than the 109's 'greenhouse', but the Perspex was more prone to scratches. The 109 pilot sat further forward over the wings, where the downwardly tilting nose from the inverted V12 engine gave him a better view ahead and downwards in combat. Setting the rudder pedals higher above the 109's cockpit floor allowed pilots to resist black outs in high 'g' manoeuvres.

In the Spitfire, the fuel tank was ahead of the pilot, so if attacked from astern, the armour plate behind the seat tended to prevent rounds hitting the fuel tank and setting it on fire. In the 109, the pilot sat above the main fuel tank, which appeared more dangerous than it was. Ordinary rounds would slow down passing through the tank and the fuel it contained, so did less damage to the pilot or the airframe. If an incendiary bullet entered the tank it would set the fuel ablaze, but even then the slipstream tended to blow the flames away from the pilot.

During its first year of full operational service, two modifications were made to the Spitfire which gave it an edge over the 109. Constant speed propellers were more efficient than the variable pitch type, which had replaced the fixed pitch airscrews originally fitted. Second, and possibly more important, was the availability of 100-octane fuel. Compared to the 87-octane supplies used by the Luftwaffe it was easier for RAF pilots to catch their opponents and shoot them down in a stern chase or escape 109 attacks by accelerating out of danger. RAF combat reports often mention the extra performance provided by these changes. Of the constant speed airscrew:

'Attacked e/a [enemy aircraft] which seemed to be above. Attempted to get on his tail, he immediately turned left to return the attack. We manoeuvred for a long while, during which he fired quite a fair amount. I got two short bursts which had no effect. After about 12–15 minutes he tried to out-climb me, I immediately went into fully fine pitch and easily caught up. The instant at which I opened fire, he rolled over and went straight down. I followed him and he started to smoke and eventually went into the sea. I steered 300° and after about 6–8 minutes made a landfall 3 miles West of Dungeness.'[1]

'The 109s split up and I picked one and gave him a short deflection burst. I did not have time to see the effect of the burst as another 109 was on my tail. I outclimbed the 109 without difficulty. When I got on his tail I gave him all the ammunition I had and saw my tracers going in. The 109 flew off very unsteadily towards the French coast.'[2]

In addition, 100-octane fuel gave pilots the option of pushing the automatic boost cut-out override tab at the forward end of the throttle quadrant. This allowed engine boost to be increased to +12psi for short periods (up to five minutes) in emergency or during combat. If this option was used, it had to be reported on returning to base and entered in the engine log book. However, Dowding noted in a memo of 1 August 1940 on the handling of Merlins fitted to Spitfires, Hurricanes and Defiants; 'Some pilots "pull the plug" on every occasion.' Pilot Officer John Freeborn of 74 Squadron used boost override the day after his CO had been rescued from Calais/Marck. 'As I broke away two Me109's got onto my tail. I dived steeply with the 2 e/a following me, one was on my tail and the other on my port quarter. As I dived to ground level I throttled back slightly and the e/a on my tail over shot me and I was able to get a three seconds burst at a range of about 50 to 100 yards. He seemed to break away slowly to the right as though he was badly hit and I think he crashed. The second Me109 then got onto my tail but I got away from it by using the boost cut out.'[3]

Two days later, Flight Lieutenant Brian Lane used the same option to escape. 'A dogfight now ensued and I fired burst at several e/a, mostly deflection shots. Three e/a attached themselves to my tail, 2 doing astern attacks whilst the third attacked from the beam. I managed to turn towards this e/a and fired a good burst in a front quarter deflection attack. The e/a then disappeared and was probably shot down. By this time I was down to sea level, and made for the English coast, taking violent evasive action. I gradually drew away from e/a using 12lbs boost which gave an air speed of 300 mph.'[4]

1. Combat report, Sergeant H. Chandler, 610 (County of Chester) Squadron, 14 August 1940.
2. Combat report, P/O Peter St John, 74 Squadron, 10 July 1940.
3. Combat report, P/O John Freeborn, 74 Squadron, 24 May 1940.
4. Combat report, F/L Brian Lane, 19 Squadron, 26 May 1940.

'At about 17.30 we were patrolling Manston at 12,000 feet when control informed us Canterbury was being dive bombed. About five miles south of the town when at about 3,000 feet a Me109, silver with black crosses, dived past my nose, flattened out about 50 feet up and headed south. I executed a steep turn, pushed in boost override, and sat on his tail. At about 50 yards, I gave him one small burst with little effect, closed to 30 yards, and gave a slightly longer burst. Black smoke poured from him as I overshot him. The a/c crashed in a field, turned over two or three times and burst into flames in a clump of trees. 70 bullets were fired from each gun.'[5]

'I then dived for sea level 10 miles from coast, saw five aircraft I thought were Hurricanes and climbed to them for protection. They proved to be Me 109's which chased me back to coast, one continuing chase after others had left me: on seeing this I went into a turn, got onto its tail closed to 70 yards and fired 2 second burst. I saw this A/C hit the sea in flames.... My Spitfire easily outdistanced Me 109's at 10lbs boost 2800 rpm.'[6]

Anything which gave a sharp increase in performance was bound be used despite official limitations. Soon, orders were being issued by commanders in the air and controllers on the ground, to use boost override to speed up interceptions or attack targets more effectively. 'I was the leader of Blue section 610 Squadron, which was ordered to take off at 0745 hrs on 12/6/40. There were two aircraft in the section as Blue 3 had trouble starting and when he did take off he failed to contact us. Immediately we were airborne we were ordered to 'vector 120' and 'gate" [push the boost override tab].[7]

This seemingly one-sided view of fighter-to-fighter combat was not the whole story. For all the RAF pilots able to catch and bring down 109s by using the performance and handling of their Spitfires to the full, others died after being surprised and attacked by 109s approaching England with height and sun behind them. Flying from northeast France to southeast England virtually guaranteed the blinding disc of the summer sun was on the German side. Because 109s could climb to high altitude before crossing the Channel they usually dived to the attack, so relative top speeds were irrelevant. RAF defenders were invariably climbing to intercept German raids and were usually outnumbered, leaving them at a real disadvantage.

Naturally, Luftwaffe pilots felt the same confidence in their 109s and were convinced they had the better bargain, especially when free to use tactics which exploited their strengths. In the final analysis the British fighters would prove good

---

5. Combat report, Sgt Jack Stokoe, 603 (City of Edinburgh) Squadron, 1 September 1940.
6. Combat report, F/L Robert Boyd, 602 (City of Glasgow) Squadron, 18 August 1940.
7. Combat report, F/L John Ellis, 610 (County of Chester) Squadron, 12 June 1940.

Flat hood of early Mark 1 Spitfire. *(Author photo)*

enough to render German aims impossible. As tactics changed to reduce bomber casualties, 109 losses increased. Ten days of the battle when fighting was hardest showed 448 Luftwaffe losses against 256 suffered by the RAF, but German losses included bombers. In terms of fighter losses alone during the entire battle, 1023 RAF fighters were lost (Spitfires and Hurricanes), compared with 873 German fighters (Bf109s and Bf110s). Further analysis suggested a ratio in favour of 109s over Spitfires and Hurricanes of 1.2 but with more Hurricanes in RAF service, the Spitfires probably achieved a rate nearer parity.[8]

From 1941 onwards both sides made mistakes. The RAF embraced offensive tactics over France which brought heavy losses and starved overseas theatres of first class fighters for trifling advantages over occupied Europe. The Germans wasted experienced pilots by failing to rest them from front-line combat so their expertise could train replacements. Aces ran up huge victory totals against inexperienced adversaries with poor tactics, but fighting tougher opponents saw the skill and experience levels of the Luftwaffe dwindling through losses.

The third highest scoring German fighter pilot of all time, the late Gunther Rall, had 275 victories, but was in no doubt of the value of British fighters. In a 1996 interview he described RAF pilots as 'Outstanding…, a well trained and highly motivated force, with good equipment and good morale … We had tremendous losses against the Royal Air Force. I had the highest respect for them…. At the

---

8. Bungay, *op.cit.*, pp. 370–2.

Bulged hood of later Spitfire. *(Will Owen photo)*

beginning of the war we flew short-range missions and encountered Spitfires which were superior.... I think that the Supermarine Spitfire was the most dangerous to us early on. I flew the Spitfire myself and it was a very, very good aircraft. It was manoeuvrable and with good climbing potential."[9]

Rall himself was a first-class pilot with excellent deflection shooting which brought him many victories. Yet, like many comrades who amassed staggering victory scores, he found limited success in the Battle of Britain. At one end of the spectrum, early aces like Galland, Oesau and Wick shot down 105 RAF fighters between them, but they were exceptions. Many future scorers like Erich Rudorffer and Heinz Bär (222 and 220 victories respectively) shot down ten adversaries apiece during the battle, while Rall himself had no confirmed victories over England in what he admitted was a challenging and demoralizing campaign.

Bär in particular had been lucky to survive the battle. He began by taking on RAF fighters in dogfights. As a result he was shot down an astonishing eighteen times, seven by Spitfires,[10] before he finally built up a fearsome reputation against less well-equipped opponents over Russia. By that time, his comrades in the West would have a more formidable 109 to restore the advantage against the RAF, which would have to be met by improvements to the Spitfire to maintain its position in the British first line of aerial defence.

---

9. *World War II online magazine,* Gunther Rall interview published September 1996.
10. Spick, *Luftwaffe Fighter Aces,* pp. 220–221.

Chapter Fourteen

Learning Curve: Training the Pilots

With German industry gearing up for a world war, selling their newest and most formidable fighter to overseas air forces hardly made sense, but the country desperately needed metallic ores, iron, chromium and copper, to make new weapons. This was the original reason for participation in the Spanish Civil War and handing over the *Legion Kondor* 109s to the Spanish Air Force. Now the Swiss and the Yugoslavs wanted the new wonder fighters.

Ten Bf109Ds were delivered to Switzerland over the winter of 1938–39 to familiarize pilots with a modern high-performance machine. They arrived without radios or weapons and with minimum instrumentation. The main contract for thirty 109Es was then followed by another for fifty more. However the last of these only arrived on Swiss airfields by 27 April 1940.

At first the Yugoslav order fell on stonier ground. Göring himself turned down their request on the grounds that the 109s were too difficult for their pilots to handle. He insisted that 'There is too much with which your pilots are unfamiliar in this new aircraft. They should convert by degrees to such fast warplanes or they will suffer many casualties, particularly during landings, and will develop a fear of really advanced fighters.'[1]

Nevertheless, the Yugoslavs were ready to pay for the fighters with iron ore, chrome ore and copper ore, an offer the resource-starved *Reich* could not refuse. An order for fifty 109Es and twenty-five spare engines was signed on 5 April 1939 and a follow-up order for another fifty fighters followed on 23 June 1939. Seemingly, the Yugoslav upgrade for their fighter force had been successful. They would have done better to follow the Swiss by buying obsolete versions of the 109 to begin with. Only seventy-three 109Es were delivered, many unflyable because of spares shortages. It seemed Göring had spoken the truth, as Yugoslav pilots found the 109s appallingly difficult to handle after their forgiving Hawker Fury biplanes. With landing accidents more common, they took Göring's advice and tried to bridge the gap between the Fury and the 109 using the Bf108 as a conversion trainer. It could carry an instructor with the pupil and had enough in common with the 109 to simplify the transition.

---

1. Green, *Augsburg Eagle*, p. 44.

RAF primary trainer – the De Havilland Tiger Moth. *(USAF)*

Surprisingly it proved a failure, with enough 109 in its pedigree to belie its reputation for easy handling, and was undeniably difficult to fly. The Yugoslavs had to switch their pilots to Hawker Hurricanes, supplied by the British. Trainees would learn on RYAF Hurricanes first, then convert to the Bf108 before tackling the 109. They were still completing the process when on 6 April 1941, Germany invaded Yugoslavia to help themselves to the raw materials they needed.

At the time the RYAF had forty-six surviving Bf109Es. On the first day they brought down ten German bombers, but the attrition rate was terrible, thanks to the pilots' lack of confidence in their demanding machines. Fourteen 109s were lost that first day with another twelve the following day and even more shot down by their own AA. By 11 April, after just five days of fighting, all surviving machines were burned to prevent them falling into German hands.

This was confirmation that flying a high performance fighter is one of the most difficult and dangerous tasks facing a military pilot. World War Two fighter pilots faced an especially lethal challenge. While their comrades learned to fly bombers or transports with an instructor sitting next to them, single seat fighters offered no such benefit. With huge differences in performance from their predecessors, flying was no longer a matter of reflexes. Events happened much faster. All too often the right action, selecting a switch or adjusting a control, done too slowly was as bad, or worse, than a wrong action done quickly.

Some learning was done with the aircraft on the ground and the engine off. The new pilot could learn the position of each button, switch or lever, how it worked and what it did. He could learn checklists and procedures, with the right order of steps to start up the engine, check it was working properly and prepare for take-off, but when all was ready to go, when the brakes were released and the throttle opened, he met a huge surge of power and a massive torque reaction trying to force him sideways off the runway. He had to counter that force and keep his machine straight as it left the ground and climbed to a safe height, clear of all hazards. And he had do it right from the very first time he tried, to survive his training. Any mistake in a complex sequence in a crowded cockpit, or in letting speed drop and risking a stall, or forgetting to lower the landing gear before reaching the ground, risked a catastrophe. At best it meant damaging a valuable and expensive fighter needed to train other pilots. At worst it could take the life of a young man possibly on the verge of reaching the standard needed to fly on operations.

This was difficult enough in peacetime. Once fighting started, rising casualties and increasing demand for replacements meant training had to be cut below what was needed. Astonishingly, most successful pilots managed to cope with nothing more than a degree of blind panic at the crucial moments, and what seemed dangerous soon became routine and predictable. But almost every air force under-estimated the true cost in terms of scarce and expensive aircraft, and young and inexperienced pilots, of maintaining squadron strengths. Spitfire pilots who followed instructions could usually make a passable take-off. With rudder and elevator trims set and the stick held back to keep the aircraft from pitching forward and hitting the propeller blade tips on the ground, they could lift the tail gently as speed increased to see the runway ahead and keep the heading straight with the rudder. As power increased, the aircraft usually took off by itself without the pilot rotating it into a flying attitude, and almost every flyer marvelled at the fighter's instinctive ability to do what was needed. Bf109 pilots on the other hand found the torque-induced swing was made worse by the high ground angle of the machine. The position of the centre of gravity made it more tail heavy than the Spitfire, and taking off was much less instinctive. The standard drill was to push the control column to the front right-hand corner of the cockpit to persuade the tail to lift as soon as speed was high enough and ensure that when it did, the swing to port could be held in check. Because of the lockable tailwheel the fighter was reasonably controllable until the tail lifted, but it was essential that the interval before the rudder took over, lateral control was kept as brief as possible. It was also vital to stop the tail rising too high and making the swing worse. The pilot also had to check the air-speed indicator for the take-off speed of 68 mph – and then do nothing. Pulling back the stick to lift the 109 off the runway was potentially lethal. The machine might leave the ground but fall back to it too soon. Usually the port wing dropped first, so the pilot instinctively pushed the control column to the right to level the wings. This usually caused the anti-stall slots to open, but in extreme cases the 109 might turn over on its back and kill its pilot.

In both machines the cooling system tried to boil before the plane lifted off. The undercarriage blocked airflow into the radiators, so coolant temperature went on rising until the pilot could raise the landing gear. Checks and adjustments had to be done as quickly as possible without errors or omissions, no mean feat for a novice pilot trying to remember everything and to do what was needed accurately and quickly. On a cold day, starting the engine for the first time, the coolant would boil in around ten minutes. In warmer conditions, seven minutes would be the limit for both aircraft.

Once safely clear of the ground, conditions were easier. Pilots of Mark 1 Spitfires had to pump the landing gear up to the retracted position using a long handle. Keeping the other hand steady on the stick to avoid the nose oscillating up and down in sync with the pump strokes took a lot of practice. It was vital to set the throttle friction adjustment to prevent it closing while the pilot's hands were busy or the fighter might sink back earthwards with wheels half raised.

In addition, propeller pitch and mixture settings had to be altered to match height and power. The Bf109 was much easier as propeller pitch was controlled by the throttle setting at the time or manually by a switch on the throttle handle. The fuel injection engine freed the pilot from needing a mixture control, and raising and lowering the undercarriage was controlled by a handle on the instrument panel or push buttons on later versions.

There was still the challenge of landing safely afterwards. Pilots of both machines had a long nose and narrow undercarriage track to deal with. The problem of landing gear masking radiator intakes also returned, so lowering wheels could only be done after flying speed had decreased and touchdown was imminent. The pilot had to fly a curved path on approach, straightening up on crossing the runway threshold so that as the nose blocked his view he knew he was correctly set up for the landing.

Unlike some wartime fighters, both machines needed to sink gently to earth in a three-point attitude, with main landing gear and tailwheel touching down at the same moment. For the Bf109, this ensured the locked tail wheel could keep the aircraft straight and avoid a ground loop. But both fighters lost flying speed relatively slowly. Only by crossing the threshold at the right speed and height (a couple of feet above the ground) could the pilot 'flare' the aircraft and hold the nose up in the landing attitude. In the case of the 109 this was steeper than the Spitfire, but had to be maintained, as the speed fell away and with it the lift generated by the wings. If done properly the aircraft would sink to the ground with minimal shock and stress. This took care and skill; flaring too high would drop the aircraft on the ground like a grand piano out of a second floor window, causing damage to itself and probably its pilot.

Of course, pilots had learned the basics of flight on simpler training aircraft. Pre-war RAF officer pilots spent two years at its college at Cranwell in Lincolnshire without very much flying, and only received their pilot's wings when serving with their first squadron. But as it became clear a new war would demand huge numbers of pilots, so Cranwell-trained regulars were joined by Auxiliary Air Force squadrons. These flying equivalents of the Territorial Army 'weekend soldiers' were trained and equipped by the RAF to provide additional pilots. In the summer of

1936 the supply was increased by forming the RAF Volunteer Reserve. Recruits between eighteen and twenty-five would learn to fly RAF aircraft without having to pay. Classed as pilot cadets or pilots under training rather than given any formal rank until fully qualified, they attended an annual fifteen-day flying course at a new flying training school. The response was immediate and enthusiastic. Within two and a half years, 2,500 pilots were training with the RAFVR.

Most new pilots began with the de Havilland Tiger Moth, familiar with flying clubs and private owners alike. Pilot and instructor sat in tandem open cockpits. With fixed landing gear, a fixed pitch propeller and no electrics, it was as simple as possible for an aircraft to be still capable of flying. With engine running, it could slow down to twenty-five knots, with no unexpected vices in stalling or spinning. It was safe for practising aerobatics, but because it yawed when eased into a bank without rudder, it exposed the mistakes of learners who found co-ordinated turns difficult.

Others began with the equally light and simple Miles Magister. This was a monoplane more in tune with the aircraft the trainee pilots were aiming to fly. It first flew in March 1937 as a spruce framed, plywood skinned low-wing design with tandem open cockpits and fixed landing gear, but flaps were standard fittings.

Trainee fighter pilots would then move to more advanced trainers, like the two-seat Miles Master. With a Rolls-Royce Kestrel, a Bristol Mercury air-cooled radial or a Pratt and Whitney Twin Wasp Junior, it had a 260 mph top speed, retractable undercarriage, a variable pitch propeller and a closed cockpit. Lend Lease replaced the Master with the North American Harvard, another two-seat tandem monoplane with enclosed cockpits and a powerful Pratt and Whitney radial, retractable undercarriage and all-metal construction. The propeller gearing meant the tips

Miles Master advanced trainer. *(Wartime photo via Wikipedia)*

North American Harvard intermediate trainer – actually a post-war US Air Force Texan, virtually identical to the version used by the wartime RAF. *(USAF)*

approached supersonic speed at high revs, with an unmistakable crackling roar. Its performance and responses were like a fighter and though simple to fly, it responded viciously if mishandled.

If recovery from a stall was clumsy or mistimed, another sharp stall would follow, and spins demanded careful recovery. The cost of bridging the gap between training and combat flying was undeniably increasing. Flying high performance aircraft, even trainers, was dangerous, and the need for replacement pilots forced trainees to fly in marginal weather conditions, with rising casualties.

Finally, trainees would have to master their first flight in a fighter, usually at their first squadron. In peacetime the unit would spare time and machines to train them quickly. After the outbreak of war, this was impossible. Squadrons taking off to intercept German raids had no chance to bring new pilots up to the level needed to survive in combat, let alone to prove an asset to the squadron by shooting down enemy machines and deflecting attacks.

Early in 1940, a new step was added. An Operational Training Unit (OTU) posting enabled trainees to learn to fly a fighter far from the combat zone. This was no rest-cure as continuing casualties might mean a move to an operational squadron in the thick of the fighting after as little as ten hours flying an elderly and worn-out front-line fighter.

For Luftwaffe pilots, the sequence was similar with one radical difference. Training in the RAF was run by a separate Training Command, but in Germany it was down to the individual *Luftflotten* – Air Fleets – so was subject to operational demands. Because the Luftwaffe was a much larger force, training new pilots seemed far less urgent than for the RAF, still trying to catch up on their opponents.

With plans for a cut in army numbers, the Luftwaffe's core of experienced pilots seemed sufficient. However, the Yugoslavs were not alone in finding the Bf109 difficult to fly, whatever its value to an experienced flyer. Take-off and landing accidents from undercarriage collapses and ground loops reached epidemic proportions.

Distinguished pilots like Johannes Steinhoff and 'Winkle' Brown spoke of as many as one Bf109 in three, or a total of 11,000 machines lost to these causes over the entire war. Other estimates suggest 1500 Bf109s lost in landing accidents between 1939 and 1941, while Len Deighton in his study of the Battle of Britain says that, 'It has been estimated that 5 per cent of all Bf109s manufactured were written off in landing accidents.'[2] If true, this equates to some 1700 aircraft lost to this cause. Whatever the true figure, this showed a haemorrhage of fighter strength. As training began to buckle under Allied attacks the problem became worse.

The trainee German pilot began with the Luftwaffe equivalent of the Tiger Moth, the Bücker 131 two-seat biplane. Others began with plywood-skinned monoplanes like the Klemm L25, virtually a dead ringer for the Miles Magister. It had low wing loading and high aspect ratio wings like a glider, with a tendency to float on final approach, making it difficult for the new pilot. The later and more numerous L35 also became a standard basic trainer.

Trainees began with the A2 qualification with three correct landings, a loop, an altitude flight to 2000 metres [6600 feet] and a 300km triangular cross-country flight. The B1 certificate covered 1800 miles of flight, including a 360 mile triangular cross-country course over nine hours, a flight to 15,000 feet, a night flight of at least half an hour, three precision landings, two night landings and at least fifty logged flights. The B2 level introduced night flights and night landings with double the distance requirement, while instructors determined what kind of flying would suit their pupils best.

Those selected for fighters went to a *Jagdfliegerschule* (Fighter Pilot School) for three or four months to fly faster machines, ending with semi-obsolescent early Bf109s. After reaching an operational squadron, their survival depended on seniors teaching them basic combat tactics before meeting the enemy. To minimize losses, the Luftwaffe emulated the RAF by introducing *Ergänzungsgruppen*, the German equivalent of OTUs, for more tactical instruction and practice. Inevitably, losses of experienced pilots meant training units were under great pressure to draft flying instructors to the Air Fleets to which their schools belonged. Shortening training time meant casualties increased. So desperate was the need for new pilots that advanced trainers, biplane fighters of the late 1930s like the Arado Ar68 and the Heinkel He51 as well as captured aircraft like the French Dewoitine D520 and Morane-Saulnier MS406, not to mention at least one I-16 Rata, were sent on operational missions.

---

2. Deighton, *op.cit.*, p. 93.

The Bucker *Jungmann* was the Luftwaffe's main primary trainer – this is a Casa E3b made under licence in Spain. *(Author photo)*

To separate training and operations, and to maximize the usefulness of the dwindling number of instructors, the *Ergänzungsgruppen* linked to individual fighter *Gruppen* were replaced in summer 1942 by three much larger units based in central France, the Rhineland and occupied Poland, renamed *Ergänzungsjagdgruppe Süd*, *West* and *Ost* respectively. Despite these changes, the time devoted to pilot training shrank further and further as instructors, then aircraft and finally fuel were taken over by combat squadrons. Since the Luftwaffe did not rest front-line aces with time to pass on their experience to replacement pilots, the lack of instructors became a worsening crisis.

One reason why the Luftwaffe had begun as the most experienced and tactically able air force in the world had been its exploitation of front line combat experience earned in Spain. Because the size of the *Legion Kondor* was so small, even compared with the pre-war Luftwaffe, the value of spreading that combat experience and tactics over as much of the service as possible had been undeniable. Pilots were regularly rotated in and out of *Legion* squadrons. On completing a *Legion* deployment they were sent to spread their experience to as many comrades as possible, while they were replaced by fellow pilots best able to profit from the experience themselves. It was a sensible system and it increased the efficiency of the Luftwaffe many times over. Had they remembered this during the later battles, the service's decline would never have happened so quickly and so disastrously.

By the end of 1943, streamlining training and emphasizing the need for replacement pilots doubled the output of newly qualified fliers. Unfortunately, casualties meant that by 1944 only *Gruppe* and *Staffel* commanders had more than six months experience and most of the pilots flying for them had only been on operational service for between a week and a month!

Between the outbreak of war and the end of 1944, almost 10,000 trainees had died in accidents. To reduce this toll, it was decided the best innovation was two-seat training versions of operational fighters. For the Bf109, this was the Bf109G-12 conversion with an instructor's cockpit crammed into the fuselage behind the existing pilot's cockpit.

'The Messerschmitt bureau had first initiated studies for a two-seat version of the basic Bf109 late in 1940 as the Bf109S (the suffix letter indicating *Schule*), but the *Jagdfliegerschulen* saw little need for such an aircraft, and the study was not resurrected until 1942 when fighter pilot training accelerated with a general lowering of standards and in consequence, a markedly higher accident rate during conversion training.... A lengthened canopy was provided, this incorporating two sections hinging to starboard. The forward hinged section, enclosing the pupil's cockpit, was identical to that of the standard single-seater, but the aft section was bulged on both sides to provide the instructor with a measure of forward vision for take-off and landing. Full dual controls were fitted and full armament was normally retained owing to the considerable change in the c.g.[centre of gravity] position that would have accompanied its removal, the reduction in weight forward being coupled with the insertion of the instructor's cockpit.'[3]

A total of almost 500 machines were modified, mostly at the Blohm & Voss plant in Hamburg, and the idea proved successful in general. However, when Eric Brown took up the chance of flying one of these machines after the war, it was very nearly the end of him: 'One of my rashest ventures was to fly the Bf109G-12 tandem two-seater from the rear cockpit with no-one in the forward cockpit. I was interested to ascertain what sort of view the instructor had for landing. The answer was none! I had to make three very frightening attempts before regaining terra firma. The periscopic sight in the rear cockpit was of no use whatsoever in the vital final stages of flare, touchdown and landing run.... I would certainly not recommend the ultimate solution that I adopted of a split-S turning dive at the runway and then a burst of power to avoid cratering the tarmac, and making tail-up contact on the main wheels. After the tail dropped it was anyone's guess as to the direction in which the aircraft was heading. I certainly had not the vaguest idea....'[4]

In spite of this, the two-seater was successful from late 1943 and had enough of the original 109's qualities for wider use when 'many saw operational service during

3. Green, *op.cit.*, p. 124.
4. Brown, *op.cit.*, p. 213.

Royal Yugoslav Air Force Hurricane. *(Photo by Mark M via Wikipedia)*

the final stages of the fighting in Europe.'[5] But as the balance shifted more in favour of the Allies, and German pilots had to cope with shorter training times, their opponents could spend more time polishing their experience and tactical sense.

Introducing new pilots to combat was difficult with the need for reinforcements. For the Germans, minimally trained replacements were sent direct to operational units where more experienced colleagues had to coach them in the skills to keep them alive long enough to contribute to the unit. For the British, a complex but effective system rotated squadrons out of the fighting to cope with losses, and finish training pilots from OTUs when time spent there was cut down to a couple of weeks with perhaps ten hours fighter flying. Before July 1940 pupil pilots would spend six months at an OTU.

Only those who could fly these demanding machines with pure instinct could pause to think about what was happening around them. And every piece of good advice from those more experienced could change the odds on survival. Peter Brothers had an instructor who had flown in the First World War. Two pieces of advice proved invaluable – one, to lean his head on his shoulder in high-g manoeuvres to restrict blood flow and delay blacking out enough to win an advantage over his opponent.

---

5. Fernandez-Sommerau, op.cit., p. 70.

The other was, when missed by tracer bullets on one side, to alter course in that direction as the other pilot would be correcting his aim in the opposite direction. Another trick to disturb the enemy pilot's aim was deliberately to de-synchronize the rudder in turns so his aircraft skidded sideways to make it harder to hit.

In German squadrons a new pilot would never be issued one of the best prepared aircraft (or even any machine at all as losses mounted) over and above more experienced men. They ended up with less and less to do, and less chance of gaining the vital experience while the brunt of the campaign fell on the increasingly stressed old hands. With leave scarce or non-existent under the pressure of combat, this helped old hands build up huge scores but also increased casualties overall. Small wonder German pilots experienced combat stress as early as the Battle of Britain, which they called *Kanalkrankheit*, or 'Channel sickness'. Even in the RAF there was no scheme at first to rest combat veterans to give new pilots advice on the tactics and skills. The longer a new pilot stayed alive, the more senior his position in the squadron and the better the machine and the ground crew he would be given, improving his chances further. But the British, and later the Americans, planned long-term campaigns with correspondingly large training targets. The RAF had a huge advantage in training many pilots in the USA and the Empire countries, where flying weather was better and with no chance of enemy attack. The Empire Air Training Scheme began in Canada in 1939, creating 360 flying schools by the end of the war. This was widened to include Australia, New Zealand, Rhodesia and South Africa. By 1945 some 80,000 RAF pilots had been trained in Empire countries with even more in the USA, compared with some 35,000 RAF pilots trained in Britain over that period.

The Allied air forces also retained combat-experienced pilots on rest tours to serve as instructors, and issued late model operational aircraft for the advanced training schools and OTUs. When Allied fighter pilots could expect to amass 200–300 flying hours during training, the worsening fuel shortage forced elementary training in Germany to be given on gliders, and total training per pilot was cut back to 150 hours in 1944 and 100 hours in 1945. Ironically, at a time when Herculean efforts had been made to produce more operational aircraft, there was less and less fuel to power them and competent pilots to fly them.

There was however, one drawback to the RAF's separation of Training Command from Fighter Command, which the Luftwaffe avoided. Twenty-five years after the Battle of Britain, the former commander of 11 Group, Air Vice Marshal Sir Keith Park, was interviewed in France about the 1940 pilot shortage, and recalled, 'I was worried daily from July to September by a chronic shortage of trained fighter pilots and it was not until the battle was nearly lost that Air Staff of the Air Ministry assisted by borrowing pilots from Bomber Command and from the Royal Navy. Incidentally, in December 1940 when I was posted to Flying Training Command, I found that the flying schools were working at only two-thirds capacity and were following peacetime routines, being quite unaware of the grave shortage of pilots in Fighter Command …'

Chapter Fifteen

The Rest of the Field

At 11 am on 5 April 1917, six Bristol Fighters of 48 Squadron of the Royal Flying Corps took off to patrol the Western Front between Douai and Cambrai at 3,000 feet. They were led by Captain William Leefe Robinson VC, the first pilot to shoot down a Zeppelin on a night raid over England. With a two-man crew the Bristol F2B was fast, compact and well-armed; its 190 horsepower Rolls-Royce Falcon engine gave it a top speed of 123 mph.

With a pilot and a gunner it followed set tactics. Flying line abreast let gunners combine their fire against enemy attacks. Unfortunately, they were being stalked by brand-new German fighters, Albatros DIII single-seaters. Even more unfortunately, they belonged to the crack fighter unit *Jasta 11*, masters of their trade. Most unfortunately of all, they were led by Manfred von Richthofen, the Red Baron.

The German fighters plunged in from different directions to divide the British fire, and exploit blind spots where British gunners risked hitting their own machines or pilots. Meanwhile, the British flyers turned to enable their gunners to aim at their

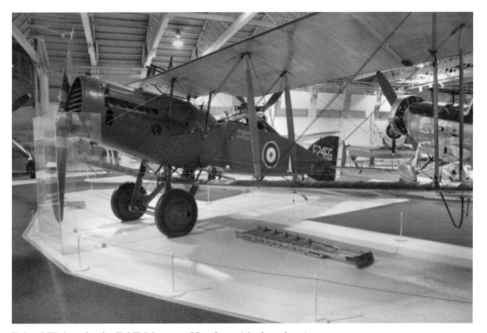

Bristol Fighter in the RAF Museum Hendon. *(Author photo)*

Bristol Fighter rear gunner's position. *(Author photo)*

Albatros fighter in Smithsonian Museum. *(Author photo)*

attackers and lost their defensive formation. One by one the gunners ran out of ammunition and the planes were shot down. Four out of the six were destroyed, and Leefe Robinson crash-landed and spent the rest of the war as a prisoner. Only one returned to base.

New tactics were clearly needed. The Bristol Fighter was thought to be fragile, so pilots flew straight and level, hoping the gunners could beat off the enemy. As this had failed, it was time the machine lived up to its name, thanks to a Vickers machine gun facing forwards and aimed by the pilot like any single-seat fighter. The transformation was amazing. Flying the F2B as a fighter let it match the performance of opponents while the gunner added a real advantage. German pilots closing in on the tail of a Bristol Fighter to deliver a fatal burst were now sitting targets for the observer's Lewis gun. Immediately after that first battle Manfred von Richthofen said in a newspaper interview that the Bristol Fighter was not fit to compare with the Albatros. Many German pilots took his words to heart and died as a result.

History seemed set to repeat itself twenty-three years later. Just after midday on 19 July 1940, three RAF fighter squadrons were sent to meet a German raid on Dover. One flew Spitfires, another Hurricanes, but the third, 141 Squadron, flew fighters similar to Hurricanes but for a large, power-operated gun-turret just behind the pilot's cockpit. These were Boulton Paul Defiants, two-seat updates of the Bristol Fighter for the age of the high-speed monoplane. It began as a different way to tackle enemy bombers in formation, protected by the crossfire of their gunners. Because British targets were beyond the range of German fighters the bombers would be unescorted, but a modern Bristol Fighter would have the speed and firepower to shoot them down.

BBMF Hurricane in flight. *(Adrian Pingstone photo via Wikipedia)*

Boulton Paul Defiant at Hendon Museum. *(Author photo)*

To meet this need, the Air Ministry issued Specification F.9/35 on 20 May 1935 for a turret-armed fighter. They wanted a top speed 20 mph faster than the latest bombers at 15,000 feet (effectively 290 mph) with a movable battery of at least four machine guns and a service ceiling of 30,000 feet. Several designs emerged, the first a modified Spitfire. The Supermarine Type 305 had a deeper and bulkier fuselage, though wings and tailplane were mainly unchanged. This provided room for a gunner sitting back-to-back with the pilot and controlling four machine guns in a streamlined barbette, set into the upper surface of the fuselage. The gunner aimed and fired the guns by remote control, and sighted his target through a transparent blister above his seat.[1]

The first design specified four Colt Browning machine guns like the Spitfire, but a later and better streamlined version would have four .303 Lewis guns. To extend his field of fire, the gunner could stand, protected by a transparent screen. His seat was geared to the barbette mounting, and he aimed through a prismatic telescope and reflector. Electric motors turned the turret and raised and lowered the guns through geared rings carrying the seat, gun mounting and sights. The Supermarine installation was almost a ton lighter than a full powered turret, suggesting a top speed of 315 mph.[2]

---

1. Price, *The Spitfire Story*, p. 25.
2. Morgan and Shacklady, *op.cit.*, p. 39.

Defiant turret and guns. *(Will Owen photo)*

On 26 March 1936 the design was shown to turret engineers Nash and Thompson, who said the mounting was flimsy and insisted the gunner should not be separated from the guns. Nevertheless the design was rejected by the Air Ministry, as was the Hawker Hotspur, since both companies were busy building single-seat fighters. Only the Boulton Paul Defiant had a fully enclosed powered turret with four .303 machine guns as used in several British warplanes. This impressed the Air Ministry enormously and initiated the specification.

Sadly it was doomed. The heavy turret crippled performance, cutting top speed, rate of climb and manoeuvrability. Unlike the Bristol Fighter, it had no forward-firing guns at all. The gunner could turn the turret until its guns faced forward, but they could not be depressed below 19 degrees above the horizontal and there was no gun sight for the pilot to aim them. Still the Air Ministry was keen. Churchill remembered the Bristol Fighter's success in the First World War and thought it the ideal future fighter. The prototype Defiant flew on 11 August 1937 and eighty-seven more were ordered followed by a 450-machine production contract. In 1938 Air Vice-Marshal Sholto Douglas, Deputy Chief of the Air Staff, ordered Dowding to form nine squadrons of Defiants for Fighter Command. Dowding realized its Achilles heel and delayed events so only two squadrons had been formed by the start of the Battle of Britain. He was absolutely right. On 12 May 1940 the first Defiant squadron, No 264, went with six Spitfires of 66 Squadron to intercept a formation of Ju88s. They shot down one German bomber. The next day, sent to attack a formation of Ju87s, they shot down four, though six Defiants were shot down by Bf109 frontal attacks. On 29 May 1940, with the Dunkirk evacuation in full swing, 264 Squadron went to intercept escorted formations of *Stuka* dive-bombers attacking British troops. In two sorties they claimed to have shot down nineteen Ju87s, nine Bf110s, eight Bf109s and one Ju88.

The German fighters seemed confused, mistaking their opponents for Hurricanes. They tried 'bouncing' them by diving from above and astern, and flew into streams of fire from the turret guns. Inevitably, combat involving turret guns involved massive overclaiming, a problem afflicting US Eighth Air Force daylight combats against massed Luftwaffe fighters. It also made using the Defiant seem far more promising than it really was.

This was the last time the Defiant deceived the Luftwaffe. Two days later, 264 lost seven machines and were retired to convoy patrol duties. When 141 Squadron entered combat on 19 July 1940 with nine aircraft to defend a convoy off Dover, the tables had turned. German 109 pilots knew exactly what to do, attacking from head-on where the Defiants could not retaliate. It was a massacre. Three Defiants escaped when the Hurricanes saw off the Germans.

The squadron was sent to defend coastal convoy 'Peewit' on 8 August and claimed twenty-one victories, eleven apparently confirmed. Flight Lieutenant Boyd was awarded five victories as German dive-bombers tried to sink as many ships as they could. It was too good to last. Two days later, five pilots were lost and two more the day after, and the squadron commander was wounded. It then moved to Drem near Edinburgh to rest and refit.

During the height of the Battle in August 1940, 264 suffered again. On 24 August it was sent to intercept a formation of Ju88s from *KG76*, exactly the target for which it was designed. Three were lost, including the squadron CO, and two days later another three were shot down with one more forced to crash-land. In four brief fights, both Defiant squadrons had been shot to pieces and withdrawn from combat. Gunners faced a specially grim fate. Bailing out meant turning the turret sideways to reach a small escape hatch. Any damage to hydraulics or electrics left them imprisoned in the event of a crash or a ditching in the Channel.

One radical revision suggested in 1940 was to replace the turret by a heavy battery of wing guns like a conventional fighter: four 20mm Hispano cannon and four machine guns, or a full dozen machineguns. The predicted top speed was 360 mph, but it never appeared as the Hurricane and Spitfire took up available production capacity. Oddly, it did appear in the American press. The September 1940 '*Popular Mechanics*' included a story under the heading '*Twenty-one gun warplane pours fire in all directions*' which read 'One of Britain's newest aerial weapons, the two-seater 'Defiant' built by the firm of Boulton Paul, fires four machine guns from a hydraulically turned turret. This enables the single-engine ship to cruise alongside an enemy bomber and pour bullets into it, a feat impossible when all guns are fixed to fire straight forward. Additional arms include fourteen machine guns in the wings close to the fuselage, two cannons synchronized with the propeller and a third cannon firing through the propeller shaft.' Sadly, it remained pure fiction, possibly to scare the Germans off. It did not work and Dowding had to retire Defiants from daytime combats.

Another role awaited them. The impossibility of using single-seat fighters against German night bombers suggested the turret-armed machine could do better than the Bristol Blenheim. Ironically the design which exposed the inadequacies of the world's fighters now lacked the speed to catch its own targets, but the Defiant's 315 mph top speed could overhaul bombers. With AI (Air Interception) Mark IV radar, the Defiant was successful enough to equip thirteen night-fighter squadrons. It could approach a bomber from below in its blind spot, or from one side or even ahead for a clear view of the target out of sight of its gunners. During the 1940–41 winter Blitz on London, Defiants shot down more enemy aircraft than any other type. It was finally replaced by faster and better-armed twin-engined machines like Beaufighters or Mosquitoes, leaving Defiants to training and target-towing.

The Spitfire's other Battle of Britain stablemate, the Hawker Hurricane, was much more practical. Designed by Sidney Camm and based on his Hawker Fury biplane fighter, it too was intended for the Goshawk but was transformed by the Merlin. It lacked the performance and advanced design of the Spitfire. Its straight taper wings were originally fabric covered, thicker than the Spitfire's and built around two spars, but with room for four machine-guns in each wing. The fuselage was a metal monocoque at the front with a fabric covered steel-tube rear framework, braced by wires and with wooden formers and stringers, less vulnerable and easier to repair. The Hurricane's wings had room for the undercarriage to retract inwards, giving a wide track for easier landings and take-offs. It could out-turn both the Spitfire and the 109, though its slower top speed made it vulnerable to 109s diving from above, but if a 109 started a turning fight with a Hurricane, it would lose. A 109 on the tail of a Hurricane would find the positions reversed in four to six turns.

Its simplicity meant the Hurricane suffered few production problems, but the first versions were more dated than the Bf109s. Wings and fuselage were covered in doped fabric, and power delivered through a two-bladed fixed-pitch propeller. There were no reflector gunsights to help aim the guns and no armour protection or self-sealing fuel tanks to avoid a fireball when hit by enemy rounds, though all these defects would be corrected in time.

The first 200 Hurricanes were complete by the end of 1938. A year later there were more than 700. Rising numbers brought improvements. Instead of simple ring and bead gunsights, Barr and Stroud of Glasgow began supplying the new GM2 reflector sight, but demand increased so quickly they had to sub-contract the order to C.P. Goerz of Vienna. The company met their obligations to the letter, even after Austria was seized by Nazi Germany in 1938. Only after four months of war did Barr and Stroud have to switch to the Salford Electrical Company.

For all its conservatism, the Hurricane had real advantages. The thick wings increased drag, but left room for wider low-pressure tyres to simplify landings and take-offs and made the Hurricane easier to handle on the ground. Placing the single radiator below the wing centre section and clear of the wheels made overheating on the ground much less likely.

Another modification made it impossible to repeat the Battle of Barking Creek. On the fourth day of war, two Hurricanes of 56 Squadron were shot down by Spitfires of 74 Squadron because their radar echoes were mistaken for hostiles. A device installed in each fighter called Identification Friend or Foe (IFF) responded to a pulse from British radar by returning a coded signal to confirm this was a friendly aircraft.

Replacing the two-blade fixed-pitch airscrew with the variable-pitch de Havilland propeller cut its 400-yard take-off run to 250 yards and the Rotol constant-speed airscrew proved even better. Ejector exhaust stacks added forward thrust. Replacing wing fabric by metal skin panels saved weight. The front screen was protected by armoured glass to deflect a head-on attack and from the spring of 1940 armoured panels protected the pilot's head and back.

Originally the Hurricane's performance was best at low altitude where it could out-turn the Bf109 and bring its guns to bear quickly, but its weakest point was the fuel tank ahead of the pilot. If hit, the resulting fireball could kill the pilot or inflict horrifying burns before he could escape. The wing tanks were self-sealing, but fitting this to the fuselage tank was more difficult. Dowding pressed for action and soon seventy-five machines a month were being converted. Problems remained, and for an equivalent number of Hurricanes and Spitfires hit by German fire, the Hurricanes were five times as likely to be brought down. For all that, they shot down twice the enemy aircraft falling to Spitfires during the battle, reflecting the relative numbers of the two types.

As the battle subsided and Spitfire numbers increased, Hurricanes were switched to other roles. Modern fighters were needed in the Middle East to protect Malta and North Africa from enemy air attacks. With Fighter Command enjoying top priority for Spitfires, only the Hurricane could fill the gap. Fitting a large Vokes air filter under the engine to prevent sand damage and hanging a pair of 40-gallon ferry tanks under the wings enabled Mark 1 Tropical Hurricanes to fly to the Mediterranean.

The Mark II Hurricane entered squadron service in September 1940 with the Merlin XX and a two-speed supercharger so the pilot could select Full Supercharge for high altitude performance or Medium Supercharge for lower down. With extra power and a roomy wing, heavier armament could down more targets. One option was an additional pair of .303 machine guns in new gun-bays on the outer side of the landing lights, making twelve in all. A better alternative was 20mm cannon, starting with a single Oerlikon drum-fed weapon in a pod slung under each wing, but this tended to jam and had a terrible effect on performance.

Like the Spitfire, fitting two 20mm Hispano cannon in each wing proved difficult. The weapon was meant to be clamped to the engine like the Bf109's Mauser for the bracing effect of a rigid mounting. Wing flexing twisted the guns, so rounds jammed in the breech. These problems were eventually solved, but heavy guns slowed the machine down. The more heavily-armed Hurricanes were Mark IIBs, and the later Mark IIC had wings modified to carry bombs or long-range tanks. These extra loads slowed the machine still more, so as soon as more Spitfires were available, Hurricanes switched from fighter duty to ground-attack. Later versions were more capable. The Mark IID had underwing pods each holding a 40mm anti-tank gun and fifteen shells. The pilot was surrounded by armour to protect himself, the engine and the cooling system. The Mark IV wing could carry 60lb air-to-ground rockets and more powerful engines like Packard-Merlins made in the USA, maintained its value for most of the war.

The Hurricane also went to sea as the Sea Hurricane Mark 1A, or 'Hurricat', a truly desperate measure to defend convoys from Focke-Wulf FW200 *Kondor* long-range maritime patrol planes which bombed ships and reported their positions to patrolling U-boats. Before escort carriers, the only way to provide fighter cover was to fit a deck catapult on a modified merchant ship to launch the Hurricane when an intruder appeared. The pilot would then shoot down the *Kondor* before bailing out close to the convoy to be picked up by an escort.

The tactic worked. Fifty war-worn machines were converted and eight actually launched, shooting down six enemy aircraft for the loss of a single pilot. Larger cargo ships carried flight decks as escort carriers where fighters could land and take off on multiple missions during a convoy's voyage. The ever-versatile Hurricane was catapult launched and stopped by arrester hook after landing. Hundreds of Sea Hurricanes served on Royal Navy carriers, still short of modern fighters.

With non-folding wings, they could only be stored below decks on the largest fleet carriers. On smaller ships they had to be worked on in the open, a daunting prospect in heavy weather. The Hurricane filled the gap well, but the Navy longed for the Spitfire, with performance to match any potential opponent. The Germans felt the same about the Bf109. Outstanding performance meant both were made into carrier-borne fleet fighters from original designs as short-range interceptors for grass airfields.

In May 1939, Richard Fairey, builder of the two-seat Fairey Fulmar naval fighter and the three-seat Fairey Swordfish torpedo biplane, was asked at an Admiralty meeting to build 500 naval Spitfires as well! Fairey insisted this would mean delays to current production. The Navy reminded him Spitfires were needed for the Fleet Air Arm with a veiled threat that failure to cooperate might mean cuts for future Fulmars. Fairey kept his nerve, offering to turn out more Swordfish instead! Unsurprisingly, replacing 350 mph monoplane fighters by more 130 mph three-seat biplanes was not what the Admiralty wanted, though they failed to press the point. Had Fairey acquiesced, the Navy might have had a seagoing Spitfire from early 1940 instead of early 1942.

The Germans faced similar problems with naval 109s. The toughest was the lack of an aircraft carrier from which they could operate! Two were being built – *Graf*

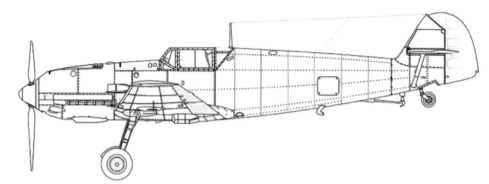

Bf109T carrier fighter drawn by Bjorn Huber. *(via Wikipedia)*

Zeppelin and *Peter Strasser* – but the *Kriegsmarine*'s low place on the Nazi priority list meant continual delays. Admittedly, they faced a huge task. With no carrier experience, they modelled their ships on obsolete British battle-cruiser conversions like *Courageous* and *Glorious*. With restricted flight decks, aircraft would have to be catapult launched and three designs were picked for carrier use; the Fieseler Fi167, a torpedo biplane like the Fairey Albacore, a navalized *Stuka* and the Messerschmitt Bf109T, for *Träger*, or carrier.

Aircraft were modified at Travemünde on the Baltic coast. The main runway was painted with the outline of a flight deck, with a working arrester wire and an electromechanical braking device so pilots could practise deck landings with moderate realism. From 1937, a test and development programme began turning the Bf109, the Ju87 and the Fi167 into capable naval machines. All three were modified and tested on the deck-landing runway rig. They were then hoisted on trolleys to be launched from a catapult on a barge moored in the estuary of the Trave river near the test airfield. The development programme with the modified aircraft completed 1500 successful arrested landings, and catapult tests involved fifteen by modified Ju87s and just four with a stopgap Bf109D.

Originally the German carrier was to be a patrolling reconnaissance vessel, with ten fighters, twenty torpedo and reconnaissance machines and thirteen dive-bombers. This was changed to copy carrier tactics in Britain, the USA and Japan. The Fieseler biplane was cancelled as obsolete, and the ship would carry twelve Ju87 dive-bombers (modified for torpedoes) and thirty Bf109s.

Since the latest 109 was the *Emil*, the RLM's *Technische Amt* asked Messerschmitt's design team to produce a carrier version of the 109E in late 1938. There were two problems; the tall, narrow undercarriage and the high ground angle of the 109 made it hard to land on a carrier deck, and the airframe needed strengthening for catapult launches, all of which meant more changes than the naval Spitfire or Seafire.

Extended outer wing panels increased the span by twenty-one inches. Thanks to uniform taper this was straightforward. A break point outboard of the gun bays allowed wing folding to save hangar space, though the flaps had to be detached first, which increased delays before launches and after recovery at the end of flights. Retractable spoilers were fitted to steepen the glide on approach, and strengthened attachment points for catapult harness and an arrester hook were fitted to the fuselage frames.

Fieseler completed the detail design and converted ten 109E airframes into 109T-0s, followed by sixty 109T-1 production fighters. Work stopped on the 109T along with the carriers, but restarted later. The first batch of 109T-1s was completed

Women assembly-line workers working on a Hurricane in a British aircraft factory in 1942, showing the conventional structure of the aircraft with its tubular engine bearers, wide track inward-retracting undercarriage and fabric-covered rear fuselage frame. *(Official wartime photo via Wikipedia)*

early in 1941, but the carrier project was suspended and the factory ordered to remove naval equipment and fit racks for bombs or long-range tanks as land-based 109T-2s. The short landing and take-off capability of the extended wing was used for small airstrips in occupied Norway.

After that, 109T fortunes rose and fell with the carriers. At the end of 1941 the T-2s were brought back from Norway and rebuilt to original T-1 specification. By the end of 1942, forty-eight Bf109T-2s had completed this process and all but two were stationed on an airfield at Pillau, on the Baltic coast of East Prussia, waiting for the carrier. Two months later, all work was cancelled and the fighters returned to normal Luftwaffe service from April 1943.

By then the *Emil* was no longer fit for front-line combat and from spring 1942 plans were produced for an improved Me155 with as many components from the current Bf109G as possible. It was powered by the same DB605 engine with the same fuselage, but with wings completely redesigned. Extra internal space let the undercarriage retract inward for a wider and safer track for carrier landings. The wings could be folded easily and quickly, and the aircraft had an engine-mounted cannon with two more cannon and a pair of 13mm machine guns mounted in the wings.

Once the carrier was cancelled, the design looked for a new role. With increasing high-altitude American bomber raids on German targets, and rumours about the spectacular performance of the new B29 being prepared to bomb the Japanese home islands, the performance and wingspan of the Me155 suggested a useful high-altitude interceptor. The engine was replaced by a DB628, a DB605 with a two-stage mechanical supercharger, the cabin was pressurized and the service ceiling predicted to be over 46,000 feet.

The new engine was fitted to a Bf109G, which reached a height of more than 50,000 feet in May 1942. The *Technische Amt* replaced the two-stage mechanical blower with a turbocharger driven from the engine's exhaust, fitted behind the pressurized cockpit. This meant lengthening the fuselage, and longer wings were made by fitting standard Bf109G wings to a simple, non-tapered centre section. The *Gustav*'s tailplane was matched with the taller fin and rudder from the experimental Me209.

Progress was disappointingly slow, and by late summer 1943 the *Technische Amt* transferred the project to Blohm and Voss as the BV155. Later Messerschmitt's role was dropped, but problems delayed the prototype's first flight to 1 September 1944. The second prototype had a redesigned cockpit canopy, cut-down rear fuselage and larger fin, but did not appear until February 1945, too late to make any difference after the B29s all went to the Pacific.

The Fleet Air Arm had the opposite predicament with several operational carriers, but no first-class fighter to fly from them. Grumman Wildcats were available under Lease-Lend, but though purposely designed as carrier fighters their performance was inadequate against land-based opponents like the Bf109G and in performance terms only the Hurricane and Spitfire would do. The Admiralty set its heart on the Spitfire, but had to join the queue for new machines.

The Navy's wait for first-class fighters ran on into early 1942. A full review of naval fighter requirements accepted the Spitfire was the best option for a day fighter with new long-term priorities; a large order for the basic version with the introduction of a folding wing version followed by an upgraded machine developed from the Spitfire Mark VIII and an ultimate Griffon-engined version.

Extra power was needed to prevent extra weapons and airframe strengthening for naval service adding weight and cutting performance. The Seafire Mark VC had stronger fuselage frames and longerons, and wings reinforced to carry bombs and fuel tanks with 25lbs of extra armour plate and heavier undercarriage components to cope with deck landings. These changes increased weight by 6 percent and cut top speed by 15 mph.

The problem of landing high performance fighters within the confines of a carrier flight deck remained. On 10 January 1942, Lieutenant Commander Peter Bramwell, CO of the Royal Navy Fighter School, landed a Mark V Spitfire fitted with an arrester hook on the deck of the fleet carrier HMS *Illustrious* anchored in the Clyde estuary. After seven take-offs, four catapult launches and eleven landings, he expressed concern over the poor view over the nose on landing. He recommended pilots fly a curved approach, but felt it might be impossible to operate the Spitfire from the tiny, exposed decks of new escort carriers.

Nevertheless, forty-eight Mark VB Spitfires were converted by fitting hooks and catapult spools into the Seafire 1B. Some were loaded aboard HMS *Furious* to provide the combat air patrol during an operation to fly Spitfire reinforcements for beleaguered Malta, then under crippling Axis air bombardment. Disappointingly, these early stopgaps were not up to the task and when launched against Ju88s watching the convoy, the German aircraft escaped relatively easily.

By October 1942, fifty Seafire IICs provided enough aircraft for four squadrons, but improvements were still needed. The standard Merlin 45 produced peak power at 15,000 feet, too high for most naval combat and German aircraft still stayed ahead of them, until the arrival of the Merlin 32 version when the Ju88 proved vulnerable at last. This had shortened supercharger blades for low-altitude performance. With this and a four-blade airscrew, the Seafire LIIC's peak power increased from 1210 to 1640 hp, a rise of more than one-third. It could climb to 20,000 feet more than two minutes faster than the normal IIC, crucial in combat. Its take-off run was shorter and when later versions had clipped wings to improve the roll rate, speed increased again. It was produced by Westland and Cunliffe-Owen Aircraft of Southampton, as Supermarine and Castle Bromwich lacked spare capacity.

Production began so quickly that both marks of Seafire joined the Navy's air strength on 15 June 1942. The reason for haste was the Royal Navy's short-term priority. Carrier-borne Seafires were needed to protect seaborne landings in North Africa in November 1942, and off Sicily and Southern Italy during 1943. Since this meant flying from small and slow escort carriers sailing further inshore than the large fleet carriers, it was vital that this worked.

On 11 September 1942, Captain Eric Brown was sent to land a Seafire on an escort carrier. At half past one in the afternoon, he approached HMS *Biter* in the Clyde estuary. After flying a circuit at 400 feet, he turned in for his landing, sideslipping to have a clear view of the deck. As he neared the round-down [the angled stern portion of the flight deck], the speed dropped close to the stall as he straightened up and touched down. Only then did he realize there was no-one on deck and the arrester wire was lying flat on the plating. When the Captain welcomed him and explained everyone was at lunch, he saw the ship was flying the 'G' flag warning him off as she was facing at 25 degrees to the prevailing wind. This created a wind of just thirteen knots over the deck, when the trial should have begun with thirty-two knots over the deck, reducing in stages to a safe limit of twenty knots!

What could have been a terrible mishap had proved the feasibility of landing a Seafire on an escort carrier. All the same, others were wary of Brown's sideslip technique in less expert hands. In fact Brown himself found a second landing on HMS *Biter* tore off his arrester hook, forcing him to run his starboard wingtip into the island to stop the aircraft in time.

To begin with, Seafire performance did not automatically fit them for carrier operations. For the Salerno landings, nine naval fighter squadrons and two additional flights, numbering 106 fighters, were loaded aboard four escort carriers, HMS *Attacker*, HMS *Battler*, HMS *Hunter* and HMS *Stalker*. With a 450-foot flight deck and a top speed of seventeen knots this proved very demanding for the pilots, and deck-landing accidents accounted for forty-two Seafires on the very first day.

Afterwards Rear Admiral Philip Vian, commanding the carriers, highlighted three main problems. He claimed the Seafire's undercarriage was too weak for deck landings, it was too slow to catch FW190 fighter-bombers over the beachhead, and its range was too short to stay on patrol for two hours, even with an extra forty-five gallon fuel tank. The slow speed of the escort carriers meant undercarriage failures were common. Many pilots tried to avoid this, with the aircraft floating over the wires to crash into the barrier. Others landed off-centre or stalled, with the same result. However, in Operation Dragoon, the Allied landings in southern France in August 1944, growing experience meant accidents were much rarer.

Two Seafire IICs tried out experimental rocket-assisted take-off equipment first at Farnborough and then on HMS *Illustrious* for sea trials. Rockets were mounted on each wing root with their blast aimed downwards. The pilot revved up the engine against the brakes and held starboard rudder to counter the swing as the brakes were released and the Seafire rushed forward. At a pre-set point he pressed the rocket-firing button and held the fighter in its normal three-point attitude until it left the deck. Once airborne he could raise the undercarriage and flaps, and drop the expended rocket tubes. Sometimes a rocket might fail to ignite, but their closeness to the centre line kept the plane controllable, but damage from rocket exhaust meant the project was dropped.

Operating from larger and faster fleet carriers further offshore in the Pacific avoided these problems, but other changes were needed to adapt the Seafire to shipboard life. Hangar storage aboard fleet carriers needed folding wings, but this proved surprisingly simple. To make Mitchell's superb wing fold without reducing strength or increasing weight, a straight fore-and-aft manual upward fold was fitted at the outboard end of the wheel wells, with a second downward fold at the wingtip joint, so it could fit below the sixteen-foot high hangar roof. The new wing, fitted to MA970, the prototype Seafire IIC, had 90 percent of the torsional rigidity of the old wing, and strengthening the main spar and fitting the joints and locks only added 125lbs weight. Seafire Mk III production began in April 1943 and it soon became the most numerous variant of all, with a total of more than 1200 in service from November 1943.

With the new Seafire, naval aviation could succeed at last. Though the other carrier navies of Japan and the USA stressed the importance of range and robustness for carrier fighters, the Seafire proved unrivalled in the closing stages of the Pacific War under the onslaught of the Japanese kamikaze suicide attacks, 'From October 1942 until August 1943, the Seafire held the crown as the fastest carrier fighter afloat, being eclipsed only by the introduction of the updated A6M5 [Zero] and the [Grumman] F6F-3 [Hellcat]. While its low to medium-level performance was respectable, the Seafire's rate of climb and acceleration were remarkable. The F.III climbed at 3250 ft/min up to 10,000 feet, some 1250 ft/min better than the Hellcat or the Vought F4U Corsair and 750 ft/min better than the Bf109G, Focke-Wulf FW190A or A6M5. For sheer acceleration the Seafire had no peer. In 1945 it was still the fastest and steepest climbing Allied naval interceptor. That turned out to be of great tactical value, because once targets had been identified on radar, the Seafire required less distance and time to reach any given altitude….'[3]

This whetted the naval appetite to convert more powerful versions of the Spitfire into faster Seafires. The first Griffon-engined Spitfire, the Mark XII, dealt with FW190 fighter bombers attacking at high speed and low level over the Channel. With extra power, clipped wings for low-level agility, and twin 20mm Hispano cannon and machine guns, they forced the raiders to switch to night operations, so production of the Mk XII was limited to 100 aircraft.

Strengthening the airframes of the Mark VIII and Mark IX Merlin-engined Spitfires would add too much weight, but the massive power of the Griffon engine offered the ideal combination for the next generation Seafire, the Mark XV. There were some problems. The Griffon turned opposite to the Merlin, so torque reaction pushed the aircraft to starboard on take-off. Using 15lbs of boost, this caused the

---

3. Article in *The Aeroplane*, August 2010, pp. 32–37 'From Torch to Iceberg' by Donald Nijboer.

fighter to hop sideways, with a danger of colliding with the island. Take-off boost was limited to 10lbs to eliminate the problem.[4]

Too late for World War II, Seafire XV production was cut back to a total of 434. However, by this time the final thirty had a cut-down rear fuselage and streamlined bubble canopy to improve visibility, and revised torsion link undercarriage suspension to withstand deck- landing shocks. Changes to catapult launches allowed heavy catapult spools to be replaced by a single hook below the centre section and a holdback point at the tail.[5] A sting-type arrester hook moved backwards when lowered into the best position to stop the fighter reliably.

Ironically, after its late introduction, the Seafire would see the final combat appearance of the Spitfire family. The 1950s Korean War involved the Seafire 47 with two-stage supercharging, fuel injection and contra-rotating propellers to eliminate torque reaction. It was armed with four 20mm cannon and up to three 500lb bombs or eight air-to-ground rockets. It carried rocket-assisted take-off boosters, allowing launches every fifteen seconds rather than the full minute needed to reload a catapult. Moving the wing fold further outboard eliminated the need for a second fold and later machines had hydraulic folding equipment. Yet enough of Mitchell's design remained to make it a superb combat machine. Ninety Mark 47s were produced up to March 1949. It retired from front-line naval service in November 1950, though lingered on in training and Reserve squadrons until 23 November 1954. By this time the last RAF Spitfires had already left production and front-line service.

4. Donald, *Air Combat Legends*, p. 120.
5. Brown, *op.cit.*, p. 138.

Chapter Sixteen

The Battle of Britain – The Opening Moves

In early summer 1940, the task facing Hitler's Luftwaffe appeared straightforward. Defeating the RAF over southern England would enable a German invasion, followed later by the defeat of the Soviet Union. With more aircraft and more trained and experienced pilots, with better weapons and better tactics, the Germans believed they would win their initial victory within weeks.

Sadly for them, events proved very different. Winning aerial supremacy over England was actually impossible, crushing the RAF was impossible, and mounting a successful invasion was impossible for several reasons. Instead the Luftwaffe experienced its first full-scale defeat at the hands of the RAF, and its losses in aircraft and experienced aircrew were blows from which it would never fully recover.

Its fate had been decided at birth. Germany lacked many essentials for modern war from iron ore to industrial capacity, bullets to ball bearings, machine guns to motor

Hurricanes of 1 Squadron at Wittering in 1940. *(Official wartime photo via Wikipedia)*

vehicles. From the Nazi takeover of January 1933, their rearmament programme strained to remedy those shortages. Propaganda was exploited to convince potential opponents that Germany's forces were so powerful and well-equipped that resistance was pointless and defeat certain. If war came, Germany had to win quickly before shortages crippled the campaign. The whole idea of *blitzkrieg* ['lightning warfare'] was for surprise and speed, rather than strength, to crush the enemy.

Yet German military invincibility was a myth. Hitler's pre-war campaigns insisted no nation could oppose the armed might of the *Reich*. The reality was different. When the *Wehrmacht* reoccupied the Rhineland, it was ready to retreat immediately had France or Britain showed signs of fight. Because the myth persuaded them into inaction, Hitler's gamble succeeded and he annexed Austria in March 1938. When German troops, tanks and motorized infantry crossed the Austrian border they found only cheering crowds, a drive in the park rather than an invasion. This was just as well given a series of worrying failures.

Germany's new tanks broke down all the way along the highway from the frontier crossings in Bavaria to the Austrian capital. General Guderian, commanding the panzer units, admitted that 30 percent, almost one third of his armour, had stalled on this unopposed jaunt. General Jodl, the *Wehrmacht's* operations chief, insisted breakdowns amounted to 70 percent of the force.[1] What might have happened against an opponent who fought back?

In fact, Germany's limited economy prevented her army from being fully mechanized. Most infantry units moved on foot and only the most elite divisions (ten panzer and six motorized) carried troops in lorries or armoured personnel carriers; the rest relied on horses. Carrying troops, towing artillery pieces and bringing up supplies required a horse for every four soldiers. Even the fast moving *blitzkrieg* which crushed Western Europe depended on a few first-line divisions with the best manpower, equipment and weapons. The remaining 118 divisions followed to occupy territory once the fighting was over. Yet the Germans knew the decisive factor was speed. Hitting enemies with a succession of heavy blows kept them off balance and prevented them realizing the scales were not tilted as heavily in Germany's favour as events and Nazi propaganda had suggested.

It worked well at first. Germany conquered Poland in just over three weeks, Norway and Denmark in two months and Holland, Belgium and France in another month and a half. With 129 divisions against Britain's 29, Britain seemed certain to be next. Figures suggested the Luftwaffe was three times larger than the RAF. What could possibly go wrong? Launching this weapon against British defences must surely knock out Germany's remaining opponent. Another quick victory would ease the pressure to replace Luftwaffe combat losses. It would reduce the need for new pilots and experienced pilots to instruct them. Meanwhile, the system to pass

---

1. William Shirer, *Rise and Fall of the Third Reich*, p. 427.

Battle of Britain Spitfires from 41 Squadron. *(Via Alfred Price)*

on knowledge and experience from veterans to less experienced comrades, which worked so well in the Spanish campaign, simply withered on the vine. As opponents caught up, the Luftwaffe's advantages quickly drained away. To make matters worse, after the French surrendered, weeks passed without orders for the next phase of the war, as service chiefs wanted to capitalize on the military advantage while it lasted. However, the *Führer* was determined to give Britain every chance to agree terms and avoid a damaging and exhausting campaign. Consequently bombing raids against Britain were sporadic and uncoordinated.

When RAF Bomber Command began raiding Germany, Göring wanted massive revenge attacks. Hitler disagreed, hoping the British would negotiate. He saw Chamberlain's presence in the government as a positive sign, even though the former Prime Minister was now convinced the dictator's word was worthless. Even so, Hitler was encouraged by the refusal of Lloyd George to join the Churchill administration. If this prominent politician pushed for peace talks, then surely popular discontent might bring down Churchill? In the meantime he offered England the same terms – a free hand for Germany in Europe, and no attacks on the British Empire. England would remain a docile puppet, preferably under Lloyd George, with Edward VIII replacing George VI on the throne.

Surprisingly, to him at least, Hitler's proposals were brusquely rejected. Invasion seemed the only way to avoid a war on two fronts should he attack Stalin's Soviet Union. But Germany had no specialized landing craft, so would have to use strings of towed barges to carry troops and supplies across the turbulent and tide-swept

Channel. Their slow speed would leave them vulnerable to air and naval attack during an initial landing and the follow-up of supplies and reinforcements. Only the Luftwaffe could break this deadlock. Food stores could be destroyed and the population terrorized by bombing to trigger an uprising. If this worked, it might bring an unopposed German landing and a peaceful occupation like that of France.

Then came Oran. In spite of Hitler's promises to leave the French fleet unmolested in its anchorage in French North Africa, Britain could not afford the risk. The Admiralty ordered Force H[2] to sail off the port and demand the French put their ships safely out of reach of German or Italian forces. The French refused, and eventually at six minutes before six in the late afternoon of 3 July 1940, the Royal Navy opened fire. Ten minutes of heavy bombardment was enough. Three French battleships were sunk or forced ashore and only the *Strasbourg* made its escape to Toulon where she was scuttled in 1942. Almost 1300 French sailors were killed and British commanders were appalled by what they had been ordered to do. Nevertheless, it proved effective. Vichy France threatened war with Britain, sending aircraft to bomb Gibraltar, but sceptics and opponents realized Churchill's rhetoric was based on grim reality. In the United States, Roosevelt was convinced Britain would endure, and deserved all-out aid. In the words of one commentator, 'If Britain was prepared to do this to her ally, what might she do to her enemy?'[3] Another lesson for German invasion planners was even more discouraging. If the Royal Navy could inflict this kind of catastrophe on heavy warships in a protected anchorage, what could it do to slow and almost unprotected barges creeping across the Channel at night?

For the moment, all the Nazis could throw at British defences was propaganda. Goebbels pressed the British people to seize Churchill and hang him in Trafalgar Square, while Hitler tried to ignore the implications of Oran. Happy that France and Britain seemed bitter enemies, he wasted two more weeks insisting sense would prevail and that the attack was merely a final desperate throw of the dice before Britain came to terms. In the meantime he sent confusing signals to his armed forces over invasion preparations.[4]

On 14 July Churchill delivered a defiant speech declaring 'We shall seek no terms, we shall tolerate no parley; we may show mercy – we shall ask for none.... Let all strive without failing in faith or in duty, and the dark curse of Hitler will be lifted from our age.'[5] On 16 July, Hitler responded by signing Führer Directive No 16 in which he announced that since England, 'Despite her hopeless military situation,

2. A heavy British naval force based at Gibraltar, at the time consisting of the battlecruiser HMS *Hood* and the battleships HMS *Valiant* and HMS *Resolution*, supported by the carrier HMS *Ark Royal*.
3. Derek Robinson, *Invasion 1940*, p. 92.
4. *Ibid*, p. 96.
5. *Ibid.*, p. 100.

still shows no sign of readiness to come to an understanding, I have decided to prepare a landing operation against England and *if necessary* [Author's emphasis] to carry it out.'

One last peace attempt remained. On 19 July 1940, Hitler spoke at the Kroll Opera House in Berlin, mixing threats with persuasion. All Britain had to do was compromise, and war and suffering could be avoided. Under appeasement this might have worked, but Britain was growing stronger and Hitler misread his opponents' mood. Within an hour the Berlin-born journalist Sefton Delmer rejected his proposals over BBC radio with blistering contempt. Even then Hitler stayed his hand. On 21 July the Luftwaffe was supposed to send more than 200 bombers to attack Southampton. He postponed the raid.

His service chiefs pulled in different directions. The *Kriegsmarine* wanted a narrow invasion front they might defend from the Royal Navy, but the Army wanted a broader-based landing area to land more troops more quickly. Meanwhile, Göring was convinced his Luftwaffe could do the job on its own. Even the level-headed Erhard Milch, promoted Field Marshal after the French surrender, argued that immediate landings by parachute troops could seize enemy airfields as the British reeled from Dunkirk. If they held them long enough, German fighters could be based there and German transport aircraft could fly in supplies and reinforcements. This possibility worried the British and was the main reason for the Local Defence Volunteers – the Home Guard. However, German shortages of men and material had already made it impossible. The plan depended on the Luftwaffe's Ju52 three-engined transports. The opening day of the western campaign, 10 May, cost one-third of these specialized aircraft, and by the French surrender just over half were left. Meanwhile the Luftwaffe was only growing slowly in spite of Hitler's orders for a five-fold expansion.

When war began, one fifth of the bomber squadrons, one third of the Ju87 squadrons and at least half the fighter squadrons were still below strength. Since then, continuing losses had not been replaced. Sixty-seven 109s had been lost over Poland and 367 fighters, most of them 109s, over France. While the RAF lost 453 Hurricanes in the French campaign, most were destroyed on the ground by surprise raids. More worrying for the Luftwaffe was the fact that only seventy-five Hurricanes had been downed in combat.

So a damaged Luftwaffe faced its first real test against a battle-ready air force with first class machines and a sophisticated defence system. It was essential German commanders had a clear set of objectives and the strategy needed to achieve them – but they did not. The RAF on the other hand had two simple aims – to defeat German attacks and to survive long enough and in sufficient numbers to continue defeating them until the Luftwaffe gave up.

The Luftwaffe's original purpose was to support the army, but tactics for a fast-moving land campaign were useless against an enemy on the far side of the notoriously rough and unpredictable Channel. Hitting targets across this formidable obstacle meant flying deep into hostile skies and facing enemy fighters with an excellent direction system. In addition, the Bf109s' limited range left them prone to running short of fuel in the heat of combat and facing the threat of ditching and the prospect of drowning.

Luftwaffe propaganda threatened massive bombing raids to disrupt morale and destroy the capability to fight: but how to deliver the reality? The delays in developing powerful aero-engines, which hindered the Bf109's progress, had an even worse effect on the Luftwaffe's bombers. There were no heavy bombers and when efforts were made to create them, engines had to be grouped in pairs to provide enough power. As with the British Avro Manchester, predecessor of the brilliant Lancaster, the double engine installations of the Heinkel He177 would be chronically unreliable.

So the Luftwaffe would have to deliver its strategic punch with medium bombers in large formations, with heavy escorts, to survive in enemy airspace. The bomber may indeed always get through, but doing so in large enough numbers depended on 109s protecting them from the RAF's Hurricanes and Spitfires.

Many Battle of Britain myths remain potent decades afterwards. The Spitfire and the Bf109 were so closely matched that victory would go to the bravest, better trained and most skilled and experienced pilots. Unsurprisingly, both sides believed in the superiority of their own men and machines. They did agree on one factor though, which proved catastrophic for the Luftwaffe. Years of pre-war propaganda stressed its immense numerical advantage over the RAF, so they saw little need to increase production and replace losses, while their own intelligence assessments consistently under-estimated RAF strength and the number of new aircraft being produced. As a result, German propaganda spurred the British into boosting aircraft production so sharply that they outstripped German output as early as May 1940, and remained in the lead thereafter. On the other hand, complacency over Fighter Command's radar-based defence system ignored what happened when Bomber Command attacked German naval installations in daylight in the opening weeks of the war, only to be cut to pieces by German fighters directed by radar.

However, the Germans failed to realize the value and vulnerability of Britain's visible coastal radar stations and their role in a rapid-response defence system. German radar was used by the navy for target finding and fire control, and for protecting its coastal bases, which is why RAF raids were intercepted. Yet the Luftwaffe used radar only for anti-aircraft gunnery and was therefore half-hearted over attacking British radar. This was because General Martini and a team of Luftwaffe signals experts had flown in a German airship up and down the English coast during summer 1939 listening for British radar signals. They searched the frequencies used by their own radars and heard nothing. At the time the relatively crude British radar used much longer wavelengths (10 metres rather than 240 centimetres), so they picked up no signals. This resonated with Hitler's short-war strategy. He saw no need for defensive radars as the British had no early warning system and were facing certain defeat.

The fact remained that contrary to the mythology, the air forces massed on opposite sides of the Channel were closely matched in fighter strength. Luftwaffe records show that earlier losses had left the Air Fleets facing the RAF with some 750 Bf109s available for operations, against some 600 – 700 Hurricanes and Spitfires. However there were complicating factors. Hurricanes were not as competitive with the Bf109 as the Spitfire and British fighters had also to hold the 1000 or so German bombers at bay.

There are many definitions of the true beginning of the Battle of Britain. For the fighter pilots, everything would depend on courage, skill and the capabilities of their machines. At last, the relative vices and virtues of the Spitfire and the Bf109 would be exposed for the world to see, but history is complex and the battle would also expose the strategy and tactics of the fighting powers. Did the Germans know how best to use their forces to defeat RAF Fighter Command? And even if they did, would a successful landing bring them victory? Or would the British devise a system to defeat this aim and keep them fighting?

Technical snobbery ensured most shot down German pilots insisted they had fallen victim to the elegant and deadly Spitfire rather than the more pedestrian Hurricane. Yet the fighters of 11 Group, covering almost 20,000 square miles of southeast England from Lowestoft in the north to the Channel in the south, and from Manston in the east to Southampton in the west, contained two Hurricanes for every Spitfire. The Germans claimed the Hurricane was easy to shoot down, but in the hands of a skilled pilot it was a killer. Czech pilot Josef Frantisek was the highest scoring RAF Battle of Britain ace with seventeen enemy aircraft destroyed. Nine of them were Bf109s. Frantisek chose to fly with 303 Squadron, a Polish Hurricane unit with an astonishing record. After spending weeks learning RAF procedures and enough English to communicate by radio, they would be declared operational on the last day of August 1940. In hard fighting during September they were credited with 126 kills, the highest total in Fighter Command. Their station commander flew with them on one patrol to verify their claims and realized they were justified. They closed with the enemy to very short range and their first-hand combat experience made them deadly opponents. Their claims to loss ratio of fourteen to one was fourth best in the whole of Fighter Command, and higher than any Luftwaffe unit.[6]

However, the Spitfire gave Bf109 pilots most food for thought. Clashes over Dunkirk underlined that the RAF would be no pushover. As the Germans closed in on British lines, the Bf109 units followed up the advance, leaving them at a disadvantage against RAF fighters crossing the Channel at almost its narrowest

6. Bungay, *op,cit.*, pp. 174–5.

point from bases in Kent and Sussex. Spitfires inflicted serious losses on German bombers over the beaches and on their escorting Bf109s. During the nine day evacuation the RAF lost 177 aircraft, mostly fighters, destroyed or damaged, but the Germans lost 240.[7] Nonetheless BEF survivors back in England often attacked men in RAF blue for their apparent failure to beat off German attacks. The Luftwaffe, on the receiving end of RAF attacks, had no such illusions.

Nevertheless, Luftwaffe commanders believed they could defeat or destroy the RAF. Göring said after Dunkirk, 'My Luftwaffe is invincible. And now we turn to England. How long will this one last – two, three weeks?' The commander of *Luftflotte 3* covering central and Northern France, General Hugo Sperrle, was even more scornful saying at about the same time, 'How long would they last in battle, they ran from Dunkirk, they deserted France completely for the safety of home. England is there for the taking.'

For all their confidence, the campaign began slowly. While waiting for Hitler's direct order the Luftwaffe began what they called the *Kanalkampf* [Channel Battle] to wear down RAF resistance by bombing coastal convoys and provoking RAF fighters to react. Either the escorting 109s would shoot them down, or they would be worn out by the time the full strength campaign began in August. It would also give Luftwaffe intelligence a better idea of RAF strength, responses and tactics.

The battle began on 9 July 1940 with skirmishes against Luftwaffe reconnaissance and weather-reporting aircraft. Perfect summer weather is another Battle of Britain myth. Often rain and low cloud ruled out operations. With prevailing winds from the west, the British knew what was coming so German bombers had to fly on weather patrols, either alone or escorted by 109s. Lone reconnaissance planes showed up well on radar and were driven off or brought down by small groups of RAF fighters. Sometimes the 109s could provoke a fight where tactics and experience gave them an advantage. On this opening day, Al Deere of 54 Squadron, the first pilot to shoot down a 109 over Calais Marck airfield back in May, spotted a German Heinkel He59 seaplane, painted white with red crosses to indicate its rescue plane status, escorted by twelve Bf109s from *JG51*. The Spitfires dived to the attack, half making for the seaplane, the others for the 109s. Deere shot down one 109 and spotted another turning towards him. Both opened fire, and Deere felt shots hitting his Spitfire as they closed at staggering speed. In seconds they had collided, leaving him to steer his crippled machine back to dry land. The 109 had scraped along the top of the Spitfire bending the propeller blades, but he managed to glide as far as Manston near the eastern tip of Kent before crash-landing. Two Spitfire pilots had been lost, and the seaplane and Deere's 109 were poor compensation.

Both sides were learning quickly. Early combats taught the RAF that bringing down German bombers was hard, even with eight machine guns. Several pilots

---

7. Professor Williamson Murray, *Luftwaffe: Strategy for Defeat*, pp. 74–75.

Air Vice-Marshal Keith Park, commanding 11 Group of Fighter Command. *(Official photo via Wikipedia)*

found Luftwaffe air gunners could hit the coolant system of an attacking fighter, turning the tables with a lucky shot. Radar was proving less effective at warning against larger attacks climbing and assembling out of range. When they revealed themselves by crossing the Channel, they needed only a third of the time required for defending fighters to meet them on level terms.

In time, radar operators would better predict the strength and intentions behind their screen displays, but height remained a problem as defending fighters had to climb quickly to reach incoming attackers. This made Spitfires ideal for Dowding and Park to use against the fast, high-flying 109s with Hurricanes going for the bombers and Bf110s. They also realized that pilots who closed in tightly on opponents would bring down most enemy aircraft. Dowding ordered RAF fighters' guns harmonized [adjusted so their fire converged] at 400 yards to give poorer shots a fighting chance of hitting their targets. Unfortunately, bullet drop and air resistance impaired accuracy. More experienced pilots reset their guns to converge at 250 yards and closed in on targets before opening fire.

Both sides agree that 10 July 1940 saw the first major battle. Rain and cloud hid a reconnaissance Dornier 17 with a Bf109 escort long enough to report a large eastbound coastal convoy codenamed 'Bread' heading for the Dover Straits. Spitfires attacked the Dornier and its escorts and a furious dogfight ensued before the Germans were driven off with no losses.

Next came a German disappointment. Twenty-four Dorniers escorted by thirty Bf110s and twenty-plus 109s appeared on British radar at ten minutes to two. Fighter Command sent three squadrons of Hurricanes and a squadron and a half of Spitfires to meet them. 111 Squadron's Hurricanes broke the bomber formation by flying head on at them, and once cohesion was lost the Spitfires fought the 109s. One damaged a German fighter and sent it home while another brought down a Dornier in the opening seconds of the scrap. One Hurricane collided with a Dornier, the only RAF casualty. Official reports credited the RAF with shooting down two Dorniers and ten escorts. Amid the fury of combat the convoy lost a single ship and the rest reached port safely. Nevertheless, Luftwaffe pilots claimed to have shot down eleven British fighters and sunk one heavy cruiser [no Royal Navy warships were present] and four merchant ships for the loss of two Dorniers, two 110s and three 109s. Both sides made differing claims, but combat reports showed the raid failed in its primary objective of sinking ships. Furthermore, German fighter losses were higher than those of the RAF. This should have been a dramatic wake-up call for the Luftwaffe, hinting that defeating British fighters might be more difficult than they had assumed.

Unsurprisingly, each side believed its own claims. The Luftwaffe thought it had downed thirty RAF fighters – they actually hit nine and destroyed one – and reported the RAF would be out of action in two to four weeks. Next day they did better. *JG27's* 109s bounced three Hurricanes protecting a convoy off Portland and shot one down. Spitfires provided reinforcements, but the 109s had climbed back to high altitude and bounced them as well, shooting down two. One small warship was sunk.

When the Luftwaffe sent in *Stukas* escorted by Bf110s, Fighter Command responded with Hurricanes which shot down four 110s and later three He111s. While Hurricanes might have difficulty with well-flown 109s, against Bf110s and Ju87s they had no trouble at all. Six Hurricanes were shot down, killing three pilots, but sixteen German aircraft were lost.

Meanwhile, 109 pilots began exploiting unpredictable British summer weather. Starting off with extra height before crossing the Channel and alerting British radar, they hid above a cloud layer to watch Spitfires sent to intercept. They then used the 109's fast dive to surprise their opponents. On 17 July, Spitfires of 64 Squadron were 'bounced' off Beachy Head, losing one Spitfire, and the 109s vanished before the rest could react. The following day it happened to the Spitfires of 610 Squadron, searching for a convoy under the cloud layer. Another Spitfire was shot down for no German losses.

Tactics were becoming decisive. On 19 July Fighter Command lost seven Defiants of 141 Squadron while the Luftwaffe lost four machines. On 20 July an eastbound convoy prompted Park to station Hurricanes above it, and Spitfires of 65 and 610 Squadrons higher still. When the Luftwaffe arrived at six in the evening, as *Stukas* escorted by 109s, the RAF fighters dived to the attack, hitting four Ju87s and shooting down two 109s. Two Hurricanes and one Spitfire were shot down, but the convoy was saved.

On 21 July Göring reformed his fighter units to enable *Experten* to protect less experienced pilots. Osterkamp was promoted to command Luftflotte *2's* fighters and replaced as commander of *JG51* by Werner Mölders. Both sides were losing experienced pilots too quickly and the outcome remained undecided. While Luftwaffe pilots were better at stealth and surprise attacks, Keith Park was emerging as a better leader than Göring.

On 24 July the Luftwaffe attacked two convoys at once and the Spitfires of 54 Squadron had to do the same to meet them. No German aircraft was shot down, but no British ships were sunk. The Luftwaffe was back at eleven with Dornier 17s escorted by Adolf Galland's 109s from *JG26*. Park sent 54 Squadron against the new

Bf109E of *JG51* Mölders with wing-mounted cannon. *(Bundesarchiv photo via Wikipedia)*

attack, but realized that in a few minutes the 109s would run short of fuel and have to dash for France and safety. To exploit their weakness he sent nine 610 Spitfires over Dover to meet the 109s heading homeward. Aware of the same problem the Luftwaffe sent more 109s from *JG52*, to clear the sky for their comrades' homeward dash. In all, five squadrons of fighters, (54 and *JG26*, 610 and *JG52*, together with six more Spitfires from 65 Squadron based at Hornchurch) joined the clash over Kent in the kind of dogfight which best suited Spitfires. For the first time the RAF were winners. Once battle was joined, no 109 could climb above the melee and bounce the RAF machines, and German pilots were forced into a turning battle which did not suit their 109s. Three Spitfires were brought down. Pilot Officer Johnnie Allen from the Calais Marck rescue of May 1940 was hit by Adolf Galland's accurate deflection shooting. With a dead engine he tried to crash-land his machine near the radar station at Foreness, but managed to restart his engine on approach. He turned towards the nearby airfield at Manston, but it cut again. He tried returning to Foreness, but his Spitfire stalled and crashed, and Allen was killed.

Another 54 Squadron machine was shot down while the pilot survived, but the Germans lost five Bf109s. *Leutnant* Schauff of *III/JG26* was hit by a burst from the Spitfire of Pilot Officer Colin Gray. He bailed out, but died when his parachute failed to

Spitfires in 'Vic' formation vulnerable to enemy attack. *(Author drawing)*

open. Another Bf109 flown by former test pilot *Oberleutnant* Werner Bartels was forced to belly-land on British soil. *Hauptmann* Erich Noack, CO of *II/JG26* was hit by 610 Squadron Spitfires and stalled his damaged machine at his base at Marquise in the Pas de Calais. Two more Bf109s from *JG52* were shot down by 610 Squadron Spitfires. Adolf Galland accused his pilots of poor discipline and tactics. He was shocked by the dedication of relatively inexperienced RAF pilots, outnumbered roughly two to one, who used their Spitfires' manoeuvrability to prevent his men protecting the bombers or disengaging from combat until their fuel ran low. He only escaped himself by a curving dive which Spitfires could not follow. From this point, many German pilots saw the RAF and the Spitfire in a different light.

Yet the next day brought a different outcome. Improving weather enabled German *Schnellboote*, high-speed motor torpedo boats, to attack westbound coastal convoy CW8 containing twenty-one small coasters, many of them colliers, heading through the Dover Straits en route to Falmouth in Cornwall. This coincided with a barrage from German heavy guns on Cap Gris Nez and a dive-bomber attack by *Stukas* which between them accounted for five ships; *Corhaven, Portslade, Leo, Polgrange,* and *Henry Moon.*

One clever tactic was for fifty escorting Bf109s to approach the convoy at sea level, strafing the ships to bring down the defenders to retaliate, leaving the Ju87s beginning their dives high above, free to pick their targets and bomb accurately. Park sent in 64 Squadron's Spitfires to tackle the 109s, reinforced by two squadrons of Hurricanes, and they fought dangerously low above the water. A Bf109 of *JG52* banked into a turn when a wingtip hit the water, smashing it to pieces. At half-past two, with the convoy creeping past Folkestone, another heavy attack was delivered by forty Ju88s escorted by fifty Bf109s, this time at higher altitude. Back came 54's Spitfires to be bounced by more 109s as they climbed to meet the bombers. Later the *Kriegsmarine* S-boats returned. The Royal Navy sent two destroyers, HMS *Brilliant* and HMS *Boreas* from Dover with two British Motor Torpedo Boats. They drove off the S-boats, but left the larger warships vulnerable to dive-bombers. Both destroyers were hit and had to be towed into Dover by tugs. Back came the S-boats

German *Schnellboote* high speed torpedo boats surrendering to the Royal Navy at the war's end. *(Official photo via Wikipedia)*

again, sinking the *Lulonga*, the *Broadhurst* and the *London Trader*. One German vessel picked up a British naval pilot, Sub-Lieutenant Dawson-Paul, whose Spitfire had been shot down by a 109. The convoy reached its destination with only two ships undamaged, and the Admiralty interrupted convoys through the Dover Straits as the attrition continued. 54 Squadron had lost a dozen Spitfires and five experienced pilots and was transferred north to recover. On 26 July bad weather brought a brief respite. Luftwaffe attacks on shipping brought out RAF fighters. One Hurricane was shot down over the sea, but three Bf109s were claimed in return. Three more RAF fighters, one Hurricane and two Spitfires, were damaged in landing accidents in wet and muddy conditions.

On 25 July clearer skies allowed *Stukas* to attack British warships off the east coast and a convoy off Swanage. One Spitfire was shot down by a 109 off Weymouth and a Hurricane lost over Dover where the Luftwaffe sank the destroyer HMS *Codrington*, in port for boiler cleaning. *Stukas* also sank a small destroyer HMS *Wren* off Suffolk and blew the bows off HMS *Montrose*, forcing tugs to tow her into Harwich. Luftwaffe losses amounted to a *Stuka* and two Bf109s, but warship losses forced the Admiralty to withdraw from Dover, imposing a heavier burden of shipping protection on the overloaded fighter squadrons.

On Sunday 28 July, two large German formations set out across the Channel before turning back to confuse the defenders. In early afternoon another massive formation, with more than sixty He111s escorted by more than forty 109s, crossed the Channel in five waves. Park called up two Hurricane squadrons to attack the bombers and two Spitfire squadrons from Hornchurch to climb and take on the 109s. One fighter was flown by *Major* Werner Mölders. In the dogfight he aimed at a Spitfire but missed another behind him. When he spotted it he made a tight turn, but the Spitfire stayed with him and fired an accurate burst which knocked him out of the fight. Nursing his machine back across the Channel he wrecked his fighter, suffered severe leg wounds and was in hospital for a month. Rumours suggested he had fallen to an RAF ace, the South African pilot Adolf 'Sailor' Malan, a former merchant seaman, and a formidable fighter pilot and tactician like Mölders. In fact, it was probably Flight Lieutenant Johnny Webster of 41 Squadron who delivered the deadly burst, but Malan would have been equally appropriate. One of the first to abandon Fighting Area Attacks, he used his experience to frame ten rules for air fighting, widely circulated, though remaining unofficial:

MY RULES FOR AIR FIGHTING

1. Wait until you see the whites of his eyes. Fire short bursts of one to two seconds only when your sights are definitely "ON".
2. Whilst shooting think of nothing else, brace the whole of your body: have both hands on the stick: concentrate on your ring sight.
3. Always keep a sharp lookout. "Keep your finger out".
4. Height gives you the initiative.
5. Always turn and face the attack.
6. Make your decisions promptly. It is better to act quickly even though your tactics are not the best.
7. Never fly straight and level for more than 30 seconds in the combat area.
8. When diving to attack always leave a proportion of your formation above to act as a top guard.
9. *INITIATIVE, AGGRESSION, AIR DISCIPLINE*, and *TEAMWORK* are words that MEAN something in Air Fighting.
10. Go in quickly – Punch hard – Get out!

Some squadrons had adopted the flexible Luftwaffe formations. One Hurricane squadron binned official tactics early in 1940, but Group Captain Johnnie Johnson, the RAF's highest scoring wartime fighter pilot, found in his old 616 Squadron logbook, 'We were still practising the dreaded Fighter Attacks in January 1941.'[8] This was a squadron from Leigh-Mallory's 12 Group where things were done by

---

8. Robinson, *op.cit.*, p. 150.

the book, even during the battle. After transfer to 11 Group at Kenley it suffered heavy losses from 109s who used more sensible tactics. When Johnson flew with them, they were under Leigh–Mallory again as head of Fighter Command. Clearly nothing had been learned in the interval.

Mölders was not alone in his narrow escape. The Luftwaffe lost five 109s and two He111s, against two Spitfires. The following day saw another dive bombing attack on Dover with more than thirty *Stukas* escorted by fifty or so 109s. Again Park sent two Hurricane squadrons against the dive-bombers with two Spitfire units to meet the fighters. This spawned a huge dogfight over the town. Eight Ju87s were shot down with two more falling to anti-aircraft fire. Seven 109s were shot down for two Hurricanes and a Spitfire.

Tuesday 30 July brought rain and low cloud. July 31 saw clearer weather and the Luftwaffe begin the invasion campaign proper. Hitler had decided the invasion could not be launched before mid-September, but then ordered Göring in *Führer* Directive No 17 that the RAF must be eliminated with a campaign ready to launch at twelve hours' notice.

How realistic was this? During July, relatively low intensity fighting over the Channel had cost the RAF sixty-seven Hurricanes and Spitfires, and the lives of forty-eight pilots. The Luftwaffe had lost 216 aircraft, though only forty-eight were fighters – seventeen fighter pilots had been killed with another fourteen missing, mostly POWs in England. So far numbers were small. More serious to German hopes of beating the RAF were losses of bombers and crews. In July alone, 168 bombers had been shot down, with the loss of 466 aircrew. The RAF's worst problem was not aircraft losses as production more than kept pace with replacements, but losing experienced pilots. Nevertheless, while Park husbanded resources, he had studied German tactics and developed counter-moves to frustrate them using his two fighter types to best advantage.

The Luftwaffe learned much less about the RAF. They knew it was more formidable than they originally assumed and the Spitfire was an adversary which even 109 pilots had to beware of, but they had no reliable idea of Fighter Command's size and strength. In a meeting on 1 August at The Hague for *Luftflotte* commanders, Göring revealed his belief that the RAF had only 500 fighters left to defend southern Britain. There was just one more week before the new campaign began on 8 August, *Adlertag*, to bring Fighter Command to its knees.

Chapter Seventeen

The Battle of Britain – The Pace Quickens

In the pre-dawn darkness of 8 August 1940, twenty-four merchant ships, escorted by nine small warships, crept towards the Dover Straits, the first convoy to run the gauntlet of 'Hellfire Corner' for two weeks. Numbered CW9 and code-named 'Peewit', it had assembled the evening before off Southend. As it hugged the coast, it headed for its destination off Swanage hoping to reach its objective with minimal damage. It would not. Heavy attacks would wreck it and RAF fighters would pay a high price beating off the Luftwaffe onslaught.

The German *'Seetakt'* naval radar on Cap Gris Nez spotted the ships first and alerted the *Kriegsmarine* high-speed diesel torpedo boats in nearby Boulogne. Just before dawn *S20, S21, S25* and *S27* slipped their moorings and headed for Beachy Head, directly in the convoy's path. With dawn breaking far to the eastward, the S-boats roared out of the darkness, forcing the convoy to scatter. Three ships were hit. The 1004-ton SS *Ouse* collided with SS *Rye* trying to avoid a torpedo from the *S20*. The 1216-ton SS *Holme Force* was sunk off Newhaven by the *S21*, and the small 367-ton motor ship *Fife Coast* was sunk by *S27* off Beachy Head with five dead.[1] Three smaller coasters were damaged by gunfire avoiding *S25's* torpedoes, before the boats turned away with all torpedoes expended.

Meanwhile, *Stukas* of *Fliegerkorps VIII* were taking off. From smaller fighter strips *JG27's* Bf 109s aimed to catch the dive-bombers up as they reached the convoy, to maximize endurance, but cloud was down to just over 2000 feet, too low for dive-bombers. As they climbed over the Channel, British radar spotted them and alerted the defences. The 109s clashed with RAF fighters, but only one ship was damaged as the convoy edged further west. Near the Isle of Wight the cloud thinned, and fifty-seven more *Stukas*, escorted by twenty Bf110s and thirty Bf109s, mounted an attack. Though RAF fighters headed them off, many pilots were inexperienced and the 109's attacks gave the *Stukas* an almost free hand. Within minutes, three more ships were sunk, the 942-ton Dutch cargo steamer SS *Ajax*, with four killed, the 1600-ton British cargo ship SS *Coquetdale* and the 989-ton SS *Empire Crusader*. She was previously the German *Leander*, captured as a prize off the Spanish coast by the Royal Navy destroyer HMS *Isis* in November 1939. Pressed into service under a new name, she was sunk by German airmen.

---

1. Ironically, the *Rye* would survive, only to be sunk by gunfire from S 27 in another convoy attacked by *Schnellboote*, this time off Cromer on the Norfolk coast on 7 March 1941.

Three new Spitfires taking off from Eastleigh airfield at the start of the war. *(via Alfred Price)*

Through the afternoon, German dive-bombers kept attacking and RAF fighters fought to hold them off. More ships sank. The 946-ton Norwegian steamer *Tres* went down off St Helens, while HMS *Borealis*, a converted 450-ton Belgian pilot cutter serving as a barrage balloon tender, foundered off St Catherine's Point, the island's southernmost tip. Survivors hid in south coast ports, so only four ships reached their destination. These effective Luftwaffe attacks underlined British failures, particularly in saving pilots from the Channel even close to dry land. In all, sixteen RAF fighters were lost with their pilots. Estimates of Luftwaffe losses varied from seventeen *Stukas,* nine Bf110s and twenty-six Bf109s claimed by the RAF to a single Bf110, three Ju87s and three Bf109s admitted by the Germans.

Restored Bf109E fitted with wing guns, seen at a US flying display. *(Jacobst photo via Wikipedia)*

The Junkers Ju87 *Stuka* was a single-engine two-seat cranked-wing dive bomber of deadly accuracy. *(Author photo)*

The biggest killer of RAF pilots remained the sea. On 11 August, fourteen pilots were killed over the Channel, with ten more missing. On 12 August, nine pilots died in the sea – one was rescued, but died from wounds. Fortunately, the Luftwaffe then switched targets from merchant ships to the mainland and the Isle of Wight. Still confused by the RAF's warning system, they decided to knock out the radar stations in time for *Adlertag*. Had they succeeded, defending fighters would have been blinded. Because coverage of adjacent stations overlapped, breaking the chain meant destroying two, three or more in the same area, and mobile back-up teams used special generator vehicles to plug gaps.

At 08.40, as the sun burned off early morning mist, sixteen Bf110s of precision bombing specialists *Erprobungsgruppe 210*, each carrying a 500 kg bomb, crossed the Channel heading west before turning northwards and splitting into four sections of four. The first attacked the station at Dunkirk, near Canterbury. Another made for Pevensey, eight miles away. A third section headed for Rye, twenty miles off, while the last aimed for the masts above Dover. Pevensey was closest and was hit hard, with several direct hits. At Rye and Dunkirk, buildings were wrecked and power cables and communications cut. Pilots reported 75 percent success, and to exploit the damage a force of *Stukas*, escorted by 109s, flew through the gap to bomb convoys in the Thames Estuary. Fortunately, the unmolested Chain Home Low radar station at Foreness reported more than sixty German aircraft approaching.

Defending fighters were too late to stop the dive-bombers and no Ju87s were hit, but damage to ships was minimal.

The next attack was spotted by radar near Littlehampton; more than 200 aircraft, including sixty-three Ju88s of *KG51 'Edelweiss'* with 120 Bf110s, themselves protected by twenty-five Bf109s. As they closed the coast, they swung westward towards the Isle of Wight, apparently aiming for Portland dockyard. Only as they passed the larger naval base at Portsmouth did forty-eight bombers turn through the balloon barrage gap over the harbour entrance, heading for the dockyard. The remaining fifteen dived at 300 mph in the opposite direction, across the eastern part of the Isle of Wight towards the vital radar station at Ventnor. With no RAF fighters present it seemed the tactic had worked. There was also no AA fire, though a battery of 40mm Bofors guns opened up after the attack began. The bombers dropped fifteen 1000-pounders into the station site, wrecking buildings and cutting communications.

This was the moment when Spitfires showed how quickly they could punish a German mistake. As the bomber leader tried to reassemble his formation to head for France, his 109 escorts waited at 10,000 feet. He ordered his own crews to climb and close in for protection, but it was already too late. Hurricanes of 213 Squadron from Exeter, followed by two squadrons of Spitfires, arrived piecemeal, so *JG53's* 109s took no notice. By the time they realized what was happening two miles below, the bomber leader had been killed by a single burst and his aircraft crashed in nearby Godshill Park.

Ten Ju88s were downed by RAF fighters and AA gunfire over the dockyard, together with one Bf110, though the Luftwaffe had succeeded in blowing another gap in Fighter Command's radar cover. The Portsmouth raid caused widespread damage and killed 100 people, and more German bombers massed over France with Hornchurch, Manston and Lympne next on the target list. Confident that British radar had suffered, *Erprobungsgruppe 210* arrived over Manston as 65 Squadron were taking off. Incredibly, all twelve Spitfires avoided the bombs bursting around them, but one had its engine stopped by the blast of a near miss. Many 109s flashed past at the bottom of their dives with no time to aim accurately as the Spitfires climbed. The Spitfires were lucky. Not so the airfield. Hangars and workshops were wrecked, and volunteers from Blenheim crews filled in bomb craters and refuelled and rearmed fighters, instead of resting before their night operations. Hawkinge was attacked by *Erprobungsgruppe 210* on their third raid of the day, wrecking two hangars, the station workshops and two Spitfires under repair, and killing two civilians and three airmen. Finally, three small raids attacked coastal towns, to test the radar coverage.

Results were disappointing. Signals traffic showed mobile teams had Rye, Dunkirk and Pevensey back in action already. Ventnor remained off the air for three days, but fake signals hid this from the Germans. Furthermore, defending fighters were as well directed as before. Knocking out British radar would be a longer and bloodier business, and many felt it was hardly worthwhile. Surely if the RAF knew a big raid was approaching, they would send more fighters to meet it, giving the Luftwaffe a chance to shoot down larger numbers? Their own reports suggested this had actually happened. German pilots claimed three RAF fighter airfields out of action and seventy fighters shot down. In fact, just twenty RAF fighters had been lost and eleven pilots killed. None of the airfields were vital and all were operating again the next day. For this partial achievement, the Luftwaffe lost thirty-one aircraft, with crews killed or missing. Delay in the Bf109s rescuing the Ju88s cost *KG51* eleven of its bombers – 17 percent of its strength. Their crews insisted the 109s should stay close for the whole attack and the flight home to protect them. Inevitably, this would rob the fighters of their performance advantages and leave them vulnerable to the enemy. In any case, switching tactics would take time, and next day saw *Adlertag* begin the carefully planned campaign to defeat the RAF.

Meanwhile their adversaries had realized the Channel fighting was claiming too many valuable pilots – 220 of them in July and August might prove even worse. Senior RAF officers showed breath-taking complacency for an island nation. They assumed Channel shipping would find downed pilots fast enough to save them, but bitter experience showed this was untrue. A single figure adrift in choppy water was hard to spot even in perfect visibility. His Mae West life-jacket was inflated by blowing into a tube, difficult when floating, possibly wounded, even if enemy gunfire had not already punctured the fabric. If a pilot went into the water, his survival chances were less than 20 percent.

Though the Germans also feared the Channel, their survival chances were better. Bf109 pilots wore life jackets which did not need inflating. Early versions had kapok-filled 'sausages' and a protective collar, but the 109's cramped cockpit made it hard to wear the jacket and fly the aircraft. Later versions had buoyancy panels inflated by a compressed air cylinder activated on contact with water. Greenish-yellow fluorescein dye would stain the surface and alert rescue patrols, while a one-man dinghy supported the pilot clear of the water.

German high-speed motor launches patrolled the Channel after any raid. Heinkel He59 seaplanes of the *Seenotflugkommando* [Air-sea Emergency Command] could land on the water beside a stranded airman and fly him to safety. These large and conspicuous twin-engine biplanes, originally introduced in 1931 as reconnaissance and torpedo bombers, were obsolete by 1939. The white-painted rescue versions wore red crosses, but the Air Ministry insisted they relayed intelligence information while on rescue missions. RAF fighters were ordered to shoot them down unless actually carrying out rescues, causing great anger in Germany. When camouflaged and armed, they remained vulnerable because of size and poor performance. They did however sometimes pick up RAF aircrew and make them POWs.

The Germans also introduced *Rettungsbojen* [Rescue Buoys], rafts anchored offshore for any pilot lucky enough to ditch near one. Each had a small four-bunk cabin with dry clothing, signalling and first aid equipment, and a stove for hot food and drinks. Rescue launches and seaplanes made regular calls. Some sources claim they saved the lives of both German and RAF pilots, but others insist most buoys were lost in bad weather.

On August 19, Park warned controllers not to send pilots over the sea. Though the Luftwaffe was switching to inland targets, the Channel battle had crippled the merchant convoys and the squadrons defending them. In mid-Channel, radar gave just five minutes warning of an attack, but defending fighters took three times as long to meet it on equal terms. This meant wasteful standing patrols, sending minimum force against each attack. Ironically, this convinced Luftwaffe intelligence the RAF had fewer fighters than they really had, which condemned them to a succession of disappointments.

The RAF had just one simple priority. Luftwaffe strength was irrelevant. As long as attacks continued, Fighter Command had to oppose every raid, though lacking height and numbers. This reinforced Luftwaffe complacency and Hitler ordered *Seelöwe* [the invasion] ready for launching by 15 September. German planners believed two weeks would see the defeat of the RAF and with heavy attacks from 5 August, this should provide enough time. *Reichsmarschall* Göring boasted in a conference at the beginning of August that 'By means of hard blows I plan to have this enemy, who has already suffered a decisive moral defeat, down on his knees in the nearest future, so that an occupation of the island by our troops can proceed without any risk.'[2] Sadly, his plans were in vain. They depended on the British being terrified to the brink of surrender and down to their last few hundred fighters. Both were untrue and typical British summer weather made things even worse, as low cloud and rain disrupted the German timetable still further.

Adlertag had first been delayed to 8 August, then 10 August and finally to 13 August. German weather predictions were unreliable as change came from the west, and poor weather caused a catastrophe. At 4.50 pm on 13 August, seventy-four Dornier 17s of *KG2* took off and headed for the English coast when a signal from *Luftflotte 2* ordered them back to wait for better weather. Their escort commander picked up the recall signal as his *ZG26* Bf110s took off to meet the Dorniers over the Channel, but the Dornier radios used a different frequency. They missed the signal and could not hear their escorts either. They droned on through heavy cloud, protecting them from RAF fighters, but *ZG26* returned to base. When the Dorniers emerged from cloud cover they saw their targets: Sheerness and Eastchurch airfield on the Isle of Sheppey. They picked up another escort en route, the Bf109s of *Erprobungsgruppe 210* which also missed the recall signal and could not communicate with the bombers either. They managed to bomb the airfield, though they lost five Dorniers when the RAF found them. The rest returned to base with another five damaged beyond repair. The Germans were consoled by the damage to both targets

---

2. Derek Robinson, *op.cit.*, pp. 176–77.

and the destruction of ten Spitfires at the Eastchurch fighter airfield. But they were wrong. Only one Spitfire was destroyed and Eastchurch was a Coastal Command station. To compound the confusion, *Lufiflotte 3* never received the recall either and sent three raids against targets further west. Airfields at Farnborough and Odiham were hit by *KG54's* Ju88s. With eighty-eight *Stukas* attacking Portland the raids were escorted by some sixty Bf110s, covered by 173 Bf109s, flying ahead of the bombers.

These attacks also failed. In dreadful visibility, the *Stukas* missed Portland and returned to base. The 109s also missed the RAF and had to leave their charges undefended. The Ju88s met RAF fighters and lost four bombers, with another eleven damaged. One Bf109 failed to return. For once the Bf110s returned to base without loss, but twenty-three escorted the Ju88s of *KG54* on another Portland attack. Again bombers and escorts could neither talk to each other nor find each other in the murk. The bombers turned back leaving the 110s unaware, so they pressed on to the target and ran into British fighters. Four Bf110s were shot down over the Channel and two more crashed back at their bases, but claimed to have shot down nine Spitfires, which raised spirits slightly.

Overall, the Luftwaffe believed they had inflicted heavy damage on their targets and shot down twenty RAF fighters. In fact the Bf110s shot down a lone Hurricane from 601 Squadron. The damaged airfields were irrelevant to the battle. Farnborough was a test and experimental airfield and Odiham was an army co-operation base with no fighters at all.

The weather cleared at lunchtime and by early afternoon *Adlertag* resumed. *Lufiflotte 3* sent fifty-eight Ju88s to bomb Boscombe Down, Worthy Down and Andover, escorted by B110s and 109s. A squadron of 109s flew across Hampshire and Dorset to clear defending fighters, and fifty-two *Stukas* attacked airfields at Warmwell and Yeovil.

This large force brought a massive RAF response. Number 10 Group threw almost every serviceable fighter into a huge dogfight. *Lufiflotte 2* attacked further east in mid-afternoon. *Erprobungsgruppe 210* pilots found Southend blotted out by cloud and returned to base while their Bf110 escorts blundered into 56 Squadron's Hurricanes in the gloom. Two *Stuka* formations from *Lufiflotte 2* went to bomb Rochester and the airfield at Detling. Rochester was under cloud, so target and attackers escaped without loss. Detling was in the clear when the Ju87s arrived at teatime, escorted by 109s. Spitfires of 65 Squadron climbed to meet the 109s of *JG26*, allowing the *Stukas* in at lower altitude. Within minutes bombs smashed the Operations Block, the airmen's mess halls crowded with personnel at the end of the working day, and aircraft parked at dispersals. It was a crushing blow. Sixty-seven men died, including the station CO, and twenty-two aircraft were destroyed on the ground before the Ju87s vanished into cloud, returning home without loss.

Fortunately for Fighter Command, when *Lufiflotte 3* tried the same tactic west of the Isle of Wight, 609's Spitfires fought the 109s of *JG53*, but spotted a *Staffel*

Dive brakes slowed down the Ju87's dive for greater accuracy. *(Will Owen photo)*

The Ju87's bomb was carried on a trapeze to swing it clear of the propeller when released in a steep dive. *(Will Owen photo)*

of Ju87s over the Dorset coast, heading for Middle Wallop. They never reached it. The Spitfires dived with the sun behind them to hit their hapless targets. It was a massacre. Six of the nine dive-bombers, together with four 109s, were hacked down by machine-gun fire and the remaining three never got near their target. Other *Stuka* formations had to give up and return to base, but the cloud which hid their targets shielded them from the RAF and they suffered no losses.

As the start of the campaign, it was a failure. Luftwaffe Intelligence claimed eighty-four RAF fighters destroyed and three airfields wrecked. In fact only thirteen RAF machines had been shot down, and ten pilots survived. None of the vital sector stations had been hit. Only Middle Wallop was a genuine fighter airfield and the only bomb to hit it was dropped by mistake. Detling was used for anti-submarine patrols and the wrecked aircraft were Avro Ansons rather than Hurricanes or Spitfires. But German losses had also been grossly over-estimated by their opponents. The RAF claimed seventy-eight enemy aircraft, whereas true losses had been forty-six, though this was bad enough for such limited results.

For several days attacks on the wrong targets in bad visibility brought losses on both sides. Patterns were emerging. If Spitfires could force 109s into a dogfight, the Germans came off badly. Yet with their advantages of height and sun, only superb vigilance could enable RAF pilots to evade German attacks, and it was increasingly clear a crisis was approaching.

On 15 August, poor weather cleared by mid-morning. Luftwaffe senior commanders were told by Göring he was worried about casualties. He wanted heavier attacks with larger formations, and fighters should stay close to their charges to keep opponents away. *Luftflotte 2* launched more than 1100 fighters and bombers at coastal targets. Dive-bombers pounded the airfields of Hawkinge and Lympne, Dorniers hit Eastchurch again, and Bf110s strafed Manston but dropped no bombs. Still more bombers attacked radar stations at Rye, Dover and Foreness. At first it seemed that greater numbers had worked. One squadron each of Spitfires and Hurricanes attacking the *Stukas* making for coastal airfields found too many to cope with, and Hawkinge and Lympne were out of action for three days. Damaged radar stations made it difficult to monitor later raids. The Bf110s of *Erprobungsgruppe 210* delivered a heavy blow on Martlesham Heath, helping dive-bombers destroy one aircraft, damage two hangars, wreck the Officers' Mess and sever communications lines and water mains.

In late afternoon, eighty Dornier 17s, escorted by 200 Bf109s attacked Rochester airfield and a factory building Short Stirling bombers, while others bombed the inevitable Eastchurch. In late afternoon, raids over the south-west resulted in German aircraft outnumbering the RAF by around six to one. Finally, some sixty Ju88s with Bf110 escorts attacked Middle Wallop and Worthy Down airfields. The Spitfires of 609 Squadron based at Middle Wallop were only scrambled as the bombers approached, but still shot down one Ju88 and three escorts.

In the day's last raid, Dornier 17s made for Biggin Hill, concealing *Erprobungsgruppe 210's* attack on the sector station of Kenley. A squadron of Hurricanes and one of

The cockpit of the formidable Junkers Ju88. *(USAF)*

Spitfires based at Biggin were sent up, and two bombers were destroyed before the Spitfires attacked the 109s. Meanwhile the bombers missed Biggin to attack the much less important airfield of West Malling, while *Erprobungsgruppe 210* missed Kenley and bombed Croydon instead.

The Germans dragged most of 11 Group into the fray. Casualties were higher: the RAF lost fifty aircraft, nine Hurricane pilots and six Spitfire pilots. Luftwaffe losses were similar at fifty-five aircraft with crews killed or captured, but German pilots had again attacked the wrong targets or missed the right ones. Instead of sector stations, they bombed naval airfields, bomber airfields and Coastal Command airfields. They missed Odiham to hit the even less important airfield at Andover. Their one success was damaging the Pobjoy works at Rochester, making Spitfire wings.

But a change in strategy by *Luftflotte 5* in Scandinavia brought disaster. With heavy attacks on southern England, they assumed northeast England had little fighter protection. They had no idea Dowding used the area for training newly assigned squadrons or restoring those needing time to recover. This meant defending fighters in plenty with experienced and capable pilots. Crossing the North Sea meant Bf110s had to fly as escorts, carrying cumbersome and unreliable 1000-litre belly tanks, and leaving rear gunners behind to save weight. With few defensive fighters, what could possibly go wrong?

The Luftwaffe force included more than a hundred He111s and Ju88s, escorted by twenty-five Bf110s. Local radar picked up the He111s and Bf110s, but underestimated their strength by more than half. Consequently the Spitfire pilots of 72 Squadron were shaken to find the raiders some 3,000 feet lower than predicted, but with no escorting 109s. The squadron commander could fly a textbook approach, circling behind his targets and diving into the attack. The Luftwaffe aircrew were given a terrifying demonstration of what Spitfires could do when untroubled by German fighters.

The Bf110s tried to jettison their belly tanks, but the escort commander's refused to release. A burst from a Spitfire struck it and blew the machine to pieces. With no rear gunners the others formed a defensive circle, abandoning the bombers. As they raced for cloud cover, more fighters arrived – the experienced Spitfire pilots of 41 Squadron and the Hurricanes of 605 Squadron. The result was a bloodbath; eight bombers and eight escorts shot down. Further south, the Ju88 formation was intercepted by two 12 Group squadrons as they attacked the bomber airfield at Driffield in Yorkshire, destroying ten obsolescent Whitleys and damaging another six. Eight Ju88s were lost, but the Whitleys were irrelevant to the fighter battle.

This was a grim truth for the Luftwaffe. In the first weeks of war Bomber Command's Wellingtons had no defence against Bf109s and 110s over German daylight targets and suffered terrible losses. Now the Luftwaffe was learning their bombers could only survive daytime raids over England when escorted by 109s. This meant daylight raids would be limited by the number of 109s available at the time, and restricted to within 109 range from their bases – the extreme south-

Captured Ju87B *Stuka* in North Africa 1941, following a forced landing. *(Official photo via Wikipedia)*

eastern part of Britain. Finally, *Erprobungsgruppe 210* had hit Croydon rather than Kenley, bombing part of London against Hitler's express orders.

The Luftwaffe returned on 16 August against the usual pointless targets of Eastchurch, Gosport and Lee-on-Solent, with commanders reassured by Göring's insistence the RAF must be down to their last 450 fighters. Two bomber formations, one of Ju87s and the other of Ju88s hit the fighter station of Tangmere. All three main hangars were wrecked with twenty people killed and fourteen aircraft destroyed. A large formation over the Thames Estuary was attacked by a squadron of Hurricanes and two of Spitfires. A fierce dogfight between Brighton and Folkestone began when another RAF force dived between German escorts and the bombers they were protecting. Total casualties included up to ten Luftwaffe machines, against five Spitfires and a single Hurricane.

In the late afternoon, two Ju88s avoided combat while defenders were refuelling and rearming, and reached the airfield at Brize Norton in Oxfordshire. 'Choosing their moment nicely, the Germans came into the circuit as if preparing to land. They even lowered their wheels in the hope of being mistaken for Blenheims'[3] [based at the airfield, with a distinct resemblance to the Ju88]. The ruse delayed the defences and allowed the raiders to hit hangars containing recently refuelled training machines and the nearby Maintenance Unit workshops. Forty-six trainers were lost and seven damaged, with eleven repaired Hurricanes also badly damaged.

Had Göring's estimate of RAF strength been true, this would have been catastrophic. But total losses were five Spitfire pilots and four Hurricane pilots killed. Furthermore, 450 fighters did not represent the remaining strength of Fighter Command, but the total of new machines produced each month, compared with 175 new 109s produced in all of August. The RAF were also shooting down more German aircraft than they were losing, though in fighter to fighter combat, losses were broadly similar. Most worrying was the toll of experienced pilots. The Luftwaffe had 1560 fighter pilots, more experienced than their 1380 RAF opponents. In less than a week Fighter Command lost almost seventy pilots killed or missing, with a similar number wounded and out of action. In half of August, only seventy inexperienced replacements arrived. Moreover, 11 Group had 245 Hurricanes, but just 80 Spitfires ready for action.

When Saturday 17 August brought almost perfect flying weather, with sunshine and clear visibility, both sides stayed away, the RAF patching up airfields and the Germans repairing bombers and dive-bombers. The lull was brief. Sunday 18 August would be the heaviest day of the battle, as Luftwaffe commanders concentrated at last on fighter airfields like Kenley and Biggin Hill, Hornchurch, North Weald, Tangmere and Northolt. One reason for poor German intelligence was Fighter

---

3. Deighton, *op.cit.*, p. 225.

Command's use of Spitfires against German reconnaissance aircraft, preventing them taking detailed pictures to show which airfields were vital to fighter defences.

Attacks began from midday, throwing an attack on Kenley out of sequence. Dive-bombing Ju88s were intended to attack from medium altitude, followed by high-level Dornier 17s, followed by a strike by nine more Dorniers flying at low-level, under radar coverage. This last attack arrived punctually, but instead of a target wrecked by dive-bombers and high-level bombers, they found Kenley alert and waiting as the other strikes had been delayed.

The Observer Corps had already reported the nine Dorniers passing low overhead. Hurricanes of 111 Squadron flew from Croydon to meet this threat while Kenley's own squadrons climbed to attack the higher-level raiders and 109 escorts. Between them the Hurricanes, the airfield's Bofors guns and rocket propelled cable launchers brought down four Dorniers, nearly half the raid, and damaged the rest.

Nonetheless, the airfield was wrecked. Hangars, offices, mess rooms and workshops suffered from small, specially fuzed bombs dropped by the low-level attackers. Soon afterwards high altitude Dorniers added more mayhem. 'Eight Hurricanes were destroyed on the ground, two hangars were wrecked and five others severely damaged, the operations room was put out of action and many other buildings, including the hospitals, were reduced to rubble. Had all the bombs exploded on impact, Kenley could have been totally destroyed, but many were released too low and hit the ground horizontally and failed to activate the warheads. Unexploded bombs were everywhere.

'But the most imminent danger was the fires, made worse because one of the bombs had exploded and fractured the aerodromes' water mains. Three of the four aircraft hangars had been destroyed, the main sector operations room lost all electricity and telephone services, and the main power cable had been severed rendering the mainframe useless. Many station buildings and the medical sick bays were destroyed as were both the officers' and the sergeants' messes. A hangar housing the station's motor transport was wrecked, and four Hurricanes and a Blenheim had been destroyed with three Hurricanes and a Spitfire badly damaged.'[4]

By the time Kenley's own fighters returned to refuel and rearm, most had to switch to Croydon or the satellite landing ground at Nutfield. Later, more German bombers tried to do the same to Biggin Hill. He111s from *KG1* were free to drop their bombs from 12,000 feet. Some eighty tons of explosive blew craters in the landing ground or a wooded area to the east, but apart from broken windows in some buildings, Biggin Hill was scarcely damaged, though one Heinkel was shot down and another hit.

Later, 109 Ju87 dive bombers of *Lufiflotte 3*, escorted by fifty-five 109s to ward off RAF fighters and 102 more as close escort, struck the radar station at Poling and

---

4. From Kenley Station Commander's Report to 11 Group.

the non-vital airfields of Gosport, Ford and Thorney Island. The Gosport attack came completely unstuck. Spitfires from Middle Wallop attacked the close escort 109s and the German pilots preoccupied with their slower and more vulnerable charges were fatally handicapped. Three were shot down before the Hurricanes of 43 and 601 Squadrons caught the twenty-eight unprotected *Stukas* and shot down ten of them as well.

Spitfires from 602 Squadron at Westhampnett caught another formation of dive-bombers as they dropped their bomb loads on Ford, and Spitfires of 152 Squadron attacked the *Stukas* targeting Thorney Island. Finally, 109s heading for home with fuel exhausted were hit by Hurricanes of 213 Squadron from Exeter. A total of seventeen *Stukas* and eight 109s were shot down for the loss of five RAF fighters and two pilots. Once again, no airfields under *Stuka* attack were fighter bases and Kenley was back in service an hour after the raid, while the German lack of hard intelligence made the sacrifice of many of their aircrew largely pointless.

The day brought both sides their heaviest losses of the entire battle. The ten days since *Adlertag* had seen 363 German aircraft shot down against 211 RAF fighters. Dowding and Park could reflect only 181 of these had been destroyed in the air, with the rest on the ground. The Germans suffered high *Stuka* losses, lacking heavy escort protection to keep Spitfires and Hurricanes at bay. *Luftflotte 2*'s dive-bombers were taken off operational orders for regular missions, causing a severe loss in accuracy and destructive power.

The Spitfire's bullet proof windscreen could absorb a direct hit from an enemy bullet. *(Via Alfred Price)*

Because combats took place mainly over land, only three RAF pilots ditched in the sea; two were missing but one was rescued. Over English soil, eight were shot down and killed, but sixteen, twice as many, bailed out and one crash landed, but all survived. Over the Channel most would probably have been lost. But RAF losses were high, with 154 pilots killed or wounded too badly to return to operations. Just sixty-five pilots were posted from training units, without the experience of those they replaced. If fighting continued at this intensity, Fighter Command might well be short of fighters and run out of competent pilots before the month was over.

Fortunately, the British had an unseen ally. Like the Russian 'General Winter' which hampered invasion attempts from Napoleon to Hitler, Britain's summer weather was another effective obstacle. The Luftwaffe was still a fair-weather air force. Its greatest triumphs had been in Continental climate zones like Poland and France, and many pilots could not cope with England's maritime conditions. The Bf109E was unusual in lacking an artificial horizon, essential for flying in cloud, so bad weather effectively grounded German fighters.

Another lull now followed. Instead of the Luftwaffe increasing its advantage by attacking RAF airfields, the defenders were given time to recover. On 19 August, 109s tried to tempt the RAF to intercept them by flying over the coast and the Thames estuary, but gained no response. Later, small German fighter formations strafed airfields from Manston to the southwest, but with no RAF reaction. Finally a Ju88 attack did the trick. One bomber and two Spitfires were lost as oil tanks at Pembroke Dock on Milford Haven were set ablaze.

When combat occurred, Park's priorities of attacking both bombers and escorts were being flouted. Sending Hurricanes against the bombers with Spitfires to keep the 109s at bay worked well in theory. However, the Spitfires were concentrating on fighter-to-fighter combat when bombers were still the most vital targets. Park advised pilots not to chase 109s but to wait until low fuel forced them home to leave the bombers exposed. He asked 12 Group to guard his airfields while 11 Group fighters tackled German raids. Leigh-Mallory insisted that forming massive 'Big Wings' were the only way to beat these raids. Park insisted combat over forward airfields left no time for Big Wings, but support for the idea was beginning to increase.

Park's pilots felt that ordering Spitfires to attack bombers first would leave them to be bounced by 109s flying top cover for raids, and would mean German bomber crews demanding even larger escorts to accompany them into RAF airspace. On the other hand, the increasingly successful Flight Lieutenant Al Deere of 54 Squadron brought a different perspective to the subject when he reasserted the value of 'The policy of using selected Spitfire squadrons to draw off the enemy escort fighters, thus enabling the remaining squadrons..., to concentrate more effectively on the bombers. Though this decision means a much tougher and unrewarding job for the Hornchurch Spitfire squadrons, I do not recall a single pilot saying other than he thought it an excellent idea. I strongly support this view, and on numerous occasions witnessed the rewards reaped when enemy bombers, shorn of the majority of their escort, were set upon by defending Hurricanes which, excellent as they were, could not have coped so effectively without the intervention of the Spitfires.'[5]

On the German side Göring wanted more raids on Bomber Command airfields to deflect the *Fuhrer's* anger for British bombs on German cities. Unfortunately, his commanders had enough trouble finding genuine Fighter Command stations to bomb. Furthermore, Bf110s were so vulnerable to RAF fighters that these also needed 109 close escorts, but bomber crews remained completely unable to talk to the escort pilots.

When operations resumed with better weather, German fighter and bomber pilots tried an effective new tactic. By climbing higher before crossing the Channel, they could trade height for speed on the way to the target or heading back to base afterwards. Even the relatively sluggish He111 could accelerate to 300 mph in a shallow dive, hard for defending fighters to intercept before it dropped its bombs. Faster bombers like the Ju88 were even more effective. Bf109s turning for home with fuel dwindling found that losing height after combat to build up speed for crossing the Channel meant less time to catch them before they reached safety.

So the contest remained deadlocked, with the outcome in doubt. Would the Luftwaffe wear down the RAF's ability to resist German attacks? Or would bad German intelligence cost them too much in casualties and resources for the results they wanted? Everything depended on the next phase of the battle, as the weather improved for 24 August.

5. *Battle of Britain Historical Society website*, 2007.

Chapter Eighteen

The Battle of Britain – The Turning Point

Weather forecasts for Saturday, 24 August 1940 were clear over southern England. Cloud to deflect German attacks would be limited to the north, leaving Göring free to resume bombing RAF fighter stations and Park needing more support from 12 Group to protect his airfields while his squadrons were in action. Leigh-Mallory, the 'intensely ambitious' 12 Group commander, convinced a Big Wing consisting of several squadrons was the only route to victory, began working hard to hound Dowding and Park out of their posts. His ultimate triumph would impose a terrible cost on Fighter Command, its pilots and the air war in general. But for the time being, his doctrine of massing a large group of squadrons for a heavy blow against each Luftwaffe raid had a superficial plausibility.

The Germans too had problems. After selecting airfields truly vital to Fighter Command, they found that only a few were within Bf109 range. The realities of air-to-air combat made it clear that fighters were needed on every mission. Even the best bombers could only survive daylight in enemy airspace with heavy escorts. The 109's short range made this worse, greatly restricting possible targets. Escorted bomber formations could only reach targets north of the Thames if relays of fighters handed over to one another to extend their time over England.

Since both air forces had broadly similar numbers of fighters, only part of the Luftwaffe's bomber strength could be used at one time, unless tactics misled the defenders. Though the RAF had to respond to every raid involving bombers because of the damage they could inflict, high-speed German fighter sweeps could safely be ignored. The campaign had developed into a battle of attrition, where victory would go to those who could fly and fight the longest.

German commanders found themselves impaled on the horns of a dilemma. They

Air Vice-Marshal Keith Park standing by his personal Hurricane OK1. *(Official photo via Wikipedia)*

Air Vice-Marshal Leigh-Mallory. *(Official photo Via Wikipedia)*

had lost men and machines attacking fighter airfields within range of their own bases: Kenley, Biggin Hill, Tangmere and even Middle Wallop. But defending fighters over home territory could always fall back to bases to the north, to a second and even a third line of defence. Air superiority over the southern half of Kent and Sussex was not an absolute, and would evaporate as soon as the Luftwaffe returned to its bases.

Attacking fighter stations north of the Thames, like Debden and North Weald, Northolt and Hornchurch, meant a more difficult choice. Previously the Luftwaffe relied on Bf110s to protect bombers beyond 109 range, but the much-vaunted *Zerstörer* squadrons could not even protect themselves. As British fighters approached, they would form a defensive circle, making them a more difficult target, but leaving their bombers to the mercy of the RAF.

Messerschmitt Bf110 – actually the night fighter version with radar array on nose. *(Author Photo)*

To help make the most of their 109s, *Luftflotte 2* moved them into the northernmost part of the Pas de Calais, where airfields were cramped and ill-equipped for an intense campaign. Meanwhile Dowding moved squadrons which had borne the brunt of the fighting out of Luftwaffe range. This was partially effective, but could not provide the experienced replacements needed to teach novices the techniques for survival. As a result, the end of each rest period would bring heavier casualties and the need for still more replacements. Dowding had already cut the time new pilots spent at an Operational Training Unit before reaching a squadron. From six months, this was down to a month during the July convoy battle, and was now reduced to a completely inadequate two weeks. Len Deighton suggested one reason there was no real attempt to introduce combat-proven pilots to fill gaps in resting squadrons was that their bitter experience might damage the morale of the new recruits.

Meanwhile, the Germans tried more ingenious tactics. From the start *Generalfeldmarschall* Albert Kesselring, commanding *Luftflotte 2,* tried to mislead his opponents. He had served in the artillery in the First World War, transferring to the still secret Luftwaffe in 1933 to become its first Chief of Staff. He was a devotee of air power, and an ingenious and resourceful commander. On 24 August he used reinforcements from *Luftflotte 3* to maintain heavy air activity over his own territory to conceal the launch of an attack. Once his forces crossed the Channel, they would split up on different courses before reassembling and making for their targets to confuse radar operators, Observer Corps posts and Fighter Command over how best to respond in the short time available.

This was potentially very dangerous, though another new tactic tilted the balance back the opposite way. To maintain the morale of the hard-pressed bomber crews, fighters were ordered to stay within plain sight all the way to and from the target. This sounded reasonable to everyone except the fighter pilots, but it would be impossible to devise any instruction better able to rob the Bf109 of its advantages. By making them fly at the same height and speed as the bombers, they gave Spitfires time to climb above them and bounce them, as an alternative to the turning dogfight where British fighters already held the advantage.

On 24 August, the Spitfires of 609 Squadron were scrambled to attack a raid heading for Portsmouth. As radar did not show the types of aircraft involved, the defences assumed this was yet another formation of Ju87 dive-bombers. They met the raid at 5000 feet to catch the *Stukas* at their most vulnerable moment as they pulled out of their dives. It was very nearly fatal. Unaware that *Stukas* were off the menu, the Spitfires saw bombers and escorts 5000 feet above. As they climbed to engage, coastal anti-aircraft batteries opened up and prevented them from breaking up the raid, letting the bombers reach Portsmouth and hit both dockyard and city, killing more than a hundred people and injuring three times as many. Yet the 109s stayed with their charges and did not bounce the Spitfires as would have been normal. Overall RAF casualties were low, suggesting the new tactics imposed on the 109s were making them vulnerable just when they were most needed.

Another fine day on 25 August reinforced this impression. Attacks on Southampton and Portsmouth cost five Hurricanes and a pair of Spitfires out of an overall total of sixteen, but the Germans lost thirty-eight fighters and fourteen bombers, with all crews killed or made prisoner. That night eighty-one RAF bombers attacked Berlin in return for Luftwaffe bombs falling on the City of London.

August 26 brought mixed weather and a reminder of the deadly efficiency of 109s when left to use their favourite bounce attacks. In mid-morning, radar revealed 150 plus raiders heading for Manston. The tattered remnants of 264 Squadron were scrambled too late, with the Spitfires of 616 Squadron from Kenley, but the 109s bounced them as they climbed. Seven Spitfires were shot down in that first pass though five pilots survived.

Back came the Luftwaffe in the afternoon with Dorniers approaching up the Thames estuary. Park sent in seven squadrons to protect Hornchurch and West Malling, arriving as the 109s left with empty tanks, and the bombers were badly mauled. Six Dorniers bombed Debden and wrecked a hangar, the main landing area, the sergeants' mess, the NAAFI and the motor transport depot. Bombs severed power cables and water mains, and the station was out of action for an hour. Park had to limit his response, sending in half his available squadrons until German intentions were clear. Then his remaining squadrons could join the fray, but vital sector stations were left unprotected in the meantime.

In late afternoon, more than fifty He111s, escorted by a hundred-plus 109s, approached Portsmouth from the southeast. In a sublime piece of close co-operation between 10 and 11 Groups, three squadrons of Spitfires and two of Hurricanes took on bombers and escorts. In worsening weather, three He111s were shot down with two badly damaged and most had to drop their bombs into the sea to get away. Once again casualties favoured the RAF. Though twenty-seven defending fighters were lost, only six aircrew were missing. The Germans lost twenty-two bombers and twenty-four fighters with most crews killed or captured.

The 109 pilots were clearly fighting at a severe disadvantage. Along with their new vulnerability to bounce attacks, they had to cruise at similar speeds to the bombers when combat began. They also lost the advantage of surprise, since RAF pilots knew where to find them. But Park was increasingly outraged by the lack of support from 12 Group. On one occasion their squadrons failed to turn up at all, on another they lost their way, and on most of the rest they arrived too late. Yet Leigh-Mallory was convinced he alone knew how to defeat the Luftwaffe with his Big Wing. He felt he should be commanding 11 Group, notwithstanding Park's experience in controlling fighters and his skill in managing the battle.

On 28 August the weather improved again, triggering attacks against Manston, Rochford and the Luftwaffe's eternal obsession of Eastchurch, where two obsolete Fairey Battles were destroyed. Park's controllers sent seven squadrons after a mixed formation of 109s and 110s on a coat-trailing mission over the Channel. The resulting battle on German terms saw five British fighters lost to bring down half a dozen Luftwaffe machines out of fourteen lost that day, evidence of the tactical tight-rope 11 Group had to follow.

So effective was this kind of German attack that even experienced British pilots fell victim. One 603 Squadron casualty, Pilot Officer Don Macdonald arrived at Hornchurch the day before with only fifteen hours flying Spitfires, barely enough to learn how to handle the machine, let alone fight in it on his first patrol. But the two other 603 victims were Flight Lieutenant Laurie Cunningham and Pilot Officer Noel Benson, each with more than 160 hours flying Spitfires in their logbooks.

The simple fact was that RAF pilots were buckling under the strain, with two British fighters mistaken for 109s and shot down by compatriots in the stress and confusion of one dogfight.

More poor weather on 29 August saw Park wrong-footed by another mixed formation of 109s and 110s, reported as 700-plus seen approaching the Kentish coast in mid-afternoon. He scrambled thirteen squadrons. By the time he realized this was a fight his pilots could not win and called them back to base, four fighters had been destroyed and two damaged with the loss of one pilot. Later, another decoy formation was spotted by Park, who sent just two squadrons of Hurricanes and one of Spitfires to meet them. One Hurricane pilot was killed, but four 109s were shot down.

Better weather on the following day brought different tactics. Kesselring sent in massed formations from mid-morning, mixing Ju88s with fighters as a trap for Park. If he ignored the attacks he risked bombers hitting his airfields where a single bomb could cause massive damage. One lucky pilot hit the main power supply to half-a-dozen radar stations, closing them down for three hours, though the Luftwaffe remained unaware of this.

Other attacks mixed He111s and Do17s with close escorts of 109s, with still more 109s flying high above, ready to bounce any British fighters attacking the bombers. When Fighter Command sent in squadrons against German formations, they found it hard to attack through the 109 fighter screen without risky tactics. One Spitfire pilot realized tackling bombers head-on avoided their fighter escort and broke up the formation, making them more vulnerable in the resulting dogfight.

For less skilled pilots or those who took on Bf110s which DID have heavy forward armament, the effects could be devastating in reverse, so Dowding tried to ban the tactic without success. But stakes were high as German formations dropped bombs on every RAF fighter station in south-east England for the first time in the campaign. As attack succeeded attack, Park fed every squadron into the battle, leaving badly damaged Kenley and Biggin Hill temporarily defenceless. He asked Leigh-Mallory to send 12 Group squadrons south to protect the airfields. From Air Vice-Marshal Brand at 10 Group this would have brought an immediate and effective response. Two squadrons were sent from 12 Group, but instead of patrolling Park's airfields they set off looking for the enemy and failed to find them. Unfortunately, the Luftwaffe were better at finding targets than 12 Group, and

Junkers Ju88 bomber version with glazed nose panels. *(USAF)*

Biggin and Kenley were bombed yet again. An incandescent Park had to pull two squadrons out of the battle to cover 12 Group's failure.

In late afternoon, Park had to commit all his fighters and again called for help from 12 Group. The response was better, but a devastating pounding of Biggin Hill by half-a-dozen Ju88s carrying 1000lb bombs put the base out of action. Thirty-nine fighters had been lost, mercifully with the loss of just six Hurricane pilots and two Spitfire pilots, but the most worrying development was the accurate bombing of fighter fields.

Fine weather on 31 August promised more of the same, and new German tactics put greater pressure on Park and 11 Group than ever before. Every raid split into smaller formations, each heading for a different airfield. The defenders could not attack over the Channel because of the dangers for pilots shot down over the sea. If they waited too long they faced multiple threats, which meant splitting into smaller formations, but half a squadron bounced by high-altitude escorts could not hope to break up an incoming raid.

Ironically, a Big Wing might have helped, but shortage of time and space made it impossible. Assembling up to five squadrons and letting them climb to take on 109s on equal terms could not be done between raids crossing the coast and reaching their targets. Fighter Command was like a boxer facing a faster and more nimble opponent, growing weaker as round succeeded round.

Debden was bombed as another Luftwaffe formation broke through. Ironically this was because 12 Group called for help for once, as forty Do17s escorted by Bf110s headed for Duxford. Park sent 111's Hurricanes to break the German formation with a head-on attack and shoot down one of the bombers before they reached their target. Given the enormous pressure on 11 Group, saving Duxford meant exposing the gap through which Debden was attacked. As some compensation, 19 Squadron's Spitfires from Duxford's satellite airfield of Fowlmere caught the fleeing bombers and shot down two for the loss of two Spitfires.

Through the day, attack succeeded attack. Biggin Hill was bombed yet again even though defending Spitfires shot down three 109s in trying to get at the bombers. So badly was the field cratered that fighters returning to refuel and rearm had to divert to Kenley and Croydon, themselves already under attack.

In the early afternoon, the Luftwaffe came back for the radar stations. Though coverage was interrupted, all were working again by the end of the day. At Biggin Hill, bombs destroyed the Operations Room, runways, workshops and repair hangars, and cut telephone links with Fighter Command. Yet the worst damage was the toll on pilots, with so many of the more experienced now victims of fatigue and loss of concentration, which could be fatal in seconds.

Fighter Command lost forty-one fighters shot down or missing over the sea. Fortunately thirty-two pilots were safe, though twenty-two had had to bail out, half suffering burns before abandoning their machines. As the pace quickened, losses mounted for both sides, but Fighter Command had to carry on fighting so long as sufficient pilots and machines remained, rather than yielding to German attacks and the possibility of invasion, however unrealistic this actually was.

For the Luftwaffe, expecting a brief campaign against an outnumbered and vulnerable foe, reality was beginning to threaten morale. The Bible says that 'Hope deferred maketh the heart sick'[1] and for German air crews the hope of meeting and defeating the last fifty Spitfires, which their intelligence insisted was the remaining strength of Fighter Command, was being deferred over so many days of fierce combat that hearts were sickening to a dangerous extent.

During August 1940, Fighter Command lost 211 Hurricanes destroyed and 44 damaged, and 86 of their pilots killed with 68 wounded. Spitfire squadrons lost 113 aircraft destroyed and 40 damaged, with 44 pilots killed and 38 wounded. This was a higher loss rate than Luftwaffe fighter casualties of 217 109s destroyed with 54 pilots killed and 91 prisoners of war, but the RAF had also been downing German bombers.

During August, non-fighter Luftwaffe losses amounted to 452 machines, and 246 crewmen killed with another 713 missing or made prisoner and 162 wounded. Perhaps the most shocking outcome from the German point of view was that their losses of fighters, on which attacks depended, represented 25 percent of their entire strength at the start of the month. Losses of bombers were slightly less serious, at almost 20 percent of initial strength, but could not be contained for long.[2]

1. Proverbs, xiii, 12.
2. Figures from Murray, *op.cit.*

Dornier 217 – a basically similar but more powerful development of the Dornier 17, which was introduced in 1941. *(Bundesarchive photo via Wikipedia)*

In a report dated 6 September 1940, Park warned, 'Contrary to general belief and official reports, the enemy's bombing attacks by day did extensive damage to five of our forward aerodromes and also to six of our seven sector stations. There was a critical period when the damage to sector stations and our ground organization was having a serious effect on the fighting efficiency of the squadrons, who could not be given the same good technical and administrative service as previously.... The absence of many essential telephone lines, the use of scratch equipment in emergency operation rooms, and the general dislocation of ground organization, was seriously felt for about a week in the handling of squadrons by day to meet the enemy's massed attacks, which were continued without the former occasional break of a day.'[3]

The sad truth was that the Luftwaffe was not Park's only opponent. Leigh-Mallory still cherished the conviction that he was a great fighter leader based on neither evidence nor experience. If Park could be forced out of his job and replaced by Leigh-Mallory, the better for 11 Group, the better for Fighter Command and of course the better for Leigh-Mallory. He stepped up his criticism of Park's tactics and fitness for command, and more and more senior officers seemed all too willing to listen.

On 30 August 12 Group's 242 Squadron was sent to protect North Weald at Park's request. Flying as a single unit they were fast enough to find an attack still in progress. Seeing Bf110s among the attackers, the Hurricanes dived on them. In

---

3. Sir Keith Park, report dated 6 September 1940, quoted on website of Battle of Britain Historical Society.

the resulting dogfight, they claimed twelve attackers shot down and Leigh-Mallory ensured the highest levels of the RAF were told. He was congratulated by Sir Cyril Newall, Chief of the Air Staff, and by Sir Archibald Sinclair, Secretary of State for Air. The truth was far less impressive. Later analysis showed the small group of Bf110s had been attacked by five of Park's squadrons and they had shot down at least nine of the eleven German aircraft destroyed over North Weald. At best, 12 Group pilots brought down two Bf110s, though more detailed analyses suggest one or both may actually have fallen victim to pilots from No 1 Squadron, also present at the time.

This was bad enough when those bearing the brunt of the fighting were under such strain, but the Duxford Wing's continuing poor performance and its absurdly overinflated victory claims made things worse. This was not the fault of the 12 Group pilots as over-claiming varied with the size of the formations involved in all campaigns and all air forces, but the inflated claims were used by Leigh-Mallory's team to prove he could do Park's job better than Park himself.

The pilots of 11 Group found their job becoming more tiring, more difficult and more dangerous. Paradoxically their drooping morale struck a chord with their German opponents who detected a falling off of the spirit with which their attacks were being resisted. Could the longed-for collapse of Fighter Command be happening at last?

The first day of September brought business as usual. The Luftwaffe wanted another chance to force the RAF to battle. In a desperate attempt to fill the gaps in squadrons cut to pieces in dogfights, other sources of pilots were combed to make up numbers. Bomber Command, Coastal Command and the Fleet Air Arm provided volunteers with plenty of flying experience, but they still needed retraining to make them fighter pilots. However, one group needing almost no retraining were the Czech and Polish pilots who had fought the Luftwaffe over their own territory in the first few weeks of war. That experience gave them reckless courage and a bitter hatred of the enemy, but limited English and unfamiliarity with RAF routines and procedures had stopped them being accepted as fully operational. Now Spitfire squadrons like 610 and Hurricane squadrons like 85 were being sent north to recuperate, it was clearly time the Poles and Czechs joined the battle.

The result was an amazing boost to Fighter Command fortunes. On the last day of August, 303 (Polish) squadron based at Northolt was finally declared operational after one pilot spotted a German bomber formation while their Hurricanes practised interceptions on RAF Blenheims. Without waiting for orders he plunged into a dive and shot down a Do17 with a perfect deflection burst from very close range. This expertise was worth diamonds and Dowding was persuaded to reverse his earlier opposition.

The remarkable record of 303 Squadron showed what a single unit of dedicated and experienced pilots could achieve against the Luftwaffe while flying Hurricanes, which German aces regarded with contempt. During the rest of the battle they shot down 126 confirmed German aircraft destroyed, both bombers and fighters. At first, they faced official scepticism, which led their station commander at Northolt to fly with them and check their claims. To his unbounded surprise he finished the mission 'And reported back, rather shaken, that "what they claimed, they did indeed get!"'[4] More remarkably, these verified claims gave 303 the highest claims-to-losses ratio of either side in the Battle of Britain. Having lost nine pilots, their ratio was fourteen-to-one when withdrawn on 7 October. The best ratio achieved by a Luftwaffe fighter unit, the 9th *Staffel* of *III/JG26*, was a claims to losses ratio of 8.7 to 1.[5] Two factors were involved. Poles and Czechs had never learned the lethal formality of the Fighter Area Attacks. Secondly, they closed to ranges shorter than even the most experienced RAF men. Though individual Czechs and Poles had flown with RAF squadrons, the exploits of these special units were a powerful tonic at a time of maximum danger.

It was just as well. The 1 September saw multiple raids heading for vital fighter airfields. Dowding and Park had to meet each incoming assault and maintain defensive patrols over each fighter airfield in case attacks reached them. As the more numerous Luftwaffe split its resources to make multiple attacks, Fighter Command risked that smaller formations might fall below the critical mass needed to deter attackers.

Kenley and Hornchurch were hit yet again with North Weald and Gravesend, but Biggin Hill suffered most with the sixth and seventh heavy raids in three days. Operations rooms were wrecked and communications severed, but repair work began as soon as the raiders left. As an insurance against a repeat performance, a replacement Operations Room was set up in the nearby village with all the necessary telephone links and power cables. Mercifully pilot losses were surprisingly low; five Hurricane pilots and a single Spitfire pilot. The Luftwaffe lost seventeen fighters, two more than the RAF, with the loss of eight bombers.

The Luftwaffe returned in strength the next day to resume pounding the Coastal Command station at Eastchurch into rubble. They detonated the bomb dump and left only one runway useable. A raid on North Weald was beaten off, but Biggin Hill was struck by a fast, low-level attack. There were also signs of a shift in priorities as German bombers struck aircraft factories at Brooklands in Surrey. Bombs fell on the Vickers plant making Wellington bombers, but the raiders missed the Hawker factory building Hurricanes. This was far more relevant to Fighter Command, though still less vital than front-line fighter airfields.

Casualties were heavy. The Luftwaffe lost ten bombers to RAF fighters and three to anti-aircraft fire. They also lost twenty-seven fighters shot down in combat and one downed by AA gunfire, but the RAF suffered fewer pilot losses with two

---

4. Quoted in Bungay, *op.cit.*, p. 175.
5. *Ibid.*, p. 347.

Hurricane pilots dead and two more posted missing. Given that most shot-down 109s took their pilots with them into captivity, even if they survived the combat, this eased the pressure slightly. Meanwhile Hornchurch's 603 Squadron demonstrated the Spitfire's growing advantage over the 109s. With one fighter forced to make an emergency landing, the rest took on the 109s in a turning dogfight, ideal for the Spitfire's manoeuvrability. They shot down four German fighters in quick succession as they tried to break out and head home over the Channel.

By lunchtime, 603 faced another massive raid with Spitfires from 72 Squadron. Flight Lieutenant Graham spotted a Dornier and moved in to attack, only to be fired on by a Bf110. He pulled hard on the stick to climb and left the German aircraft to pass beneath him before swinging around in a tight turn to starboard to close on his assailant. Despite the German pilot's evasive action, Graham stayed on his tail. After a series of short bursts, the 110 emitted thick smoke, its gunner bailed out and the pilot crash-landed three miles west of Dover. Another German fighter, a 109 from *JG54*, was hit by an accurate burst from one of 603's Spitfires and crashed near Chilham, west of Canterbury, killing its pilot. His body was found in the wreckage almost forty years later and given a belated military funeral.

Park and Dowding were now having to deploy fighters to prevent the Luftwaffe disrupting aircraft production, but when these units were dragged into combat, a gap opened up for the bombers. German morale was boosted by seeing the damage they inflicted, and the desperate RAF attempts to deflect their attacks. Finally they knocked out their cherished target of Eastchurch, declared non-operational for part of the day.

Kesselring wanted to force the RAF to wear itself out against heavy German attacks. However, the more sceptical Sperrle, commanding *Luftflotte 3*, was convinced the RAF was far from beaten and felt more airfield attacks would reduce its strength and ability to resist. For the time being that was exactly what was happening. On 3 September a big formation over the Thames split into three, heading for Hornchurch, North Weald and Debden. Park sent in 603's Spitfires with two Hurricane squadrons to tackle the raid. Bombs damaged the hangars and operations block at North Weald, and started fires. For once, help arrived from 12 Group as 310 Squadron's Hurricanes shot down four Bf110s. This was a bad day for the German long-range fighters, exposed as Park postponed interceptions to force the 109s back to base with dwindling fuel reserves. Two more 110s collided and crashed, and another two were shot down by Spitfires. More would have been lost had not 19 Squadron's cannon-armed Spitfires suffered the usual stoppages after a perfect diving attack.

Hornchurch remained operational. Biggin Hill was bombed again, but the raids tailed off during the afternoon. Göring was insisting in The Hague that more attacks were needed on Fighter Command airfields, as even Biggin Hill was still operating. Still heavier attacks would defeat the RAF before turning on London in preparation for the invasion. Ironically he suggested the Bf110s should be used more ambitiously as long-range escorts so more targets could be hit. But his Luftwaffe was wearing out almost as quickly as Fighter Command. German losses on 3 September were eight bombers and seventeen fighters, while the RAF lost twenty fighters and two Blenheims shot down by mistake.

More fine weather on 4 September brought more airfield raids. Lympne, Rochford and the Coastal Command airfield at Eastchurch were hit, together with the Short Stirling plant at Rochester and the Wellington factory at Brooklands. Once again the Germans had the Hurricane plant at their mercy but missed it in the confusion of combat. It was a brutally expensive raid for the Luftwaffe with six Bf110s shot down on their way to the targets and another nine on the way back to their bases.

Park scrambled a dozen squadrons, six Hurricanes and six Spitfires, to meet the attacks. The Spitfires of 234 Squadron were called in from St Eval in Cornwall to patrol over Tangmere. They spotted Bf110s below and their Australian flight commander Pat Hughes attacked the smaller of two formations some fifteen strong with one section, while the other three sections tackled a larger group of around fifty crossing the coast. When the German pilots spotted Hughes and two of his section approaching, they formed their usual circle. Hughes attacked the leading 110 head-on. His first burst hit the pilot and the stricken machine reared upwards and burst into flames. He manoeuvred to catch another 110 in his sights and delivered another burst, sending it to explode on hitting the ground.

By now four more 110s were queuing up to shoot at Hughes. He jinxed sharply to avoid their fire, loosing off a burst at each one. As one dived away from the fight it became a potential target and Hughes plunged after it. As its image expanded in his sights, he fired the rest of his ammunition, sending it earthwards with both engines on fire. Meanwhile his other pilots claimed fourteen more 110s between them, with a Dornier 17 for one Spitfire slightly damaged.

It was the worst day in the campaign for the German long-range escorts, and underlined what a formidable opponent the RAF remained in spite of German hopes its strength was waning. The Luftwaffe lost fourteen bombers and six 109s while the RAF lost five Hurricanes and their pilots and six Spitfires with four pilots.

That night Hitler spoke at Berlin's *Sportpalast*. In a rabble-rousing speech he promised the British would suffer massive retaliation for attacking German cities. In the sinister world of Nazi Germany, this was a clear signal that Luftwaffe chiefs had better review their priorities. Both Göring and Kesselring decided to crush the RAF's remaining squadrons by attacking the most important target within range: the city of London itself.

This would take time to organize and more settled weather on 5 September brought more attacks on 11 Group airfields – North Weald, Kenley and Biggin Hill – together with Detling, Lympne and once again the Coastal Command field at Eastchurch. Both Biggin Hill and Eastchurch were temporarily unusable, and the Hawker factory at Brooklands was damaged. One Spitfire squadron, 41, flew from Catterick to the battered landing ground at Manston. They were sent to meet the first raid over the Thames Estuary where one pilot bounced an escorting 109 caught attacking a Hurricane, forcing it to crash-land near the village of Aldington, between Ashford and Folkestone. Three more 109s were shot down by Spitfires.

It was a discouraging day for both air forces. For the RAF, settled weather meant no let-up in the strain on pilots, machines and the defence system. Tired pilots could not make out the tiny specks of approaching aircraft nor react quickly enough if bounced by the enemy. Even the seasoned fliers of 41 Squadron were vulnerable. Two Spitfires collided over Essex, killing both pilots. Each casualty added to a growing list and reserves were dwindling. In time, replacements would appear, but time itself was in the shortest supply. Increasing attacks on aircraft factories could cut replacement fighters needed to keep squadrons fighting. As with any attrition battle, the smallest faltering would give the enemy the best tonic. If 11 Group could no longer intercept German raids, the Luftwaffe might rightly assume their battle was won.

The fighting had discouraged the Germans too. They had lost twenty-three aircraft compared with twenty RAF machines. In spite of their numerical advantage the 109 pilots had to fly several demanding missions a day with the ever-present threat of the Channel to end every one. Now they faced an apparently hopeless struggle against a tenacious and determined foe. How much longer to defeat the RAF, and did they have the resources?

On 6 September the morning dawned clear and bright, bringing three large raids over the Thames Estuary against Kenley and Biggin Hill, and a switch to Heston from the usual bombing of Eastchurch. More attacks were delivered against the Hawker works at Brooklands and oil tanks at Thameshaven. Brooklands was damaged, but oil fires could be seen for miles and were hard to extinguish. Yet the day's operations hit the Luftwaffe twice as hard as it did the RAF (seven bombers and thirty-seven fighters destroyed, compared with twenty-two fighters, with eight pilots).

Dowding and Park knew Bomber Command's raids on Berlin would provoke a violent reaction from the Luftwaffe. In Fighter Command, many squadrons had already been cut from the full establishment of twenty-six pilots to sixteen or less. Park tried to raid the under-employed Fairey Battle squadrons for much-needed pilots, but Air Vice Marshal Sholto Douglas, a bomber man to the soles of his boots, refused, saying these slow and obsolete bombers would be needed to drive the enemy back if the invasion was launched. They had completely failed to do this in the May *blitzkrieg* in France in spite of appalling casualties.

It was also becoming clearer to the Germans that despite all their exertions and losses, they were no closer to crushing 11 Group and Fighter Command than they had been a month before. While Dowding's figures showed he was 200 pilots short of his planned establishment, he still had almost 200 pilots more than at the start of the Channel campaign in early July.

Meanwhile, the Luftwaffe was suffering severe shortages of its own. Their replacement pilots were equally vulnerable in combat and their aircraft shortage was worse than the RAF. Normally a fighter *Gruppe* would contain between thirty-five and forty aircraft, but average strength had now fallen to eighteen fighters. With some 75 percent serviceability, this was no way to equip an air force for an attrition campaign. This would not have mattered so much had the RAF really been limited to its last 200 fighters. The fact that German Intelligence did not know the truth made the situation worse at the moment when the campaign was about to take a dramatic shift, to take all bets off the table.

Chapter Nineteen

The Battle of Britain – Deliverance

Saturday, 7 September 1940 saw settled weather over southern England, with conditions ideal for attackers approaching over the Kentish coast or up the Thames estuary, before heading for airfields or aircraft factories. After three weeks' casualties, RAF reserves had dwindled to danger levels and it was vital to time the interception of incoming raids exactly.

Park repeated his warning to pilots to obey ground controllers' orders. With mounting losses from high-flying 109s bouncing them out of the sun, Spitfire pilots used their faster climb performance to add several thousand feet to heights given by controllers for a double advantage over the enemy. Diving quickly maximized surprise. Better shots could turn a dive into a kill, but even less skilful pilots could let the combat evolve into a turning dogfight. This improved their survival chances, but made it harder to destroy the bombers before reaching their targets.

For the moment at least, there was no sign of the Germans. At 8.30am radar screens remained blank. A single reconnaissance aircraft was shot down by Spitfires over the Dutch coast. The fighters went home and the sinister calm resumed. Hour after hour passed with no change. It was almost midday when radars at Dover and Pevensey spotted a small formation over the Channel. Spitfires from Kenley found

Junkers Ju88 bomber. *(USAF)*

Bf109s and attacked, but two Spitfires were forced down on English soil leaving the strange quiet to resume. Might the invasion already be under way?

In fact, Göring was enjoying the fine weather. With Kesselring and his First World War comrade Bruno Loerzer, he toured airfields before hosting a picnic on the cliffs above Cap Blanc Nez. There, the largest Luftwaffe formation ever sent against England, more than 1000 aircraft, finally shattered the peace of the autumn afternoon. This massive attack would restore his standing with Hitler and allow him to boast that by taking personal command, he could bring final victory.

At twenty minutes to four, radar screens at Dover, Rye and Foreness showed massed aircraft over the Pas de Calais. Fighter Command scrambled twenty-three squadrons to protect the battered fighter stations and the Thameshaven oil storage tanks. As always the Duxford Wing took far too long. The original plan to hit the German formation before Maidstone was missed by twenty minutes. They were attacked on the climb by Bf109s and found their targets well past Maidstone, where two 11 Group squadrons had had to take on the Luftwaffe on their own.

Everyone soon realized something was different. The Observer Corps reported 300 bombers and 200 Bf110s, with some 600 Bf109s as the actual escorts, but this huge attack was not aimed at fighter airfields. One part headed past Biggin Hill and Kenley without attacking, and crossed the Thames between Windsor and Maidenhead, finally heading for the wharves of east London.

A second big formation crossed the coast near Beachy Head and the Thames near the Isle of Sheppey to follow the river towards London. Finally, more fighters and

Pilot Officer David Glaser and his 65 Squadron Battle of Britain Spitfire – on the cowling is one of the baffles designed to deflect the exhaust glare during the abortive (and hideously expensive) attempts to search for German night raiders. *(Via Alfred Price)*

bombers flew from the Kent coast to the East End. For once, co-ordination worked and the groups recombined into a single huge formation. As RAF fighters reached the Luftwaffe's position they saw aircraft massed 8,000 feet deep and covering some 800 square miles. One Spitfire pilot, looking up at layers of 109s over the German bombers and diving to attack, said it was 'Like looking up the down escalator at Piccadilly Circus.' Their biggest problem was breaking into the formation to deflect the bombers from their targets. Each time they approached, the 109s would 'bounce' them into taking evasive action.

Park insisted bombers were attacked first, but this was self-preservation. The 109 pilots were experts. If the RAF attacked from the formation's starboard side, the top cover escorts would dive and force them away from the bombers, leaving the port side escorts to take over the top spot ready for the next attack. For Park, the vital question was what had happened to the Duxford wing? Once again it arrived as the bombers were heading for their French bases, helped by a stronger than usual tailwind, while another 200 bombers headed for the capital with no RAF fighters to stop them.

Fortunately, other German formations met more trouble; the northbound group was met by Spitfires, but before they could attack, the Germans turned for the docks. With extra height the Spitfires plunged through the 109s and shot down a pair of Dorniers, a pair of Bf110s and a single 109 without suffering any casualties, but the main German formation continued to its target. Another formation hit by both Spitfires and Hurricanes still managed to bomb.

By 5.30pm, bombers were hitting targets from Woolwich Arsenal to the factories and docks of the East End, and the communities between them. Only when the 109s had to return to refuel and rearm could the RAF reach the bombers in a fierce dogfight before they too had to replenish fuel and ammunition. To make things worse, Hornchurch was blotted out, not from Luftwaffe attacks, but by smoke from the blazing docks.

At dusk more bombers crossed into English airspace. With almost no RAF night fighters, they could range freely over darkened cities. When they returned and reported London's East End in flames, they convinced Göring this was the winning formula to knock the RAF out of the fight. Park thought differently. Although large German formations had ranged over British airspace for half a day, no attacks at all had been made on 11 Group's airfields. He said afterwards, 'Between the 8 September and the 6 October when Göring concentrated his day bombers on London, the London Docks, the aircraft factories and other cities in the south of England …, that was really his major tactical error because we think now that if he had carried on for another week or ten days hammering my fighter airfields he might have had them out of action, in which case we could have lost the battle, because we could not have operated effectively from aerodromes north of London.'[1]

---

1. BBC interview, recorded 1 January 1961.

Not that this worried Göring. He sent more and more bombers to fan the fires blazing over the East End. They mixed incendiaries with the latest high capacity bombs containing a ton-and-a-half of high explosive apiece, to kill firemen trying to control the flames. Huge stacks of timber and bonded warehouses crammed with spirits or stores of sugar, molasses and paint continued burning fiercely.

The ordeal of city and people continued for seven interminable hours until the morning all-clear. More than 300 tons of bombs killed almost 500 civilians and injured 1000 more and this was merely the beginning. Because powerful bombardment might mean invasion, the 'Landings imminent' alarm was issued during the night. Roadblocks were set up, the Army and Home Guard were called out and preparations made for blowing up vital bridges and strong points. Only with daylight and a temporary lull in the bombing was it clear no landings had been made.

After this first concentrated assault on London, and the first night of round-the-clock bombing, Fighter Command had lost eight Spitfires destroyed and sixteen damaged, and sixteen Hurricanes lost and eleven damaged. However, six Spitfire pilots and eight Hurricane pilots had been killed, many of them experienced leaders. The Germans suffered too. RAF pilots claimed fifty enemy aircraft, seventy-three 'probables' and thirty-nine 'damaged'. German records confirm the loss of nineteen bombers and Bf110s, but no less than sixteen Bf109s, with a damaging effect on future operations.

With larger formations, over-claiming increased in proportion. For example, 19 Squadron's Spitfires from the Duxford Big Wing were led by Squadron Leader Brian Lane. His report explained, 'A [Bf] 110 dived in front of me and I led 'A' Flight after it. Two Hurricanes were also attacking it. I fired a short burst as well as the other aircraft. Two bailed out, one parachute failing to open. Enemy aircraft crashed one mile east of Hornchurch and one crewman landed nearby and was taken prisoner.' The problem was that all three pilots would have fired at the enemy and seen the crew bail out and their machine crash. All three would claim the Bf110 in good faith and if these were not cross-checked, the victory total would have been multiplied by three.[2] So Big Wing operations invariably meant inflated claims.

Meanwhile, the Luftwaffe continued to mislead Fighter Command. Formations assembled behind the French coast within radar range. Many would not attack, but if Park sent up fighters there was a danger they might have to refuel when the real raid arrived. He could only afford to react as raiders crossed the English coast, leaving insufficient time for warning 12 Group.

10 Group had even less time, but sent squadrons when ready. 609's Spitfires reached aircraft factories at Brooklands and Weybridge in time to catch the attackers. One pilot reported, 'I went for the nearest bomber and opened fire from

---

2. Combat report, Squadron Leader B.J.E.Lane, 19 Squadron, 7 September 1940.

about 400 yards, meanwhile experiencing heavy return cross fire from the bomber formation. After about 12 seconds, smoke started to come from the port motor and it left the formation. I then waited for it to go down to 3000 feet and then dived vertically on to it and fired off the rest of my ammunition.'[3] The bomber crashed into the estuary, but the chance for the Spitfire pilot to concentrate on its destruction probably meant German fighters had already left.

However, darkness tilted the balance firmly the other way. RAF fighters could not find and shoot down German bombers at night. Even flying a patrol and finding one's base to land safely was difficult, and lives were lost in trying. The solution would depend on two-seat night fighters, Defiants or Blenheims or the later radar-equipped, faster and heavily armed Beaufighters, but the necessary technology was in its infancy.

Early on 8 September, Park flew his Hurricane over the East End and the centre of the previous night's attack. It made a grim picture with fires stretching seven miles from Tower Bridge to Woolwich and beyond. The main line railway stations at London Bridge, Waterloo and Victoria had been crippled. Fires still burned at Woolwich Arsenal with the threat of major ammunition stores being detonated by the heat, and flames threatening Beckton gas works, the largest in Europe. 'So widespread were the fires that the Fire Officer at the Surrey Commercial Docks on the south bank sent the often-quoted message to "Send all the pumps you've got …. The whole bloody world's on fire…."'[4]

The Rotherhithe tunnel was closed by debris. Shops, houses and pubs were reduced to piles of rubble, with mains services cut off and roads blocked. But Park found this a blessed deliverance. If the Germans had tried so hard to smash British resistance, how could they give up now? 'It was burning all down the river' he said afterwards. 'It was a horrid sight. But I looked down and said "Thank God for that", because I knew that the Nazis had switched their attacks from the fighter stations thinking that they were knocked out. They weren't, but they were pretty groggy.'[5]

So they were, but the main reason for the German change of strategy was the cost of daylight attacks. Like the RAF earlier and the USAAF later, they realized the terrible cost of daylight bombing against a tough and well-commanded fighter force. They would increasingly turn their bombers to night raids and use more single engine fighter-bombers, faster and more difficult to intercept and shoot down in daylight. Meanwhile, Park could move his hardest pressed squadrons out of the battle and replace them with fresher units.

Then the Luftwaffe returned, sending twenty Dornier 17s and thirty Bf109s across the Kent coast near Deal. Park's Spitfires were beaten off by the 109s and their fuel

---

3. Combat report, Flight Lieutenant McArthur, 609 Squadron, 7 September 1940.
4. Quoted in '*Wartime Britain, 1939–1945*', Juliet Gardiner, p. 290.
5. Quoted in Battle of Britain Historical Society website.

was running low when reinforcements arrived. One Spitfire and four Hurricanes were shot down with the deaths of two pilots, but the attack was deflected short of London. However, night raids could clearly not be deflected. Sirens sounded at 7.30pm as bombers approached, safe under the blanket of darkness. Two hundred and fifty bombers unloaded tons of incendiaries and high explosive, re-igniting many earlier fires, killing more than 400 people, and injuring almost twice as many.

On 9 September, cloudy conditions kept things quiet until late afternoon. Park ignored a sweep of 109s, waiting for the main attack. As a hundred bombers with sixty escorts followed, two Spitfire squadrons were sent to patrol over Kenley and Biggin Hill. More squadrons formed a second defence line, with a request for 12 Group to fill a gap north of the Thames between North Weald and Hornchurch.

Douglas Bader commanded the Duxford Wing and fought the battle as he wanted. As the sun moved to the west he climbed to 22,000 feet, 2,000 feet higher than ordered, to attack out of its glare. Meanwhile, 607 Squadron's Hurricanes attacked bombers heading for the Surrey aircraft factories before the 109s could spoil their aim. The German raid was beaten off and turned away for France.

The Duxford Wing met more German bombers approaching from the east after they had been intercepted by Spitfires from Hornchurch. As they attacked, 19 Squadron had to scatter before the 109s, but for once the Wing was in the thick of the fighting, generating claims for twenty-one enemy aircraft destroyed. RAF casualties were three Spitfires destroyed and four damaged, and twelve Hurricanes destroyed and three damaged. Pilot casualties were lower at six Hurricane pilots, and both attacks had been repelled.

Leigh-Mallory decided that what was needed was more of the same. He wanted a still bigger Big Wing with more faster-climbing Spitfires. He ordered 74 Squadron to move from Wittering to Duxford and moved 611's Spitfires from Digby in Lincolnshire to boost the Wing to a potential five squadron formation, three of them Spitfires, with a total of some sixty fighters.[6] This meant two problems. Five squadrons would take even longer to assemble than three, and would mean more overclaiming. On one occasion, the Wing claimed to have destroyed fifty-seven German bombers and fighters, when Luftwaffe records showed only eight machines were lost.[7] [All air forces suffered from overclaiming enemy casualties but were usually accurate over their own since these were easier to verify.]

Nevertheless Leigh-Mallory continued to pass uncorrected claims to senior RAF commanders to bolster his reputation, and no one queried his figures. One factor helped. Because of delays, the Big Wing usually arrived as the Germans were turning for home, a coincidence easily explained as the Big Wing forcing the Luftwaffe to abandon its attack. Park had sent most of his squadrons to attack the

---

6. Bungay, *op.cit.*, pp. 313–4.
7. Deighton, *op.cit.*, p. 240.

approaching Germans, leaving Hornchurch and North Weald unprotected. He asked 12 Group to cover the airfields, but the Big Wing went searching for Germans instead, leaving them completely undefended. Fortunately the Luftwaffe did not capitalize on Leigh-Mallory's behaviour.

Unfortunately, before his weaknesses were revealed, Dowding and Park found their other flank exposed as German night attacks continued. On 8 and 9 September, 200 bombers attacked the City of London around St Paul's Cathedral, lighting fires on both sides of Ludgate Hill, around the Bank of England and the Guildhall, providing markers for each successive wave of bombers. Bombs fell on Somerset House, the Law Courts, St Thomas' Hospital and Buckingham Palace, killing 782 people, including 73 buried under a school used as a refuge, which collapsed under a direct hit. However, the daytime respite gave Fighter Command time to replace pilot losses and restore damaged airfields.

Using day fighters to attack night bombers was useless and expensive. Dowding needed time to develop radar controlled night fighting, but was attacked for taking too long. His solution was right, but by then he had been forced to retire. Park was equally handicapped. Had he done as Leigh-Mallory and his supporters wanted, and called 12 Group while the Germans were assembling over France, he would have to guess their intended targets and risk squadrons being sent to the wrong place or left on the ground when the real attack arrived. Only by waiting until an attack was committed could he deflect German formations from their targets. On the other hand, had Leigh-Mallory compromised over massed formations, individual squadrons could reach the battle quickly enough to affect the outcome.

Even 12 Group pilots found the Big Wing cumbersome and dangerous. Two Spitfire pilots from 19 Squadron thought it made little or no difference as the complex formation broke up as soon as a fight began. On two occasions the Wing's arrival alarmed other RAF pilots at the approach of such a large formation. When they closed to investigate, Wing pilots circled defensively until fuel reserves dwindled and then returned to base.

For the beleaguered pilots of 11 Group, the Big Wing seemed a waste of time. Pilot Officer Tom Neil spoke for many. 'All too frequently, when returning to North Weald in a semi-exhausted condition, all we saw of 12 Group's contribution to the engagement was a vast formation of Hurricanes in neat vics of three, steaming comfortably over our heads in pursuit of an enemy who had long since disappeared in the direction of France....'[8]

With all these failings, it defies explanation that senior officers insisted on awarding game, set and match to Leigh-Mallory and his flawed ideas. At a high-level Air Ministry meeting on 14 October to discuss tactics, he insisted they agree the minimum fighter unit to meet large enemy formations should be a wing of three

---

8. Neil, *Gun Button to Fire*, p. 120, quoted in Bungay, *op.cit.*, p. 358.

squadrons, and criticized all of Park's decisions which had kept the Luftwaffe at bay so effectively. Nevertheless he had the overwhelming support of top Air Ministry commanders with their unerring preference for the wrong decision. One exception was Air Marshal Charles Portal, soon to be Chief of the Air Staff, who suggested large formations risked leaving local bases undefended. Leigh-Mallory said plans had been drawn up to avoid that, and everyone agreed he had dealt with the problem. This was similar to the dogma that bombers alone could defend a country from air attack.

The reality was, inevitably, different. Park tried hard to give 12 Group every chance to help. He reported its contribution over late October when the fighting had passed its peak. In ten sorties, the Wing made one interception and shot down a single German aircraft.[9]

Sadly, it was too late. It was decided that Leigh-Mallory would replace Park commanding 11 Group where his theories would hand victory to the Luftwaffe over occupied France in the months following the battle. Park was switched to Training Command before being sent to Malta in 1942 so his tactics could defeat the Luftwaffe all over again.

Worsening weather brought another respite on 10 September, but it would improve later. Nevertheless, German commanders realized Fighter Command was far from defeated and postponed the planned invasion date for a final all-out assault. This began with German formations attacking during the early afternoon. Fortunately Park was ready. At 2.45 pm radar plots warned attacks were on their way, and requests went to 12 Group to send squadrons from Duxford and Coltishall in Norfolk. Within thirty minutes two large formations containing 300 bombers and 200 escorts headed for the Kent coast and the Thames Estuary, while others approached Southampton and Portsmouth. Park repaid 10 Group's help by sending squadrons from Tangmere and Westhampnett to meet this attack, leaving his Group again at full stretch.

Park's fighters, together with those released by 12 Group, met the Luftwaffe over the Thames Estuary. With Bf109s running short of fuel, some RAF fighters shot down Dorniers and Heinkels, but others were caught as they tried to climb. The Bf110s formed defensive circles, causing heavier German bomber losses, though some were left to hit targets in docklands and central London. Intense combat further west caused more RAF losses, though attacks were blunted and damage to ports avoided while fighter airfields were left unmolested. Again the Luftwaffe's main effort was made during darkness, with 180 bombers pounding London, and others bombing Liverpool and laying mines along the southern and eastern coasts.

This was encouraging, but Fighter Command paid a high price. Six Spitfires and nineteen Hurricanes were lost with another six badly damaged. Twelve RAF

---

9. Air Ministry Correspondence, PRO 2/5246, quoted in Bungay, *op. cit.*, p. 359.

pilots were killed and another four badly injured, an unsustainable casualty rate. Fortunately, rain and low cloud kept the next three days quiet. On the afternoon of 14 September, small bomber formations escorted by masses of 109s tried unsuccessfully to tempt British fighters to fight at a disadvantage and incur heavy losses.

This reflected a deepening German conviction that decision time for the invasion was near. The *Kriegsmarine* opposed the enterprise unless the RAF was defeated first. Hitler more or less agreed, but seemed to believe landings were still possible, given five days of clear weather. No one mentioned this same condition had been made by Göring a month before.

Some Germans still believed Fighter Command was buckling under the strain. Because raids were sporadic, Park responded carefully, in case feints left fighters refuelling and rearming for the main event. The RAF was now better prepared with radar and communications restored, and most squadrons were back to full strength though many pilots still lacked experience. Finally, Bomber Command's night raids on invasion barges in French and Belgian harbours had sunk more than 30 percent. The problem remained fighter losses. On one quiet day eleven, RAF machines were lost for eight German fighters, though only four RAF pilots died.

Park's reading of combat reports led him to consider head-on attacks to scatter enemy formations. He also insisted RAF pilots should not waste time and ammunition on damaged and probably doomed enemy aircraft. 'If you still have

Present-day author photo of the site of Ventnor Radar station.

ammunition left, help your mates in the main engagement. It is unfortunate not to have the pleasure of seeing your victims crash, but this serves no useful purpose.'[10]

Now the Luftwaffe would make a last attempt for the invasion, to end this dispiriting drain of losses. Decision day was imminent, and improved weather suggested it should be Sunday, 15 September. Park would nearly be caught out by an early German raid, but ironically the Duxford Big Wing would show the enemy how far from defeat the RAF remained after weeks of combat. It would also see the highest number of victory claims made by the defenders. Even the disappointing reality would strike a blow to German morale.

From the start, both sides resolved to strike hard. Göring issued orders to bomber and fighter units for a powerful attack using two promising new tactics. New methods of jamming British radar were only moderately effective. British radar cover was poor above 20,000 feet, though until now the Luftwaffe misunderstood the reason why high-flying squadrons could fly closer to their targets before meeting defending fighters. They simply assumed defenders took longer to reach these higher altitudes, failing to spot the inherent radar weakness.

Park expected heavy attacks with clearer weather before German invasion hopes vanished with autumn storms. He repositioned squadrons to deflect attacks on London, should the Luftwaffe maintain its usual routine. When radar reported German aircraft around 9.30 am, this was earlier than usual. Park placed two squadrons at each airfield on stand-by, with pilots strapped in and engines warmed up, ready for take-off and a climb to interception height. Then followed a series of feints and responses. Two Luftwaffe formations closed in on the Kent coast and the Thames Estuary. Park scrambled squadrons from Gravesend, Hornchurch and Croydon, but as they reached combat height the Germans headed back to base. By 11 am there were reports of more than 200 bombers with heavy escorts heading from Calais to cross the coast near Dungeness and head for the Thames Estuary. More and more squadrons came to 'Stand By' status and within five minutes two Spitfire squadrons took off from Biggin Hill. Ten minutes later, six squadrons of Hurricanes were sent after them. Five minutes later still, at 11.20 am, Park sent off two more Hurricane squadrons with two squadrons of Spitfires. Facing a single German attack, for the time being Park could order squadron commanders to concentrate on their principal objectives; Hurricanes to break through to the bombers and Spitfires to climb high enough to keep the 109s at bay. This was no easy task, as the Luftwaffe mounted high-altitude attacks at heights from 15,000 feet up to 26,000 feet, across a two-mile front. Most were Dorniers and He111s, with some Ju88s, and with Bf110s flying close escort and 109s providing top cover, to pounce on defenders approaching too close. They met over the Kent coast at approximately 11.30. Already, earlier losses showed in the threadbare German

---

10. Fighter Command Papers 16/735, PRO Kew, quoted in Bungay, *op.cit.*, p. 317.

Captured Heinkel He111 in North Africa wearing RAF markings. *(Official photo via Wikipedia)*

bomber formations. *III/KG76* only had nineteen Dorniers out of an establishment of thirty, so were reinforced by *I/KG76* with just eight machines left. They took off on schedule at 10.10 from bases at Beauvais and Cormeilles-en-Vexin to fly the 100 miles to Cap Gris Nez to meet their escorts. By 11.04 am, British radar knew they were on their way at last, but not whether this was yet another fighter sweep or a genuine raid. As a precaution, Park ordered two Biggin Hill Spitfire squadrons to climb to 25,000 feet to meet the highest German escorts. This was a measured response to radar information. Once it was clear the raid had no follow-up waves, Park sent reinforcements. At 11.15 he ordered up squadrons from Northolt, Kenley and Debden, and five minutes later from North Weald and Hornchurch. He now had two squadrons high over Canterbury, four over Biggin Hill and Maidstone at 15,000 feet to meet the bombers, reinforced by two over Chelmsford. Spitfires were sent over Dover at 20,000 feet to arrive as combat descended to that altitude, giving them the chance to cut their way through to the bombers. Two squadrons from North Weald went to Maidstone to prevent bombers making for London. He asked for a squadron from 10 Group to protect his flank and the aircraft factories at Brooklands and Weybridge.

At 11.30 his leading squadrons were closing the Kentish coast just ahead of their opponents, a straggling formation two miles across. RAF fighters dived into the

Glassed-in cockpit of He111. *(Author photo)*

fray with Park's pairing of Spitfire and Hurricane squadrons working properly at last. The Spitfires were high enough to bounce the Bf109s and deter them from protecting the bombers, themselves threatened by Hurricanes striving to break their formation by head-on attacks.

Park's plan was that the leading squadrons would slow or deflect the onslaught, but reinforcements were needed so they could return to base for more fuel and ammunition. He ordered four more Spitfire squadrons and seven Hurricane squadrons to join the battle, leaving airfields undefended. His only remaining option was to call the Duxford wing to protect those airfields by reaching them before the Germans. Fortunately, on a day when 11 Group's impeccable tactics were frustrating the Luftwaffe's plans, his opposite numbers at 12 Group did the right thing, thanks for once to an element of luck.

The Big Wing's five squadrons were ordered to patrol at 20,000 feet over Hornchurch so they could bounce the Germans out of the sun as they approached their targets. Already the Luftwaffe had severe problems. Formations had to close up to beat off head-on Hurricane attacks. This meant slowing down, so that 'Some of the Bf109s had to fly with their flaps down in order to hold position, which made them sitting ducks.'[11] To make things worse still, weather records show attackers

11. Bungay, *op.cit.*, p. 321.

facing an unusually strong headwind, a jet stream slowing progress over the ground to little more than a crawl.

Rather than follow orders, the Duxford Wing made for Gravesend on the opposite bank of the Thames where bombers were turning for base after bombing South London. With 11 Group squadrons already attacking, the intoxicating sight of RAF fighters outnumbering the Germans forced the Big Wing to wait their turn. The effect on Luftwaffe morale was catastrophic. After weeks believing RAF reserves were almost exhausted, here were fighters in plenty pressing home attacks with remorseless determination.

They had one last hope. They knew a still larger second wave was forming up to catch RAF fighters on the ground as they took on fuel and ammunition. By now Park knew this too, and prepared to meet the blow. Kesselring scraped together a large force by amalgamating his depleted units. His 114 Dorniers and Heinkels were reinforced by 26 Bf110s and a mass of 340 Bf109s flying above the bombers, with only one Gruppe of *JG54* contributing to the close escort. Ahead of them, the defenders' first wave was refuelling and rearming.

Fortunately, Park posted a sentry to watch for Germans and report their strength. A 92 Squadron Spitfire from Hawkinge was ordered off at 1.45 pm, with the first radar plots showing the attackers forming three groups. After twenty minutes the pilot watched the Luftwaffe's approach from 26,000 feet. Probably realizing his significance, half-a-dozen Bf109s climbed to attack. He decided to bounce the 109s, but speed built up in the dive to stiffen his controls too much to allow him to aim accurately. He left the controls trimmed for level flight so that after flashing past the German fighters he could ease pressure on the stick. The Spitfire pulled out of the dive automatically and he blacked out, recovering to find it flying normally so he could land at Biggin Hill to pass on his information.

By then Park had sent up pairs of squadrons over Sheerness and Chelmsford at 20,000 feet, one over Hornchurch at 15,000 feet and two more at 20,000 feet. He also asked for the Big Wing at either 2.05 pm or 2.15 pm [sources differ] and realizing this raid too was not being followed by a second wave, he committed his last reserves. At 2.15 pm three Hornchurch Spitfire squadrons bounced the German escorts from above and up-sun, bringing down fourteen 109s for the loss of a single Spitfire.

More and more squadrons arrived as Park sent his second line into the attack. Some were so inexperienced that 66 Squadron's Spitfires were ordered to use the old Area Attacks. On this one occasion with the escorts beaten off, they worked. The formation of now unescorted bombers was shot to pieces, causing heavy German losses and plummeting morale. Nevertheless, as Park brought all his fighters to bear on deflecting the bombers from the capital, the situation was dangerous. Soon fighters would run short of fuel and ammunition, and have to disengage. Would the Big Wing arrive in time? And in the right place?

By 3.00 pm the Wing was still climbing to its designated height, making a nonsense of Leigh-Mallory's claims they could reach their height and target area within half an hour of being called. At 16,000 feet they found 109s of *JG26* above them and ready to pounce. The Germans made a single diving attack, bringing down two Hurricanes and a Spitfire, and then swept away towards their bases with fuel reserves running out.

The unescorted bombers carried on into a bank of thickening cloud. In the end the British weather did what the fighters had narrowly failed to do and defeated the attack. The Dorniers turned away and followed their escorts. The Big Wing reported to a delighted Leigh-Mallory that they had beaten off the Luftwaffe. They claimed they had been warned too late but had destroyed twenty-six German aircraft and probably destroyed another eight for two pilots killed and two wounded.

The day's actions ended with two afternoon raids. An attack on Portsmouth by twenty-seven He111s from *KG27* was caught by Spitfires from 152 Squadron at Warmwell as it headed back to base, losing one bomber with another damaged. *Erprobungsgruppe 210* sent thirteen Bf110s at extremely low altitude under the radar. They showed up well enough to attract three squadrons from 10 Group, with four from 11 Group, to defend the Supermarine factories at Southampton. The Bf110s hit the wrong factory, but the fighters flew at normal height and missed the 110s far below.

As the battles ended, both sides took stock. The hugely inaccurate claims of 12 Group (one-third of the total in the morning and a quarter for the afternoon) meant the RAF total was greatly inflated. The total claimed was 185, a figure so dramatic it sent British morale sky-high and persuaded the Americans to celebrate. The Germans claimed they lost eighteen aircraft in the morning combats and another thirty-eight in the afternoon, a total of fifty-six against British losses of half this number. Aircrew losses made the biggest difference. The Luftwaffe lost eighty-one killed, thirty-one wounded and sixty-three prisoners against twelve RAF pilots killed, fourteen wounded and one fished out of the Channel by German rescuers.[12]

The figures were bad for both sides in the long run. The Luftwaffe's dwindling morale convinced them daytime attacks in British airspace were impossible. Future daylight raids would use Bf109s and 110s in high-speed fighter-bomber missions. The RAF's inflated figures reinforced Leigh-Mallory's claim that only he knew how to command RAF fighters in battle. His jubilant report to Dowding reached a delighted Air Ministry a week later. In it he claimed that four Big Wing operations in early September had destroyed 105 German aircraft and probably destroyed another forty for the loss of fourteen aircraft and six pilots. On 15 September the Wing had taken off before noon and destroyed all the German bombers they could see. 'In the afternoon, the Wing produced satisfactory results, but could not break

---

12. Figures from Bungay, *op.cit.*, p. 333.

up the bomber formation because they had not had enough time to reach patrol height and so were attacked by Bf109s while trying to get into position.'

The clear implication that the poorer performance in the second combat was due to Park calling the Wing too late was seized by his high-level supporters. The Deputy Chief of the Air Staff, Air Vice-Marshal Sholto Douglas, had asserted that, 'It does not matter where the enemy is when shot down [in other words, before or after he attacked his targets] as long as he is shot down in large numbers.'[13] The difference was all too obvious to those actually trying to defeat the attacks.

For the time being however, all was well. The historian John Terraine cited the words of the official history, *The Defence of the United Kingdom*, which summed up the situation with clarity: "If 15th August showed the German High Command that air supremacy was not to be won within a brief space, 15th September went far to convince them that it would not be won at all."'[14] Clearly, Dowding and Park had convinced the Luftwaffe, if not their colleagues, that Fighter Command had the measure of its opponents across the Channel. Faced with the escalating casualties of daylight bombing, *Seelöwe* was postponed to 1941, for better weather and hopefully better ideas for turning the tables. It would never be resurrected, as the following year would see the Luftwaffe sent east to find new victories, like Napoleon's *Grande Armée* following the equally catastrophic defeat of Trafalgar.

---

13. John Terraine, *The Right of the Line*, p. 202.
14. *Ibid*, pp. 210–11.

Chapter Twenty

Battle of Britain – Decline and Fall

Both sides paused for breath after 15 September 1940. While RAF victory claims were clearly over-optimistic, Fighter Command was still so far from defeat, the Germans knew their efforts had failed. The RAF had survived this bitter campaign of attrition, and the system developed by Dowding made that possible, together with tactics devised by Air-Vice Marshal Keith Park commanding 11 Group. Within limitations of height, distance, radar coverage and aircraft speed, he met incoming raids without defending patrols having to land and refuel as follow-up attacks arrived.

However, their other opponents wore RAF uniforms. During its brief history the Royal Air Force suffered particularly badly from senior officers' mistakes. The useless Fighting Area Attacks and the unremitting pressure of the bomber lobby were two prime examples. Even with Fighter Command's clear and hard-won victory, the door of the Air Ministry remained ever open for anyone suggesting it

Statue of Air Vice-Marshal Keith Park, commanding 11 Group. *(Author photo)*

would have been better to use different methods. Those with no direct experience of commanding a fighter group against a stronger and more experienced opponent could usually ensure their views triumphed in high-level meetings.

In a BBC Radio interview more than twenty years after the battle, Park summed up his massive responsibilities: 'Göring was attacking my fighter aerodromes with the object of knocking Fighter Command out of action as a preliminary to invading England. So my tasks were really three and they were quite clear to me, and to Dowding, we never even questioned or discussed them. First I had to avoid my pilots and fighter aircraft [being] destroyed on the ground, the warning was so short in the forward aerodromes that there wasn't time to mess about dispatching two, three, four or five squadrons before the bombs would be falling.

Now my second was to prevent my fighter aerodromes with their vital Operations Rooms, hangars, workshops, messes and all their ground organisations from being bombed out of action as was done in Poland, and as was done in France. Thirdly, I had to inflict the maximum casualties on the German Air Force in order to wear them down so that we could prevent them from obtaining air superiority over any part of England.

'You see that to win my battle, we were fighting at 11 Group with our backs to the wall, this wasn't a little game, we were not in a position to try out tactical theories of this or that, Big Wings or Little Wings, we were fighting for our existence and the existence of London and the Empire, and I say now that after all these years that had I tried to adopt Bader's theories of the Big Wings of five squadrons, I say now that I would have lost the Battle of Britain, and…, might well have lost the war for the Allies.'[1]

Park had had a varied military career. Serving at Gallipoli in 1915, he was blown off his horse by a German shell and sent to England to recuperate. Reassigned to the Somme, he transferred to the RFC to fly Bristol Fighters with 48 Squadron. While commanding the squadron he was awarded the MC and Bar, the DFC and the Croix de Guerre, and shot down five enemy aircraft, with another fourteen probably destroyed, to become an ace. In peacetime he commanded 25 Squadron, another fighter unit, before staff and training roles led to his posting to Fighter Command as Senior Air Staff Officer to Dowding in 1938. He was finally appointed AOC 11 Group in April 1940.

Leigh-Mallory left the law to join the infantry in 1914 and was wounded at Ypres a year later. He too transferred to the RFC and trained as a pilot, but specialized in reconnaissance, bombing and army co-operation. He served in peacetime staff and training commands in Britain and overseas, but had never flown in a fighter squadron let alone commanded one. His staff appointments taught him the value

---

1. BBC radio interview with Air Chief Marshal Sir Keith Park, 1 January 1961.

of high level contacts in the RAF and Air Ministry, and in November 1937 he was appointed AOC 12 Group, more than two years before Park was given a similar post.

Park worried that German losses on 15 September were less than he had hoped. At fifty-six German aircraft to twenty-eight British, German aircrew losses were much higher, but Park's paired squadrons were not always sharing their tasks properly. He had wanted Spitfires to go for the German fighters, leaving Hurricanes to attack the bombers, but combat reports showed Spitfires often went after the bombers first, leaving Hurricanes to take on 109s at a disadvantage. Park now ordered Spitfires to tackle the topmost German fighter screen first. Spitfires from forward airfields like Hornchurch and Biggin Hill would assemble in pairs at height if visibility was clear. With high cloud they would assemble below the cloud layer before climbing to interception height to target 109s properly. Those fighters based further back at airfields like Northolt or Tangmere could even be formed up in three squadron wings to react to earlier warnings of second and third wave attacks.

However, carefully planned tactics were already out of date. As Park prepared to meet criticism from Douglas and Leigh-Mallory, Luftwaffe commanders were absorbing their lesson from 15 September, that daylight bombing attacks would never bring victory and merely increased losses. Meanwhile the weather was worsening. On 16 September cloud and heavy rain restricted the Luftwaffe to feints and fighter sweeps. One Hurricane and one Spitfire were lost, but both pilots were saved and there were no RAF casualties that day. The Luftwaffe lost ten machines during day and night operations – seven Ju88s, two He115 seaplanes and a He111. Bombs were dropped on the docks and the East End, and on more distant targets like Liverpool and Manchester, Coventry, Bristol and Birmingham.

New German tactics now made life more difficult for the defenders. The uneasy cooperation between bombers and escorts had left Bf109s much worse off and given Spitfires the best options. When the Luftwaffe switched to large mixed formations of fighters and bombers, the true numbers were harder to assess. With Bf109s operating as fighters or fighter-bombers, defending fighters often did not know what they faced. A fighter sweep could safely be ignored, but some machines might then drop bombs with disconcerting accuracy. Yet Spitfires taking on these

Air Marshal Sir William Sholto Douglas, Dowding's chief critic, took over his job after the Battle of Britain. *(Official photo via Wikipedia)*

Air Vice-Marshal Trafford Leigh-Mallory, commander of 12 Group. *(Official photo via Wikipedia)*

Wrecked Supermarine Woolston works after German bombing by Heinkel 111s of *KG55* on 26 September 1940. Both this and the nearby Itchen plant were abandoned with machine tools, and production workers shifted to the dispersed facilities in the area to avoid future raids.

(Via Alfred Price)

formations could find themselves attacked by fighters rather than *Jabos*, and 109s free from escort duties to bounce their opponents were formidable as ever.

Still, poorer weather meant lower casualties and fewer machines lost by either side. On 18 September the Germans switched tactics again. Using 109 fighters, *Jabos* and Ju88s, they split into small fast formations, each heading for a different high-value target. Afterwards the RAF claimed twenty-three bombers and ten fighters destroyed, mostly by the Duxford Wing, with twelve unescorted Dorniers claimed by 242's Hurricanes alone. Leigh-Mallory passed the uncorrected figures upwards for congratulations from the Chief of the Air Staff and the Secretary of State for Air. Though based on unproven evidence, no other participants received such praise, and confirmed Leigh-Mallory's campaign to replace Park at the head of 11 Group was gathering strength.

Ironically his claim was less exaggerated than usual. A whole bomber formation found itself vulnerable to a wing of experienced pilots whose arrival for once was well timed. Because this happened as German tactics were changing it would never reoccur. However it increased Park's difficulties just as Leigh-Mallory's claims undermined his case. The weakness in those figures would be exposed four months later at the expense of RAF fighter pilots.

With better weather and faster German formations, Park needed earlier warning of raids. He and Dowding sent up 421 Flight after the first radar alarms. These were Hurricanes which, crossing the coast at high altitude, radioed back details of incoming German raids.

Occasionally the battle flared up. On 23 September, radar showed four approaching formations, all Bf109s and too fast for RAF fighters to intercept with the advantage of height. Even so, casualties remained similar on both sides. Eleven British fighters were brought down and three Spitfire pilots killed, but the Germans lost ten Bf109s and a single Bf110.

Raids on 24 September struck south coast ports and naval bases, and the Supermarine factory at Woolston. Bf 109s bounced Spitfires near Dover, shooting down two, but both pilots survived. Twenty minutes later Bf110s from *Erprobungsgruppe 210* and *ZG76* flew under the radar to wreck Supermarine's Woolston factory and bomb the Portsmouth suburbs. Three machines were downed by anti-aircraft fire and another was damaged but returned to its base. Six British fighters and two pilots were lost.

Next day the Luftwaffe attacked the Bristol plant at Filton at midday. Assembly hangars and aircraft were destroyed with 132 plant workers killed and more than 300 injured. The Germans lost six aircraft and eighteen crewmen against four British fighters and one pilot. Five hours later, twenty-four bombers escorted by Bf110s attacked Plymouth and lost a single bomber. London was bombed again at night and mines were laid in the Thames Estuary. The next day 100 bombers hit Southampton and the Supermarine plant again. Ten RAF fighters were lost with three pilots, for sixteen German bombers and sixteen fighters.

It seemed Park had the measure of his opponents as successful interceptions were forcing bombs to be jettisoned over the sea or open country. On 27 September, the Duxford Wing caught the Bf109 escorts as they headed for home, claiming thirteen destroyed for five of their own machines and three pilots. However they headed straight for the fighting instead of protecting Park's airfields. Park made an official complaint about 12 Group cherry-picking targets instead of following orders. Leigh-Mallory ignored it. He had other priorities.

The victory was less decisive than it could have been as assembling the Wing still took too long, and two squadrons almost missed the action. Nevertheless, Leigh-Mallory insisted that 'The Wing is airborne in six minutes and over Tilbury less than eighteen minutes after take-off. Our record proves it! If we were only given enough warning, we could have the Wing in position to meet a raid while your squadrons are still climbing.'[2] As this was a reply to Park's complaint that the Wing had taken almost an hour to reach Sheerness, it was clear neither side now believed the other. Leigh-Mallory continued demanding earlier warnings from Park, who insisted it was impossible to be sure of the composition of a raid until it was almost upon them. Wild guesses risked wrong-footing the defences.

Next day the Germans returned in force, but all three raids were beaten off for nineteen fighters shot down and eleven pilots killed, against six 109s. On 29 September, the Luftwaffe sent in a late afternoon fighter sweep, but Park refused to be tempted. Finally, the last day of September would be most discouraging for the Luftwaffe. As if worried by time passing, they sent in four heavy raids at 9 and 10 am and 2 and 4 pm. The first was turned back by low cloud, and others suffered similarly. One used the German *Knickebein* radio navigation system to try to find its target. The bombers followed a radio beam, until they picked up a second crossbeam which triggered the dropping of their bombs.

This would be successful in the devastating raid on Coventry in November 1940, but this time the bomber commander missed the secondary beam and followed the primary beam under heavy RAF attack until his escorts began running short of fuel, without a common radio frequency to warn him. The Luftwaffe lost fourteen bombers, a single Bf110 and no less than twenty-eight Bf109s, their heaviest fighter loss of the whole campaign, underlining the impossibility of a decisive victory. The Luftwaffe lost 433 aircraft in September compared with 242 RAF losses, but British aircraft factories were out-producing the Germans over this period. Göring faced increasing discontent from pilots. The *Jabos'* bomb load reduced performance, increased fuel consumption and cut endurance over Britain. However fighter-bombers now approached above 25,000 feet so that only Spitfires could catch them and shoot them down.

---

2. Quoted in Michael G Burns, *Bader: The Man and his Men*, p. 119.

Luftwaffe tactics occasionally failed completely. On 7 October, twenty-five Ju88s approached Portsmouth, and the aircraft factories of Westland at Yeovil and Bristol at Filton. Beyond 109 range, their escorts were Bf110s instead. The AOC 10 Group sent two Spitfire squadrons to tackle the 110s and a pair of Hurricane units to attack the bombers. The Spitfires dived at high speed through the Bf110s to reach the Ju88s direct, forcing them into evasive action while the 110s tried to catch up, but were attacked by Hurricanes in a reversal of Park's usual tactics. Without 109s, both RAF fighter types hit bombers and escorts hard. Only as they made for their home bases did more 109s arrive to cover their withdrawal.

Meanwhile, Park was trying new ways to exploit Spitfire qualities. With faster and more powerful Mark IIs, squadrons could climb above 109s and force them towards Hurricanes waiting lower down. On 2 October it had been the turn of *8/JG53* to try operating as *Jabos*, with other units from the *Geschwader* as escorts. These escorting 109s crossed London above 30,000 feet and met no RAF planes at all. The *Jabos* on the other hand did.

Unable to reach the fighters supposed to protect them, they suffered casualties. Spitfires from Hornchurch bounced the *Jabos*, shooting down four in a single pass and making the others jettison their bombs and turn for home. Their escorts followed them back, having seen nothing. German pilots became even warier of *Jabo* missions, but by the end of October Göring ordered every fighter *Geschwader* to denote one *Staffel* in each *Gruppe* as a specialist fighter-bomber unit.

Meanwhile, one night's raid on London killed more than 500 civilians with 1000 more injured. With no answer to night bombing, Dowding and Park were heavily criticized. During daytime attacks, casualties were limited only by worsening weather. On 19 October sixty bombers and fighter-bombers were turned back for the loss of two British fighters and one pilot. On the twentieth the Luftwaffe lost one Bf110 and six Bf109s against RAF losses of three fighters with all pilots safe. After fog cleared, the Germans sent in fighter sweeps, shooting down six British fighters for three of their own. On 25 October the Italians joined in with twin-engined Fiat BR20 medium bombers escorted by Fiat CR42 open cockpit, fixed undercarriage biplanes, attacking east coast ports from bases in Belgium. After a month of mounting losses they returned to the Mediterranean.

The Luftwaffe continued switching tactics. On 25 October they shot down ten British fighters for fourteen of their own. Two days later, they sent in mixed fighters and bombers, losing six fighters and four bombers, but shooting down eight British fighters and killing four pilots. Next day two Ju88s and a pair of Bf109s were lost for no British losses. Finally, on 29 October, the Luftwaffe returned to earlier tactics with bomber formations escorted by Bf109s. That morning 12 Group fighters took more than twenty minutes to assemble, again arriving after the raiders had gone home.

An afternoon raid by more than 100 bomb-carrying 109s was bounced by nine RAF squadrons, with eight attackers shot down in minutes. German losses for the entire day were two Dorniers, three Bf110s and no less than twenty-two Bf109s. The RAF lost seven fighters, but five pilots survived. The last official day of the campaign, 30 October, brought more losses still with one He111 bomber and eight Bf109s shot down, and five British fighters lost with all save one of their pilots. Bad weather on the following day meant no losses for either side.

After this, the battle petered out into Luftwaffe nuisance raids with Bf109 and Bf110 fighter-bombers flying at high speed and altitude, while 12 Group's leisurely approach was no help. As weather worsened, Spitfire and 109 pilots faced high-altitude temperatures far below freezing. Canopies and windscreens misted up or became coated with ice, blotting out visibility, and Park told squadrons to climb to at least 25,000 feet before attacking.

High altitude front-line combats depended on Spitfire Mark IIs, as Mark I aircraft were limited to 27,000 feet, and only the more powerful engines and superchargers of the Mark IIs let them fight higher. Until Park's high-altitude watchers could spot incoming 109s and decide which were fighters and which were *Jabos*, Spitfires risked attacking fighter-bombers as ordered, only to find fighters operating at their ideal altitude. Often 109s would fly over at 35,000 feet while Spitfires stayed at 27,000 feet. If the 109s came down there would be a fight. If not, then not. During October 1940 the RAF lost 186 aircraft, an average of six per day, with 120 pilots, or an average of four a day. The Luftwaffe lost double this total.

Meanwhile, Dowding wanted airborne radar to guide night-fighters to their targets. Sholto Douglas insisted more day fighters be used at night, but without navigational aids and approach lighting, day fighter pilots often crashed, landing after an abortive patrol. These increasing and profitless casualties had no effect on German bombing, but like most suggestions from senior officers, it risked losing the day battle, the most vital of all.

The fact remains that Douglas and Leigh-Mallory were much better at convincing those without experience than either was at dealing with the problem. So those who planned, organized and controlled the defences under heavy attack were criticized by those keen to replace them, but who lacked any means for improvement. The increasingly isolated Dowding, whose sober assessment of the threat and the development of the defence system which had triumphed over the Germans, fell victim to opponents within the RAF.

Less than four weeks after the official end of the battle, Dowding was replaced as head of Fighter Command by none other than Sholto Douglas. Already Douglas's decisions showed appalling misjudgement. As Chief of the Air Staff in 1938, he ordered Dowding to form nine squadrons of Defiant two-seat turret-armed fighters. 'For work over enemy territory', he insisted, 'a two-seater is best.' His staunch supporter, Air Vice Marshal Stevenson, mistakenly insisted the Defiant could repeat the success of the Bristol Fighter in the previous war, ignoring the fact that it left the pilot weaponless and compelled to fly so his gunner could aim at a target outside his own range of vision. He went even further. In spite of its carrying half a ton extra weight in gunner and turret, he insisted it was 'slightly faster' than the Hurricane with the same engine. The truth was the Hurricane was 26 mph faster, and could reach 15,000 feet, two and a half minutes quicker, while the Spitfire could do even better. Dowding was convinced Defiants would be a liability and formed only two squadrons, which suffered appalling casualties and had to be withdrawn. Later, when the shortage of trained pilots threatened defeat, Dowding asked Douglas to release pilots saddled with the useless Fairey Battle which had killed so many crews during the fighting in France. Now these men were eager to take part in the air fighting. Douglas predictably refused.

Meanwhile, Keith Park was replaced by Leigh-Mallory. At an Air Ministry meeting on 17 October 1940 chaired by Douglas, a planned coup took effect. It was packed with their supporters and the objective was to criticize every aspect of Dowding and Park's control of the battle, which had worked, and replace it with Leigh-Mallory's tactics, which had not. In an atmosphere more reminiscent of Alice in Wonderland than senior officers in a fighting service, Douglas insisted day fighter pilots fly standing patrols in the darkness, while Dowding and Park stressed the appalling casualties this policy would cause.

In his post-war autobiography, Douglas claimed that Dowding had become 'A little blinded' to the 'more simple hit or miss, trial and error, use of single-engined fighters' and that 'The effort had to be made.'[3] 'Hit or miss, trial and error' were vague phrases for a process for which the fighter pilots involved would pay with their lives at a time they could least be spared.

Finally it would prove unnecessary. Dowding proposed a radar-equipped fighter flown on instruments and controlled from the ground like his day fighters. His critics poured scorn on the idea though it closely resembled the German system for attacking British bombers. Fitting airborne interception radar into a Beaufighter finally did the trick. The first successful radar-guided night bomber interception was completed by Flight Lieutenant John Cunningham as early as 19 November 1940, and others quickly followed. Dowding had been right, but by then he had been sacked.

It is possible the critics were right and their limitations hampered Fighter Command's responses to German attacks. They were accused of not using 12 Group's squadrons, preferring to concentrate on the beleaguered units of 11 Group which bore most of the fighting. The idea of a vendetta between Park and Leigh-Mallory fails to convince as Park's limitations were due to geography and technology. The positions of German air bases and the distances to Park's airfields, to London and other high-priority Luftwaffe targets would always limit warnings to defenders.

---

3. Douglas, *Years of Command*, p. 354.

An approaching raid was detected when radar spotted German formations over France. To know the essential details, raiders had to approach the coast for the Observer Corps to estimate heights, numbers and types of bombers and escorts. Only then could Park respond, wary of a second or even third wave catching his fighters after they landed to refuel and rearm. Squadron-sized formations could respond quickly and meet each new threat as it emerged. An especially heavy attack would be met by multiple squadrons, but reserves were still needed to protect sector airfields while the first waves of defenders topped up fuel and ammunition. This imposed constant pressure on front line airfields, but the alternative was risking squadrons caught on the ground or airfields undefended by a second attack.

Park's opposite number commanding 10 Group to the west, Air Vice-Marshal Sir Quintin Brand, faced the same problems and coped equally well. Both men respected one another, and covered one another's territory when German attacks crossed inter-Group boundaries. Requests for extra squadrons at vital moments were met effectively. With 12 Group relations were very different. Leigh-Mallory's lack of experience commanding fighters did not stop him obsessing the only way to meet heavy German attacks was to form massive fighter formations. Nevertheless, what Park most needed from 12 Group were individual squadrons to cover these bases and intercept Luftwaffe attacks.

Leigh-Mallory simply refused to comply. He insisted his Big Wing was the only battle-winning tactic developed for the air war and anyone who denied this must be shunted out of the way. Yet insisting on a Big Wing virtually guaranteed failure. Even if Park asked for the Wing to be sent south as the Germans appeared on radar, the timings would still not have worked. Stephen Bungay explained in his masterly study of the battle, *The Most Dangerous Enemy*, 'It would have done no good even if he had called them when the potential raiders were still over Calais. Assuming it climbed at its optimum speed, a Spitfire would take about 23 minutes to cover the 70 miles from Duxford to Canterbury. It could take off about five minutes after the first radar warning of a build-up over Calais. A bomber travelling at 180 mph would reach Canterbury in fourteen minutes. Even without adding in time for forming up, and even assuming the Wing used Spitfires rather than the slower Hurricanes which would take longer, a German bomber formation would be over Canterbury in half the time it would take a single fighter from Duxford to get there. The whole thing was absurd.'[4]

Even more absurd was the idea this was the best use for the Big Wing. If the incoming raid was a feint, or a high-speed formation of Bf109 fighters or *Jabos* or a mixture of the two, the Wing would be ineffectual. Both Douglas and Leigh-Mallory thought it better to shoot down large numbers of German aircraft after

---

4. Bungay, *op.cit.*, p.356.

they had bombed than to strike before they attacked. Given the time it took the Big Wing to reach the enemy, this was usually the only choice.

They insisted the enemy would be so discouraged by the resulting defeat that he would deliver no more attacks. However, hitting 11 Group's airfields while those ordered to defend them were taking off, assembling and heading for a target somewhere else was exactly what Göring had been praying for. Leigh-Mallory, Douglas and their supporters risked handing him victory on a plate.

North Weald was bombed on 24 August, Debden on 26 August and Biggin Hill on 30 August, all as a result of 12 Group failing to appear. Yet Park was placed on the defensive under a storm of criticism for not making better use of Leigh-Mallory's squadrons. Finally he was left to carry the blame which had been the intention throughout. At the same time, Leigh-Mallory's report that no less than 105 enemy aircraft had been destroyed in one day by his Big Wing, at a cost of fourteen British fighters and six pilots, was accepted uncritically.

Was he truly the only man for the job? There were signs even before the war that he and Douglas lacked a real grasp of what was involved. In October 1938, Leigh-Mallory proposed his 12 Group take over twenty-nine of the nation's forty-one fighter squadrons, leaving a dozen to defend the capital and none at all for anywhere else. Dowding told Park it showed Leigh-Mallory, 'Failed to understand the basic ideas of fighter defence.' Major exercises in summer 1939 showed 11 Group intercepted 60 percent of raids, which was classed as 'reasonable', but concerns were expressed over Leigh-Mallory's actions. He used too many fighters on standing patrols defending airfields, leaving too few guarding high priority industrial targets.

After the Battle of Britain, Leigh-Mallory's confident claim that he could get a Wing of five squadrons into the air in six minutes, and over Hornchurch at 20,000 feet in twenty-five minutes, should have been challenged. On 29 October the wing would take seventeen minutes to leave the ground and another twenty before it even set course from base – after two months' constant practice. Unfortunately, Leigh-Mallory looked and sounded impressive because his views were simple and appealed most to those who knew least about the battle.

For Park, like Dowding, it was now too late. Opponents were queuing up to attack his tactics and performance and trash his reputation. In early 1941, with Douglas and Leigh-Mallory firmly in charge, the Air Ministry would publish its official history of the battle [without mentioning Park or Dowding by name] and the Big Wing controversy remains bitterly divisive. Many insist they were ill-served by an Air Ministry which ignored their role in winning the air battle over south-east England in the summer and autumn of 1940. They were left victim to a shabby and self-serving intrigue, and replaced by officers in thrall to a misconceived and unsuccessful doctrine imposing a terrible price in casualties and a malevolent effect on Allied air operations. In particular, the Big Wing concept squandered the Spitfire's advantages in the wrong campaign with the wrong tactics, denying other theatres where it could have made a dramatic difference to the air war.

Descendant of 'Britain First' – the night fighter version of the Bristol Blenheim. *(Author photo at Wroughton Air Display)*

Others insist Dowding and Park deserved to be ejected from their commands and replaced by their opponents. Ignoring personalities, did their removal help Fighter Command build on its hard-won victory by taking the fight to the enemy and enable the RAF to achieve its strategic and tactical aims? In Churchill's opinion the Air Ministry was 'a most cumbersome and ill-working administrative machine, where jealousies and cliquism were rife.'

Leigh-Mallory failed to show up to pay the usual courtesies on taking over Park's command, so Park had to hand over to his deputy. Once safely at Park's desk, Leigh-Mallory suggested to Douglas they use huge nine-squadron formations, but the problems they faced when trying out this tactic were greater than those faced by the Luftwaffe against Park and Dowding.

Wanting a chance to demonstrate his Big Wing theories on a wider stage, Leigh Mallory set up a paper exercise to re-run the battles of 6 September 1940. On that day, the Luftwaffe had attacked 11 Group fighter airfields at Biggin Hill, Kenley and Hornchurch, and he wanted to show how much more effectively the Big Wing would have coped with the threat than Park's tactics had done. In fact it did nothing of the kind. It merely underlined the catastrophic limitations of the Big Wing yet again.

Independent umpires decided the Germans bombed Kenley and Biggin Hill while Leigh-Mallory's fighters were still on the ground. Ironically, for someone who blamed Park for insufficient warning of German attacks, Leigh-Mallory did far worse. In the real battle, Park's Group shot down seven German bombers and

Massive firepower of the Bristol Beaufighter which became the first truly successful night fighter rather than the intended Spitfire replacement. *(Author photo)*

thirty-seven fighters, a toll twice that suffered by the RAF. This did nothing to dent Leigh-Mallory's sublime self-confidence. He maintained that if the problem recurred, he would let the Germans bomb the airfields and hit them hard on the way back to base so they could launch no more raids. He discounted the losses of men and machines caught by heavy bombing, a trait he shared with Douglas.

Later analysis showed every criticism Douglas and Leigh-Mallory levelled at their predecessors was unjustified. Having effectively proved the Big Wing theory failed in defence, Leigh-Mallory and Douglas then proved its shortcomings in attack as well. They decided their new priority was to 'lean towards France', and Big Wings would be ideal to do it. It was another monumental misjudgement. To follow the Luftwaffe's defeat by making the same mistakes and paying an even higher price was criminal. All the advantages enjoyed by the RAF in fighting over its own territory with an effective defensive system were handed to Göring, but the Germans had extra factors to make the contest even more unbalanced. Fighting over an occupied country, they could give battle when conditions were favourable and decline it when they were not. Within months of the start of this new campaign, the Luftwaffe turned to face the invasion of Russia the following summer. Fighting over France became an irrelevance. Thanks to Fighter Command's new masters it was an increasingly expensive irrelevance.

They began with 'Rhubarbs', fighter sweeps over northern France. As losses of fighters and pilots climbed for little worthwhile return, large formations made over-claiming inevitable, claims which convinced Leigh-Mallory the Luftwaffe was suffering. Again he believed his own figures, but in truth German aircraft losses were much less than those of the RAF. Operating over occupied territory, pilots bailing out were back in action within hours.

When the Germans failed to react, the RAF followed failed Luftwaffe tactics a step further, using fighters to escort day bombers against coastal targets. These 'Circus' operations brought a sharper German reaction and gave the 109s an additional advantage. No need to match speed and endurance flying close escort to their own bombers; now they could climb to their best performance height and strike when the RAF were guarding their slower charges. Limited range now worried RAF pilots like those flying 109s over England, and the Luftwaffe also had a better radar warning system to decide where and how to intercept.

Squadron Leader Johnnie Johnson, one of the RAF's best pilots, found Wings hard to control in action as pilots got in one another's way. His experience confirmed Park was right and two squadrons made the ideal combat formation. During the first half of 1941, Fighter Command lost fifty-one pilots and their machines, most of them Spitfires. They claimed to have shot down forty-four German aircraft, a poor exchange, but efforts were stepped up to deter the Luftwaffe from transferring fighters against Britain's new ally, Stalin's Russia. RAF losses climbed higher still with no effect on German actions.

Between 14 June and 3 September 1941 Fighter Command lost another 194 pilots killed or taken prisoner. Leigh-Mallory claimed 11 Group had shot down 437 German aircraft with another 182 probably destroyed. His figures remained pure fantasy as the Germans only kept some 200 fighters in the area. Their records showed 128 destroyed and seventy-six damaged, with many pilots safe. From mid-June to the end of 1941, Fighter Command claimed to have shot down 731 enemy aircraft for a loss of 411 RAF fighters. These claims too were ludicrously inflated. True German losses were 114 aircraft. In the words of one expert, 'It was for such results as these…, that a force of seventy-five day fighter squadrons [were] retained in this country throughout the latter part of 1941. Whether this was a wise allocation of resources at a time when there were only thirty-four fighter squadrons to sustain our cause in the whole of the Middle and Far East is perhaps, an open question.'[5]

To take Britain's best fighter and waste it in a fight against Germany's equivalent with every advantage on the enemy side and to no real purpose, was criminal. But to do so when even a handful of Spitfires with experienced pilots could have made a vast difference to the Desert War, to the skies over Malta and even the Japanese attack on Malaya and Singapore, seems sufficient insanity to persuade the Air Ministry to reconsider. Sadly, it did not. Even Douglas realized something was wrong by February 1941, but nothing could stop Leigh-Mallory's Big Wings from failing. On the Dieppe operation of 19 August 1942, Leigh-Mallory's deputy, Group Captain

---

5. Denis Richards, author of *The Royal Air Force 1939–45*, quoted in Terraine, *op.cit.*, p. 285.

Harry Broadhurst, argued that individual squadrons would cope better than large and cumbersome Big Wings, but was overruled. The RAF lost ninety-six aircraft on the first day, but claimed to have shot down ninety-five enemy fighters. The Germans actually lost forty-eight. The RAF lost forty-seven pilots killed and seventeen made prisoners of war. Broadhurst flew over the operational area three times to see what was happening, and saw small groups of German fighters bounce the huge Wings time and again.

All this reflected badly neither on the Spitfire nor on those flying it, but followed the disparity in the opposite direction during the Battle of Britain. Even the best fighters available to either side could only succeed with the right tactics, the right objectives and the most professional commanders. Fortunately, changes were on the way. Later that year Douglas was moved to the Middle East, but in spite of his huge casualties, Leigh-Mallory was appointed AOC Fighter Command. In Air Ministry eyes, the Dieppe debacle made him ideal to command air forces for the Normandy invasion.

He was finally posted to command the Far East Air Force but en route to the appointment, he overruled his subordinates once too often. He and his wife and party left Northolt in a four-engined Avro York in worsening weather on November 14 1944. His crew asked to wait till cloud cleared over the French Alps, but he insisted they fly on. The aircraft hit a ridge near Grenoble and all on board were killed.

After Training Command, Park returned to front line duty, commanding the beleaguered island of Malta where he fought another defensive battle against heavy odds and once again triumphed. In a final irony, he was appointed Allied Air Force Commander in Chief at South-East Asia Command, to replace Leigh-Mallory.

Chapter Twenty-One

The Friedrich and the Spitfire V

The 22 October 1940 brought the first signs of approaching winter, fog and low cloud, blotting out visibility for Luftwaffe raiders. Only as the fog blanket frayed were fighter-bombers tempted to attack a coastal convoy instead. RAF fighters met them, and dogfights developed over Kent and the Channel. Finally, late afternoon saw eighteen more 109s cross the coast at Dungeness and head for Hornchurch before turning south for Biggin Hill.

On the way they met Hurricanes from 46 and 257 Squadrons. Their pilots noticed one German machine had rounded wing-tips and a much larger spinner, though still clearly a Bf109. It was flown by the Luftwaffe's leading ace, *Oberstleutnant* Werner Mölders, commanding *JG51*. Within minutes, he had shot down three Hurricanes. For Mölders these routine victories appeared in his log as claims number 63, 64 and 65. His total would reach 115 by the time he died in a crash in 1941 in a He111, flying to Ernst Udet's funeral.

Now he was flying the first Bf109F *'Friedrich'*, a new version of the Luftwaffe's principal fighter. More than any other military aircraft, fighters have to keep on improving to avoid obsolescence, and the Spitfire and 109 were no exceptions. The race was now on to steal an advantage over the opposition. First one side would edge ahead, then the other, striving for more engine power, less drag and heavier armament.

Replacing the original Merlin of the Mark I Spitfire with the Merlin XII created the Spitfire Mark II. The engine ran on 100-octane fuel at up to 12lbs of boost for longer periods while a higher supercharger gear ratio lifted peak power to 1175 hp. It also had cartridge starting where an explosive round turned over the engine when the starter was pressed. Other improvements included a constant speed propeller and hydraulics to raise and lower the undercarriage without having to pump a handle up and down.

The Spitfire IIA had the same eight-gun armament as the Mark I, and first appeared in August 1940 during the Battle of Britain. It was soon joined by the Mark IIB with a pair of Hispano 20mm cannon and four Brownings. However, while extra performance was welcome, the Mark II's significance was complex. By the time it appeared, Air Ministry determination to replace the Spitfire with one of a range of inadequate substitutes was threatening Fighter Command's hopes of defeating the Luftwaffe. Both the Bristol Beaufighter and the Westland Whirlwind promised much but delivered little, being too far behind in development and too

Adolf Galland and Werner Mölders at Theo Osterkamp's birthday party. *(Photo via Bundesarchiv – Wikipedia)*

cumbersome in flight to survive fighter-to-fighter combat. Furthermore, switching priceless factory capacity to make unsuitable Westland Lysanders or obsolete designs like the Gloster Fighter were insane options for a nation at war.[1] The idiocy of this policy is hard to exaggerate. The 'Gloster Fighter' was a single seat all-metal monoplane powered by an 840 horsepower Bristol Mercury IX air-cooled radial engine, and armed with eight Browning machine guns in the wings. Intended to replace both the Spitfire and the Hurricane, it was slower than either. Its top speed was only 282 mph (by the Air Ministry's own figures), and 302 mph (claimed by its designer, H.P. Folland). It actually managed 316 mph at 16,000 feet when the second prototype was tested, but when this flew at the end of 1937 it was already obsolete, and potentially ideal prey for the Bf109.

Fortunately for national survival, Mitchell's successor at Supermarine, the skilled engineer Joe Smith, now had the chance he needed to produce his own Mark I Spitfire replacement. The Mark II was his reworking of Mitchell's design to make it suitable for production at the huge Castle Bromwich plant and remain competitive with the Bf109, and the first examples emerged as welcome assets in the battle's closing stages.

---

1. Quill, *Spitfire: A Test Pilot's Story*, p. 151.

Then it was Messerschmitt's turn. His Bf109E *Emil* would give way to an improved design as the 'ultimate standard fighter for the Luftwaffe', with stronger and more efficient wings and fuselage to reduce weight and drag, and provide higher performance with heavier armament. Like the Spitfire II, the 109F needed a more powerful engine. In this case the new DB601E left bore and stroke unchanged, but detail improvements raised power output. Better breathing let it run faster, while revised valve timings and cam profiles kept valves open for more of each revolution. Inlet ports were widened and polished to smooth out the mixture flow on the induction stroke. Ignition timing was advanced by 10 degrees and boost pressure raised slightly[2] from 1.3 to 1.42 times atmospheric pressure by adding another four blades to the twelve on the supercharger impeller. Stronger valve springs enabled the engine's top speed to rise from 2400 to 2700 rpm and its power from 1100 to 1350 hp.

Of course there was a price to pay. Greater high-speed power meant poorer low-speed output and a lumpy tick-over, possibly cutting out altogether when the throttle was closed. This problem was solved by a parallel induction system, bypassing the main throttle and letting fuel into the cylinders, even with the throttle closed, through a secondary throttle with final closure controlled by the pilot.

The new engine was 18 inches longer than the DB601A, so the cowling had to be reshaped. It fitted closely around a larger spinner, and the tighter fit reduced turbulence and made the engine more accessible for repairs and maintenance on operational airfields. The cowling was split longitudinally into two panels, each lifted by releasing three latches and propping it open on a strut. A lower panel carrying the oil cooler dropped down when released and swung to starboard for access to the engine from below. Finally, the drag of the tailplane struts was eliminated by mounting the tailplane higher than before and strengthening the rear fuselage structure with a smaller fin and rudder. The ailerons were narrower but deeper in chord and no longer linked to the flaps. Each wing had a new radiator, twice as wide but half as deep, to reduce drag with a double flap behind it. Turbulent air flowing under the wing passed over the radiators to absorb excess heat, before flowing out through the inboard sections of the split flaps. On the *Emil* both flaps and ailerons were simple control surfaces with completely separate radiator flaps. Now the outer sections of the wing flaps could be lowered to reduce speed and add lift during the landing approach, but inboard sections were double layered. The bottom layer worked as a conventional flap with the outboard flap, controlled by a wheel on the cockpit wall. The upper layer was controlled by a thermostat regulated by coolant temperature.

The wing's internal structure was similar to the 109E, with a two-foot shorter span to improve the roll rate. Leading edge slots were reduced in span but increased

2. Engine information from Report No EA. M/116 from RAE Farnborough, 1943.

in chord. More efficient ejector exhausts were fitted like those of the Spitfire and the supercharger air intake extended further into the slipstream over the nose. Other changes included a slightly smaller light alloy, three-bladed propeller with an electrical pitch control and a constant speed unit, controlled by a throttle-mounted button. Auxiliary drop tanks increased the range from the *Emil's* 410 miles to more than 1000 miles.

Combat capability was improved by fitting the nose cannon originally meant for the 109E. Sited between the twin cylinder banks and firing through the hollow spinner, it had a higher muzzle velocity and rate of fire than the wing-mounted 109E cannon, delivering a heavier punch. Removing the wing guns also improved its aerodynamics and agility, and eliminated their ammunition boxes and the bulges in wing panelling which covered them.

Many pilots believed the closer grouping of nose cannon and cowling machine guns made it easier to hit targets. Others wanted genuinely heavier firepower as the new setup cut the weight of a given burst. The aircraft could carry bombs in the fighter-bomber role, and would later have extra instruments and armour and self-sealing fuel tanks. The cooling system was redesigned with tanks on opposite sides of the engine. Coolant was a 50:50 mixture of water and ethylene glycol with a dash (1.5 percent) of *Schützol 39* rust inhibitor running in a closed circuit. Later in 1941, each would have a cut-off valve to stop flow from a tank damaged by enemy fire from running the engine dry and causing it to seize.

The cockpit was protected from below and behind by armour fixed to the floor and the frame between the cockpit and fuel tank. The windscreen normally carried extra armoured glass panels on a metal support frame. The forward panels of the sides of the screen could be slid open and the hood opened to starboard carrying with it the pilot's rear armour. The undercarriage hydraulics had an emergency lowering facility. In spite of widespread claims, its track was actually a foot wider than the Spitfire at 2.10 metres, or 6.8 feet. The Spitfire Mark 1 had a track of 5 feet 8.5 inches, but the Bf109's poor ground handling was due more to a high ground angle and awkward steering geometry than merely track width.

This was the basic design of the *Friedrich* and its eventual successors, the Bf109G *Gustav* with the larger and more powerful DB605 engine, and the late-war Bf109K *Karl* or *Kurfürst* [Elector]. However, they suffered from a typically Messerschmitt problem. His enduring obsession with reducing weight to the minimum led to a series of accidents when the 109F entered service. The project might even have died at birth like the original Bf109.

A German Air Ministry contract late in 1939 specified fifteen Bf109F-0 pre-production machines between November 1939 and spring 1940. In an echo of the Spitfire's slow beginning, only three had been completed by the end of January 1940 when Göring cancelled the whole project. Fortunately for Messerschmitt and for the Luftwaffe, he was too late. Production plants were set up for series assembly, and the first two prototypes, V22 and V23, had already flown. V22 was a compromise powered by the DB601A which first flew on 26 January 1939, followed soon afterwards by V23, the first genuine *Friedrich*, powered by the DB601E. Two more semi-prototypes were modified from old Bf109D airframes, but already the programme had met trouble.

In April 1940 the first production machine was being assembled, but its first flight was delayed from July to October. Later, changing priorities revived the 109F programme, but changes were still needed. The most serious problem was shortage of specialized raw materials needed to produce DB601E engines.

As a stopgap, the DB601N *'Nordpol'* [North Pole] version was used instead. This had redesigned cylinder heads and higher compression. It burned the same 100-octane fuel as the Spitfire II and produced 1200 hp at take-off or 1270 hp for up to a minute at a time at a height of 16,400 feet (5000m).[3] The 109F went into series production early in 1940 with a top speed of 321 mph at sea level and 373 mph at around 20,000 feet, almost identical to the Mark II Spitfire. By 1940 Daimler-Benz had turned out more than 7000 DB601A engines, but only around a dozen DB601Es and some forty DB601Ns. In fact the *Nordpol's* lower power proved a blessing as Messerschmitt had again edged uncomfortably close to structural weakness in his modifications. Wing failures caused several early 109Fs to crash or make forced landings. Tests showed wrinkles or cracks in skin panels covering the wings and even the most experienced pilots could fall victim.

Wilhelm Balthasar, commanding officer of *JG2 'Richthofen'* with forty-seven victories to his credit and an ace of the Spanish Civil War (see Chapter Twelve) was attacked by Spitfires over the Channel on 3 July 1941 and did what most 109 pilots did when surprised. He pushed the stick forward into a vertical dive to escape. His engine's fuel injection helped him leave his attackers behind, but stress caused a wing to collapse and he died in the ensuing crash. Rumours suggested lucky hits from RAF fighters had damaged the 109's wing, but structural weakness was a more probable cause. Strengthening the main spar and fitting thicker skin panels eliminated the problem.

A second weakness suggested removing the tailplane struts was another improvement too far. A series of tailplane failures was only explained when one pilot sensed a vibration and made an emergency landing. They found signs of fatigue at the section carrying the tailplane, but riveting a pair of four-inch metal strips on each side of the fuselage across the weak point solved the problem.

Once production models were modified the weakness never returned. The cause was almost certainly high-frequency resonance in the tailplane main spar at certain engine power settings, making the structure fail at the rearmost fin to fuselage connection. Ironically, if more DB601E engines had been available early in

---

3. Power and performance figures from Green, *Augsburg Eagle*, p. 85.

the *Friedrich's* operational career, greater power would almost certainly have caused more casualties.

Clipping the wings to improve manoeuvrability worsened the 109's already unforgiving handling, so pilots with insufficient expertise found it almost impossible to fly. This played havoc with combat effectiveness. If a pilot was fighting his own aircraft rather than opponents, what use was extra manoeuvrability? New semi-elliptical curved wingtips were added to restore its handling. Later Bf109s kept the rounded wingtips unlike clipped-wing Spitfires, where the roll rate was improved without marring handling.

Armament problems caused more delays. The designers wanted guns grouped in the nose to avoid the 109E's cumbersome wing guns. Because mounting the MG/FF cannon in the nose caused vibration, a stopgap was needed. First they fitted the 15mm MG151/15 heavy machine gun designed for nose mounting, but this cut firepower. The 7.92mm Rheinmetall-Borsig MG17 cowling machine guns had 500 rounds apiece and the Mauser heavy machine gun in the nose had 150 rounds, belt-fed from a box in the port wing. Spent links and cartridge cases fell down chutes to bins in the cowling and under the cockpit.

There were now two engine types and two armament configurations, making four versions in all: 'light' versions with the nose-mounted heavy machine gun, and 'heavy' with the MG FF cannon once this could be fired without too much vibration. Each would originally have the DB601N and later the DB601E once enough were produced. The 'heavy' versions with the N engine became the 109F-1, and with the E engine the 109F-3. Similarly the 'light' version with the N engine would be the 109F-2 and with the E engine the 109F-4.

Strenuous efforts were made to cut production time and costs. The F-1 and F-2 versions were priced at RM90,000 each, and the F-3 and F-4s at RM70,000 apiece. Where each 109E had taken almost 20,000 man-hours to produce at the start, dropping to 9,000 for the final batches, the 109F began at 16,000 man-hours, falling to less than 7,000 at the end.

Production began at Regensburg from October 1940 and by spring 1941, 226 109Fs had been produced, but problems continued. Lubrication and electrical defects caused unreliability, and bad weather limited deliveries to front-line units. October 1940 also saw 109F-3 production start, but this least successful variant comprised just fifteen examples. Main production concentrated on the 109F-2 until May 1941 when F-4 production started, continuing until mid-1942.

F-4s carried a new MG151/20 modified by fitting a long 20mm barrel to fire 800 explosive rounds per minute at higher muzzle velocity than the original MG FF. It was belt-fed and effective against light armoured vehicles as well as enemy aircraft. Where the MG FF of the earlier 109s fired 530 rounds per minute at just under 2,000 feet per second, the new cannon fired at 2,600 feet per second, an increase in firepower per weapon of more than one-third. However, using a single cannon decreased the total weight of rounds in a burst of fire.

Leading edge slots and trailing edge double flaps on Bf109G. *(Author photo)*

More attempts were made to increase performance and firepower against American heavy bombers. The GM-1 boost system injected nitrous oxide into the engine, raising top speed by 12 mph at almost 20,000 feet, making it faster at high altitude than all adversaries at the time. The resulting 109F-4/Z had a tank of nitrous oxide in each wing, and a larger oil cooler as later fitted to the 109G. The propellers of GM-1 machines had broader blades and were standard on most 109F-4s.

Separate modifications increased the 109F's armament with extra guns, bombs and even rockets. One of the first added a pair of 15mm machine guns slung in underwing gondolas. As these were outside the propeller arc they could fire at their maximum rate, but this additional punch imposed extra drag.[4]

The rack for the auxiliary fuel tank could also carry different types of bomb, and various cameras were fitted for reconnaissance missions. Finally, the underwing machine guns could be replaced by the old MG/FF low-velocity cannon of the Bf109E or the 20mm version of the MG151 cannon. This reduced performance slightly, but greatly increased firepower with three unsynchronized 20mm cannon and two cowling machine guns.

---

4. Recently revealed information from German sources showed factory tests of a 109G with mock-ups of the underwing gondolas suggested only a drop in maximum speed of 3.6 mph.

After the Battle of Britain, the range of the later Bf109s was transformed by long-range tanks.
(Author photo)

Ironically these cannon-armed 109Fs suffered even more problems than early cannon-armed Spitfires. From spring 1942, when the prototype three-cannon 109F-4 went for trials to Tarnewitz, the armaments proving ground on the north German coast, the weapons proved unreliable from the start. Guns and mountings had been sent by rail because of bad weather, and set up for ground tests, whereupon they suffered successive failures.

The weapons were dismantled and modified before fitting to the aircraft. The first air firing tests on 24 March 1942 were followed by sixty more flight tests, with little improvement. The aircraft went back to the makers with detailed criticisms, and in May 1942 four cannon-armed 109F-4s were sent to Tarnewitz from the Messerschmitt plant at Wiener-Neustadt in Austria. Over a month they completed twenty-seven air test firings. More than half were failures: the port cannon failed sixteen times. On six tests the gun jammed before 100 rounds had been fired. The starboard cannon suffered seventeen failures, five before a mere eleven rounds had been discharged.

Once again the machines went back to the factory. Later in May, four 109F-4s went from *JG54* on the Russian front to Tarnewitz for fitting with the latest weapons. Three ground firings worked well enough, but three air tests were failures. The guns were dismantled and reassembled, and used for a dozen air tests with a single failure and two reloads to clear jams. Finally, the cannon installation was passed as satisfactory, but ammunition had to be loaded slowly and carefully to limit failures, and an attempt to fit heavy 13mm MG131 cowling guns was dropped.

As faults were identified and cured, the 109F began to live up to its promise, but production remained disappointing. The first 208 109F-1s had the DB601N and the stopgap MG FF/M nose cannon, while the 109F-2 with the same engine and the nose mounted 15mm heavy machine gun numbered another 1380. The F-3 with the new DB601E and the old MG FF/M cannon was the rarest with fifteen machines produced before switching to the 109F-4 with the Mauser nose gun and the DB601E engine, which comprised the rest of *Friedrich* production. This included 240 109F-4/R1s with underwing gun mountings, 544 109F-4/Zs with GM-1 injection for high-altitude combat and another 576 109F-4/*trop* tropicalized versions, vital in the air fighting over Malta and North Africa.

Several later examples served as prototypes for experimental upgrades. Three had revised landing gear for Me209 and Me309 replacements. The inner wing sections were rebuilt for inward-retracting main wheels with a retractable nose wheel for a tricycle undercarriage. A fourth prototype was used for trials with wing-mounted air-to-air missiles and another replaced a Bf109E fitted with skis. More examples were used to test modifications for the '*Gustav*', so the introduction of the most numerous 109 version was greatly simplified.

Though the *Friedrich* suffered development headaches, it was deadly in the hands of a master. The first Bf109F-1s included the example issued to *Oberstleutnant* Mölders in autumn 1940. A pilot of his ability could best use the machine even with sub-standard engine and armament, and after his initial success he flew it again on 25 October 1940, with two *JG51* comrades. *Hauptmann* Hans Asmus flew Mölders' old *Emil*.

Over Kent at 30,000 feet they spotted Spitfires sent to intercept, but still climbing. Mölders had already shot down two RAF fighters, but he and his comrades dived on the rest and prepared to open fire. Then Asmus yelled a warning that they were being bounced by more Spitfires behind them. For once, Mölders had to run for home. Pushing his throttle to the stop, he forced his 109's nose down and headed for France at top speed. Lacking fuel injection, the Spitfires fell back. Before they could catch up, Mölders reached his airfield at Wissant on the Pas de Calais.

Asmus was not so lucky. After shouting the warning, his *Emil* was hit by a bomb jettisoned from another 109 as the Spitfires closed in and he crashed on British territory. Because he was flying Mölders' old machine the RAF assumed for the second time they had captured Germany's leading ace. Their first mistake resulted from the capture of his younger brother *Oberleutnant* Victor Mölders, shot down by a Spitfire on 8 October.

Even with its stopgap armament, Mölders had shot down five British fighters, three Hurricanes and two Spitfires, on two successive sorties, but he was an exceptional pilot. Originally crippled by debilitating airsickness, after taming his nausea through sheer willpower, he developed the courage and ability to close in on

Spitfire Mark Vs on the assembly line at the massive Castle Bromwich factory. *(Via Alfred Price)*

an opponent and inflict lethal damage with two synchronized machine guns without even firing the cannon – the two weapons had different firing buttons on the Bf109.

It was clear the 109F rewarded skilful pilots even more than the *Emil*. Limited by Messerschmitt's wing, the armament was grouped in the nose. With the fragile and unstable undercarriage, this made a machine with limited appeal for those who handled it clumsily. For those *Experten* able to close in before firing, it proved a precision weapon, lethal in fighter to fighter combat.

In the Mediterranean and North African campaigns, tropicalized 109Fs proved deadly against RAF Hurricanes. The 109Fs had a compact filter over the supercharger air intake to prevent damage from sand being sucked into the engine. A vital aid to pilot comfort when waiting at cockpit readiness was a pair of clips on the fuselage sides for umbrella sunshades to protect them from the glare of the open sun. However, against different opponents, its armament proved deficient as the US Eighth Air Force entered the air war over Germany. Massed formations of heavy bombers from airfields in Britain took a much greater weight of fire to bring them down, and closing in to short range to guarantee hits made lethal damage from the massed crossfire of a bomber formation almost inevitable.

More than 3400 109Fs were built and its career lasted up to summer 1942. At its peak, pilots thought it the best handling variant and it comfortably re-established Luftwaffe fortunes against the Spitfire IIs equipping RAF fighter squadrons as they took the fight back across the Channel to France. The 109F's improved high altitude performance made it difficult for Spitfires to attack and almost impossible for Hurricanes.

The next contest between the fighters would raise stakes further. While the Air Ministry continued its desperate search for a Spitfire replacement, the Supermarine team remained convinced the best fighter to replace the Spitfire was an improved Spitfire. And just as the switch from the 109E to the 109F was based on a more powerful DB601, the successor to the Spitfire II would need a more powerful Merlin.

Time was pressing. The Hawker Tornado's performance was inadequate, the Beaufighter was late and like the Westland Whirlwind, needed two engines for each machine, with the limited manoeuvrability inherent in most twin-engine designs. The Typhoon was more promising, apart from its immensely powerful but cripplingly complex 24-cylinder Napier Sabre. As with the *Friedrich*, increased power exposed lethal airframe weaknesses. When it entered service in 1941 (long after failing to replace the Spitfire as a front line fighter) more were lost through engine or airframe failures than enemy action in its first nine months of operations.[5]

Joe Smith and his Supermarine team had one great advantage over rivals and opponents; Rolls-Royce had a more powerful Merlin engine ready. The Merlin 45 was a Merlin XX with a two-speed supercharger, with drives switching at 10,000 feet. As this made the engine more complicated, the lower altitude drive was removed to give it the improved high-altitude performance of the Merlin XX, but able to fit the cowling of the Spitfire Mark I and Mark II. It produced 1515 hp at 11,000 feet and 1210 hp with 3lbs of boost at 18,000 feet, while flight tests confirmed a 370 mph top speed at 20,000 feet.

The Merlin 45 used 100-octane fuel and fitting it into a Spitfire fuselage was simple, so the first new Spitfire Mark VAs were simply 180 re-engined Mark Is and IIs. Because the engine was heavier, the undercarriage was strengthened and the wheels set two inches further forward to reduce a tendency to nose over on the ground. The Mark VA retained eight machine guns, but the VB used the Mark IIB wing which held two Hispano 20mm cannon and four machine guns, while the VC had a stronger wing for a second pair of Hispano cannon to be fitted alongside the originals.

Of course the first Mark Vs went to feed the constant need for replacements arising from the expensive tactics of Douglas and Leigh-Mallory over Occupied France. Ironically the Germans had already learned this lesson well. Because Bf109s were scarcer than Spitfires, they had to be careful over where and how they were used. Amazingly, they held Fighter Command at bay over France with just two fighter *Gruppen*, *JG2* and *JG26*, with the rest on the Eastern Front.

---

5. McKinstry, *op.cit.*, p. 256.

Spitfire Mark V being readied for loading on to USS *Wasp* at Port Glasgow. *(Via Alfred Price)*

Meanwhile, Spitfires were urgently needed over North Africa where Rommel threatened the Suez Canal and Middle East oil. Vital to the desert war were the bombers and torpedo bombers based on Malta. With just a couple of dozen Blenheims and Beauforts, eight US-built Baltimores and thirty Wellingtons, they imposed catastrophic casualties on Axis shipping supplying the *Afrika Korps*, and the Germans tried to bomb them into oblivion. An average of one bomber in three was lost on each raid, but in seven months of 1941 they sank the majority of German and Italian merchant shipping in the entire Mediterranean, and the supplies which might have got Rommel to Cairo.

The Luftwaffe needed no temptation to fight. Constant raids and heavy losses of fighters caught refuelling and rearming emphasized the need for a high-performance airfield defence fighter: the latest version of the Spitfire capable of coping with harsh desert conditions. Fortunately, the growing Mediterranean crisis provoked a more effective response.

The RAF sent tropicalized Spitfire Vs to Malta and North Africa. Large Vokes filters mounted under the nose lopped 11 mph off the top speed, and increased the time needed to reach 20,000 feet by almost a minute and a half. Fortunately, a maintenance unit at Aboukir in the Nile Delta[6] designed a more compact filter

6. No. 103 MU. Details from *Air Combat Legends, Vol. 1.*, p. 31.

which cut drag and weight. This became the standard fitting and could be left in place in kinder environments by removing the filter element.

Persuading Fighter Command to ease its grip on Spitfire Mark Vs helped make life harder for German fighters in Sicily and North Africa, but they still had to fly far beyond Spitfire range, even with auxiliary tanks. The quickest way across the gap was to launch Spitfires from an aircraft carrier as close to Malta or Egypt as possible, to be flown the rest of the way by pilots who would fly them in combat.

Getting fighters to Malta was fairly straightforward; keeping them there was harder. Furious air battles over the island ebbed and flowed with each side gaining an advantage before losing it again. Malta had been a British naval base since the Napoleonic Wars and the loss of Rommel's supplies provoked a sharp reaction. Göring sent a major formation, *Fliegerkorps II*, under *Generalfeldmarschall* Albert Kesselring to Sicily in late 1941 to beat the RAF into submission. First results were promising, with Bf110s and Ju88s catching and shooting down a third of RAF reconnaissance and bomber aircraft in the first two months.

Kesselring stepped up raids, devoting one in three to vital airfields defended by Hurricanes. In January 1942, eight RAF fighters were shot down in combat, but fifty were destroyed on the ground, leaving twenty-eight serviceable. Air and naval commanders pressed for Spitfires as the only option, and fifteen Mark VBs flew from the carrier HMS *Eagle* to Malta on 7 March 1942. They carried a new 'slipper' long-range ninety-gallon tank to more than double the fighter's range. Its bulky but streamlined shape was slung below the centre section, making it possible to fly 660 miles to Malta from the Algerian coast.

The pilots made a short and anxious take-off from the flight deck and two more delivery flights saw another sixteen Spitfires reach Malta by the end of the month. Unfortunately, constant German raids kept the defences on the back foot and fighter reserves dwindled back to danger level. A direct appeal from Churchill to Roosevelt brought the American fleet carrier USS *Wasp* to dock at Glasgow and take aboard forty-seven more Spitfires. These were launched on 20 April, followed almost three weeks later by another sixty-four from *Wasp* and *Eagle*.

Spitfires showed they could make a fight of it, shooting down Bf109Fs and bombers, but attrition remained high and regular German reconnaissance flights revealed the arrival of reinforcements, triggering heavy raids to destroy as many as possible on the ground. By dawn on 21 April, only twenty-seven Spitfires remained and by the end of the day numbers had fallen to seventeen. Nevertheless times were changing. Hurricanes could target bombers, leaving the faster-climbing Spitfires to take on 109s. However, continuing losses to bombers and torpedo bombers reduced the toll on Rommel's supply ships with nearly all cargoes reaching their destinations during April.

By early June, *Eagle* made three more ferry runs, delivering another seventy-six Spitfires. These tilted the balance back in the RAF's favour, while increasing losses left Kesselring short of aircraft. Axis pressure maintained the siege and shortages of

Spitfire VB in flight. *(Wartime print via Wikipedia)*

food, water and fuel sank to danger levels. Two British supply convoys sailed during June. One was turned back by enemy action after losing a cruiser, three destroyers and eleven merchant ships, but the other made it to Malta, with the loss of fifteen out of seventeen merchantmen.

In August, another large convoy fought its way through to the island, losing nine out of fourteen merchant ships. By now Air Vice-Marshal Hugh Lloyd, AOC Malta, had been replaced by Kesselring's old opponent from 1940, Air Vice-Marshal Keith Park. Fresh from Training Command, he possessed unrivalled experience commanding fighter squadrons in battle and he arrived on a flying boat in the middle of an air raid on 14 July 1942.

He faced a sadly familiar situation. He had more than 100 serviceable Spitfires, backed up by Hurricanes able to bring down bombers. He had better radar, and faster-responding squadrons, reaching the same height as incoming raids in a matter of minutes, but he found the baleful doctrine of the Big Wing still reigned supreme. Raids occurred three or four times day, every day, but time was wasted by defending fighters turning away to assemble into big wings, allowing the enemy to bomb docks and airfields. Fortunately, he knew exactly what to do. He said later 'I immediately changed the tactics to the same as 11 Group, that is I dispatched the squadrons singly or in wings of two squadrons. I sent them forward to intercept the German bombers over the sea before they could drop their bombs and broke up their formations before they ever reached Malta, and I boast that within two weeks of arriving in Malta, my squadrons completely put a stop to daylight raids over Malta without any additional strength and that was just by an alteration of tactics.'[7]

---

7. Park interview with BBC, 1 January 1961.

Once again Park's approach was vindicated. Sending smaller units meant a quicker response. Instead of climbing southwards to interception height, the faster Mark Vs could hit the Germans before they reached their targets. Park usually sent three squadrons, one high enough to bounce the 109s out of the sun, a second to hit the close escort or the bombers and a third to attack the bombers head-on and break their formation. It worked almost immediately. The enemy abandoned heavy daylight raids within a week and *Stukas* were withdrawn altogether. Kesselring ordered his fighters to approach the island at even higher altitude, but Park kept his Spitfires at 20,000 feet, their best performance height. Geography was now in his favour with the sun behind his formations, making it impossible for the Luftwaffe to bounce the Spitfires by surprise. They could only attack by descending to meet them, where the Mark Vs had the advantage in a turning dogfight.

August 1942 left Rommel trying to force the Alamein defence line before Cairo. By the end of the month, Malta-based aircraft and submarines had sunk a third of his cargoes and two-fifths of his fuel. In September, only a quarter of his supplies survived the Mediterranean crossing and had to be hauled hundreds of miles from Tripoli to the front. In October supplies fell further below what was needed with fuel supplies dwindling almost to zero.

More Spitfires arrived on Malta. Out of 385 flown off carriers (HMS *Eagle*, USS *Wasp* and HMS *Furious*), 367 landed safely. From the last week of October, smaller numbers flew the entire 1100 miles from Gibraltar in a single five-hour flight thanks to a larger oil tank in the nose, a twenty-nine gallon fuel tank in the rear fuselage and a 170-gallon tank slung beneath the centre section. This did little for top speed, but more than trebled the range, and by now the Luftwaffe lacked the means to attack incoming reinforcements. To save weight, all guns were removed except a pair of machine guns. When they landed, tanks were removed and guns replaced.

By mid-October, Kesselring's forces were defending Rommel's army retreating all the way to Tunisia. Combat took place mainly at low level with Spitfires carrying a 250lb bomb under each wing or a 500lb bomb under the fuselage. To make the most of low altitude performance, the LF Mark V used a Merlin 45M or the almost identical 50M with supercharger impeller blades reduced in diameter, or 'cropped'. Most fighters also had wingtips clipped by 4 feet 4 inches to improve roll rate and manoeuvrability. Performance was better below 12,000 feet, though the ever-cynical pilots complained the tired aircraft were 'clipped, cropped and clapped' Spitfires.

At the same time, and in the same area, other Mark Vs were modified to the opposite extreme against high-altitude Junkers 86P intruders from airfields on Crete. These bombers, fitted with twin turbocharged Junkers Jumo opposed-piston diesel engines, wings lengthened by 10 feet and a pressure cabin fuselage for a two-man crew, could fly higher than 42,000 feet. Over Britain they proved immune from defending fighters and dropped single 500lb bombs without harassment.

From May 1942 they appeared over the Mediterranean. The Aboukir maintenance unit fitted a Spitfire V with a more powerful Merlin and a four-blade propeller.

The armour and four machine guns were removed, and the machine intercepted an intruder north of Cairo on 24 August 1942. As it closed in on the Junkers at around 37,000 feet the Ju86P began climbing, but the Spitfire closed to some 150 yards before firing, hitting the target's starboard engine and making it dive to safety.

Later changes included a lighter battery and less fuel, and the removal of the radio and the aerial mast. This left the 'Striker' aircraft with no communication with ground control so an unmodified Spitfire V acted as a 'Marker'. Both aircraft took off together with the 'Marker' homed in by ground control on the approaching Junkers. The pilot of the 'Striker' stayed in formation until he spotted the target, then climbed and intercepted it. This combination brought down two and possibly three high-flying Junkers, though one modified Spitfire had to ditch in the sea having run out of fuel chasing the target, with its pilot swimming safely ashore. The Junkers flights ceased soon afterwards, as their limited range kept them from the fighting far to the west.

The Mark V proved successful for pilots from all the Commonwealth and Allied countries including both the United States and Soviet Russia, together with non-participants like Portugal, Turkey and Egypt. It served in all major theatres of war from the Eastern Front to the Pacific, including Burma and Australia. In most cases it fought well against every opponent, though in others its short range and sophisticated design created unexpected drawbacks. High over northern Australia for example, the air was cold enough to cause the propeller constant speed unit to fail, shifting it into fine pitch, making the engine over-speed and forcing the pilot to bail out or crash land.

Spitfire Mark V seaplane. *(Via Alfred Price)*

Soldiers, sailors and airmen collaborate at replenishing a tropical Spitfire VC at Takali, Malta, 1942. *(Via Alfred Price)*

Spitfire V being hoisted by crane on to USS *Wasp*: note Vokes filter and absence of wingtips. *(Via Alfred Price)*

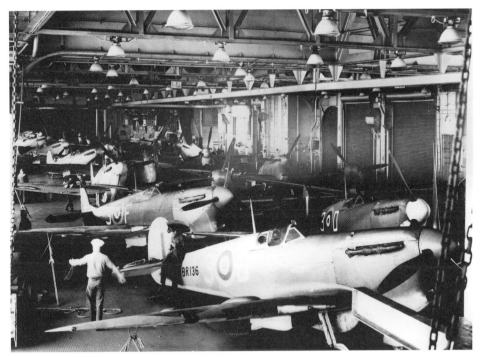

Spitfires in hangar aboard USS *Wasp*. *(Via Alfred Price)*

It also provided the basis for specialized photoreconnaissance Spitfires. Mark Vs of the South African Air Force in the desert had oblique F.24 cameras in the port side rear fuselage for tactical reconnaissance missions with the advancing Eighth Army after the Alamein victory. There was another version harking back to Mitchell's racing seaplanes, when three Mark Vs were fitted with floats designed by Supermarine. As lack of directional stability proved such a problem with the racing machines, these had modified fins above and below the tail plane to remedy this.

The intention was to hide them in the Dodecanese Islands for surprise attacks on German supply planes. The extra weight and drag of the floats slowed them down, but a 324 mph top speed was more than enough to threaten slow Ju52s. However, when the enemy captured the islands of Leros and Kos in autumn 1943, the opportunity vanished and they were converted back to landplanes.

Finally, another radical modification was carried out by the Germans. A Spitfire Mk VB operating with 131 Squadron was hit by light anti-aircraft fire returning from a sortie over Normandy on 18 November 1942. The pilot, Sous-Lieutenant Bernard Scheidhauer of the Free French Air Force, brought it down in a turnip field on occupied Jersey. The Germans took him prisoner, repaired the Spitfire, painted it in German colours and markings, and flew it to the Luftwaffe test centre at Rechlin.

Spitfire V with large ferry tank running up on USS *Wasp's* flight deck. *(Via Alfred Price)*

Spitfire V taking off from USS *Wasp*. *(Photo from US Navy Historic Center)*

It was then sent to the Daimler-Benz plant at Stuttgart to be fitted with a new DB605 engine with a Bf109G air intake, electrical system and propeller. Tests showed the new engine improved performance and raised the ceiling by more than 5,000 feet. Pilots liked the easier handling of the Spitfire, both on the ground and in the air, and its improved visibility compared with the Bf109. It became a popular machine for test flights and joy rides until wrecked in an American bombing raid on 14 August 1944 and broken up for scrap. Scheidhauer himself went to *Stalag Luft III* in Poland and took part in the Great Escape. Recaptured by the Germans, he was murdered with other escapees on Hitler's orders.

Ironically, it was over the Channel and Northern France that the Mark V was most outclassed by the *Friedrich*. Here the Luftwaffe had experienced pilots, the freedom to ignore attacks and the sun mainly at their backs.

Meanwhile a new and deadlier surprise was about to strike Fighter Command. For the first time this would not be a new version of Messerschmitt's fighter, but a new design from another company which threatened RAF morale until another, still more formidable Spitfire could turn the tables yet again.

Chapter Twenty-Two

The New Generation – The FW190 and the P51

O*berleutnant* Armin Faber was adjutant of *III Gruppe* of *JG2 'Richthofen'* based at Morlaix in Brittany. On 23 June 1942 he was cleared to fly a combat mission with the *7th Staffel* of *JG2*. That afternoon, their base had been bombed by six twin-engined Douglas Bostons in a 'Ramrod' operation. Twelve more Bostons had attacked Dunkirk with a complex covering operation involving Spitfires from as far afield as Cornwall and Yorkshire.

JG2 had been sent to chase the RAF back across the Channel and had had a reasonably successful day. Though the Bostons all returned safely, the Luftwaffe airmen claimed seven Spitfires shot down, for the loss of three of their own. *Leutnant* Walter Goring died when his plane crashed near Pont Audemer in Normandy, though this might have been an accident. *Unteroffizier* Willi Reuschling was posted missing after colliding with the Spitfire flown by Wing Commander Alois Vasatko, CO of the Czechoslovak Fighter Wing. Vasatko died crashing into the Channel, but Reuschling was taken prisoner.

Faber too had been successful – so far. After shooting down one Spitfire and crossing the English coast, he had become separated from his comrades. Then he was spotted by Spitfire pilot Warrant Officer Frantisek Trejtnar of 310 Squadron. Faber had only one cannon working, so he automatically turned northwards towards Exeter as the Spitfire climbed after him. Finally, he soared upwards in an Immelmann turn, rolling back upright at the top of the half-loop. Both fighters met head-on, the pilots firing as they closed. Faber's cannon shells badly damaged the Spitfire. Trejtnar had to bail out, landing heavily and breaking a leg.

Faber was short of fuel, and urgently needed to make for base. Trying to collect his bearings after the violent manoeuvres of combat, he saw the sea ahead and

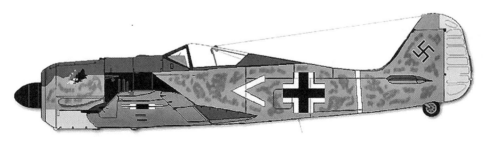

Armin Faber's FW190. *(Author drawing)*

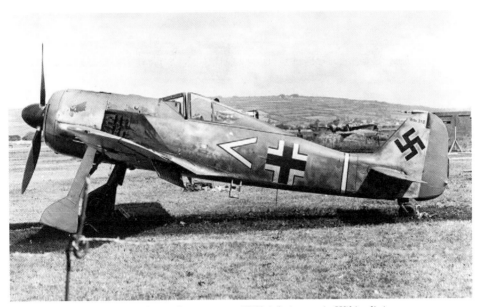

Faber's FW190 at RAF Pembrey after capture. *(Official picture via Wikipedia)*

turned towards it to cross the Channel. As land loomed up ahead once more, he looked for the nearest airfield in what he believed was Occupied France, where he could land and replenish his tanks before returning to Morlaix. He spotted runways ahead, and reassured by his safe return, he completed a series of victory rolls before turning into a landing approach. Ahead, he could see airmen wearing the blue grey uniforms of the Luftwaffe running out to greet him.

On that same afternoon, Sergeant Jeffreys was Duty Pilot at the airfield at RAF Pembrey on the South Wales coast. This was not a demanding job. The airfield had been busy during the Battle of Britain and its aftermath, hosting Spitfires and Hurricanes, but had been switched to Training Command a year earlier for training air gunners and testing new weapons. So the Sergeant carried nothing more lethal than a single shot Very signal pistol, and had little to do between routine landings and take offs.

Then, as the sun began sinking just after 8.30 pm, everything changed. He heard the roar of a single-engine fighter approaching at speed. As it crossed the airfield, it completed a series of victory rolls, and as it flew inverted on the last of these, the pilot lowered the landing wheels. By now the astonished Jeffreys, watching through binoculars, had spotted large black crosses on wings and fuselage, and a swastika on its tail fin. Though this identified it as the enemy, it was no familiar Bf109. Its large radial engine, wide-track undercarriage and long sloping hood meant he had no idea what type of aircraft it was. Then to his even greater amazement, the enemy pilot orbited the airfield, turned on to final approach, lowered his flaps and landed smoothly. Jeffreys jumped into action. Ordering his men to signal their visitor to

Faber's FW190 in RAF markings being test flown at Farnborough. *(Official picture via Wikipedia)*

park at dispersal, he clattered down the control tower steps and ran over to the aircraft. As it stopped, he climbed on to the wing. The pilot pushed back the canopy, and stared down the muzzle of Jeffreys' signal pistol as the sergeant told him he was now a prisoner of war. The news was an appalling shock. Not only had he presented the Luftwaffe's newest fighter to the RAF, but his war was now over after a catastrophic misjudgement. Later, he would make a failed suicide attempt before seeing out the war in a POW camp.[1]

The astonished Faber had made the same mistake as dozens of pilots on both sides after surviving intense high speed combat. Instead of flying south to return to base, he had flown a reciprocal heading to the north instead. He assumed the Bristol Channel was the strip of water he had to cross to reach his base and that the South Wales coast was actually France. Failing to spot specific landmarks, he landed at the first airfield he spotted to refuel and ask for directions.

Group Captain David Atcherley, Commanding Officer of the nearby RAF station of Fairwood Common near Swansea, was sent to Pembrey in an official car to pick up the German pilot and take him back to be entertained in the Mess. The journey must have added to Faber's distress. Atcherley kept his hand on his Service revolver in case of any attempt to escape, but when the car hit a particularly bad pothole on the country road, he inadvertently squeezed the trigger and fired, the bullet passing through the door within inches of the German. He was later taken by train to London for questioning before going to a POW camp.

---

1. Brown, *op.cit.*, pp. 78–81.

FW190 in flight. *(USAF)*

For the RAF, this capture of a German fighter was a huge bonus. Ever since the previous autumn, pilots had been reporting new German fighters quite different in appearance from the Bf109s they knew well. Their radial engines persuaded intelligence officers they were Curtiss Hawks supplied to the French *Armée de l'Air* before the outbreak of war, or even French Bloch 151s, pressed into service to replace combat losses. Neither was close to the truth. The mystery aircraft's speed and armament ruled out both possibilities. Camera-gun footage from a Mark V Spitfire of 129 Squadron shot on 13 October 1941 revealed this as an entirely new fighter and a very formidable one.

Eventually it was identified as the Focke-Wulf (FW) 190, a new design by Kurt Tank using the BMW801 radial engine originally proposed as an alternative power unit for the Bf109. Key to its performance was the way in which the contours of the radial engine had been faired into the narrow fuselage to reduce aerodynamic drag to the minimum. Pilot reports suggested it was well armed with machine guns in the wing roots synchronized to fire through the propeller arc, and a pair of cannon further out towards the wing-tips.

At the time it was assumed that a radial-engined machine would suffer from the bulk and drag of its power unit, though its greater resistance to combat damage was a major advantage. Even a single bullet through a radiator or coolant pipe could cause a liquid–cooled engine to seize, but radial engines sometimes survived even with one or more cylinders shattered by cannon shells.

On the other hand, with 109 output set to peak early in the war, it seemed sensible to assume that Daimler Benz engine production would be required for Messerschmitt's needs for the foreseeable future. Any alternative fighter would need a different engine to enter series production. Moreover, the Luftwaffe Operations Staff denied that another fighter was needed in the first place, as they felt the 109 could cope unaided.

Whatever the truth of that prediction, the FW190 would depress Spitfire pilots meeting it in combat. Later it emerged that German pilots had suffered from the engine's tendency to catch fire without warning, leading to a ban on flying over water. By the time these weaknesses had been eliminated and *JG26 'Schlageter'*, followed by *JG2 'Richthofen'*, had been equipped with the new fighter, it had become clear the Mark V Spitfire was seriously outclassed. Now the RAF desperately needed one of the new aircraft to find its strengths and any weaknesses they could hope to exploit.

One desperate expedient was proposed by an officer in No. 12 Commando, Captain Philip Pinckney, for two men to be landed by motor gun–boat (MGB) on the French coast. One would be a Commando, for which he volunteered, and the other an experienced test pilot for whom he suggested his friend Jeffrey Quill, Supermarine's test pilot. To Quill's horror, the plan involved them making their way alone to a German fighter airfield in Northern France, to steal one of the new fighters. Quill would fly it back to England, while Pinckney would make his own way home using a light folding boat to rendezvous with the MGB.[2]

The operation was initially aimed at Abbeville airfield and then at Maupertus near Cherbourg, (one of the bases of *JG2*) both within easy reach of the sea, with hiding places nearby. There they would wait for Luftwaffe mechanics to warm up a FW190 whereupon they would approach, shoot the mechanics and seize the aircraft before the defences could react. The plan reeked of desperation. It was named 'Operation Airthief' to hint to the Germans what was planned, and the chances of success would have been lethally slim. Now Faber had saved them the trouble and given them precisely what they wanted.

This incredible stroke of good luck still concealed major worries. The FW190 was taken by road to Farnborough, where it was dismantled, analyzed and reassembled, painted in RAF colours and given a new identity as MP499 before its first test flight. Ten days later it was sent to the Air Fighting Development Unit (AFDU) at RAF Duxford to be compared against Allied fighters, where the results proved ominous. Compared with the Mark V Spitfire, the FW190 was faster at all altitudes up to 25,000 feet by between 20 and 30 mph. It matched the Spitfire V in the climb,

---

2. Quill, *op.cit.*, pp. 222–7.

but ascended at a steeper angle. This was a valuable advantage, particularly when pulling up from high cruising speed or after pulling out from a dive.

The FW190 was also faster in diving than the Spitfire, particularly at the start. In a tail chase the 190 could usually pull up into a climb and easily evade its pursuer. With similar fuel injection to the Bf109 it could pitch forward into the dive without any loss of speed or power. It also had the highest roll rate in the world. The Spitfire's only real advantage was its tight turning circle, priceless in a dogfight. Even then FW190 pilots had two options. A flick roll into the opposite bank and a diving turn when 'bounced' by a Spitfire left the British pilot with a very difficult deflection shot before his target pulled out of range. The other was to emulate Bf109 pilots and limit themselves to dive and zoom attacks to avoid a close encounter dogfight.

What could a Spitfire pilot do to escape a FW190 attack? Mock combat tests showed the Mark V was most vulnerable cruising at low speed and would usually be forced into a series of tight turns. When cruising at higher speed, a shallow dive could force the attacking 190 into a long and unwelcome tail chase, provided the enemy was spotted in time. Diving steeply was never recommended as the 190 would soon close the range and open fire with an easy zero-deflection shot.[3]

The FW190 had only two real limitations. Engine performance dwindled sharply with height, due to supercharger limitations compared with the Bf109 or the Spitfire, so it was limited to low and medium altitude combat. It lacked any stall warning for the pilot like the Handley Page slots of the 109 wing or the Spitfire's curved leading edge. Consequently an inexperienced pilot or one under combat stress might enter a stall, causing the aircraft to flick into the opposite bank followed by a vicious spin. Only brave or experienced pilots were willing to risk a stall, which occurred at higher speeds than the 109 or the Spitfire.

Its approach speed for landing was relatively high. If this fell too low it would stall without warning. Its air-cooled engine took a long time to warm up and early examples overheated. Originally it had neither an artificial horizon nor a vertical speed indicator, limiting it to fair weather operations. Nevertheless, its good control harmony, wide track undercarriage and well-thought-out controls meant it was popular with its pilots.

There were some positive implications for the RAF. Provided they only approached FW190s at maximum cruising speed and retained the initiative, they had a chance of holding their own. But just over three weeks after Faber's arrival, the Commander in Chief of Fighter Command, Air Chief Marshal Sir William Sholto Douglas, wrote an anxious letter to the Under Secretary of State for Air, in which he pointed out, 'At the beginning of the war our fighters possessed technical superiority over those of the enemy. We have gradually lost this lead and we are now in a position of inferiority. It is essential that this position should be remedied before

---

3. Alfred Price, *The FW190 at War*, pp. 46–47.

next spring when it is anticipated that intensive air fighting will take place.' After criticizing what he called the 'spirit of complacency' of the Ministry of Aircraft Production, he complained that they 'find it difficult to believe that we really have lost our lead in fighter performance. There is however no doubt in my mind, nor in the minds of my fighter pilots, that the FW190 is the best all-round fighter in the world today.'[4]

Fortunately this was unduly pessimistic. Plans already existed for a re-engined Spitfire, the higher performance Mark IX (Chapter 23) to meet the FW190 on more equal terms and experienced pilots could already use tactics against this new threat. For those without, the future seemed bleak. On 1 June 1942 FW190s from *JG2* and *JG26* inflicted serious damage on RAF fighters. Seven squadrons of Spitfire Vs from Hornchurch and Biggin Hill wings escorted eight Hurricane fighter-bombers against a target in Northern France, with more Spitfires from the Debden Wing in close support. Nevertheless, the Luftwaffe ace 'Pips' Priller led forty FW190s from *JG26* in a copybook attack out of the sun to shoot down eight Spitfires and send five more home with battle damage, without loss or damage to themselves. On the very next day over the Pas de Calais the relatively inexperienced 403 Canadian squadron, led by the veteran New Zealander Al Deere, were bounced by 190s from *I/JG26* and *II/JG26*. Here again tactics and experience proved decisive. The Germans waited until the RAF fighters turned for home and used cloud cover to manoeuvre into position for attack, distracting Deere's attention with a single *Staffel* in plain view. Deere ordered his pilots to break towards the enemy to deflect their attack. Unfortunately this split 403 from the rest of the Wing as two more *Staffeln* of FW190s burst from cloud cover at odds of three to one. Within minutes seven Spitfires, more than half 403's strength, were shot down, with six pilots killed. One crashed later on landing, but the Germans suffered no losses at all.[5]

After shocks like this the FW190 appeared invincible. However, some pilots had the ingenuity to use the Spitfires' inherent qualities against these new opponents. For example, 64 Squadron based at Southend, was led by its new CO, Squadron Leader Duncan Smith, who had first met FW190s with his previous unit. On the afternoon of 13 May 1942, his Mark V Spitfires formed part of a large fighter sweep over the Pas de Calais. Attacked by fifty FW190s as they crossed the French coast at 10,000 feet over Gravelines, Duncan Smith and his wingman forced one of the German fighters into a turning fight and shot it down in minutes, after which they lost height in a series of defensive turns before heading back to base. What could have been a disaster was encouraging, with two FW190s destroyed, three probably destroyed and another damaged for the loss of one Spitfire and its pilot.

---

4. Alfred Price, *ibid.*, pp. 38–9.
5. Spick, *op.cit.*, pp. 121–3.

By now, the pilots of 64 Squadron flew in 'finger-four' formation like their opponents. By stepping port and starboard sections slightly higher and lower than the centre section, it was impossible for the enemy to approach without being spotted. In addition, they took care to fly at the highest feasible cruising speed, to retain the option of quickly diving on to German aircraft spotted below, or climbing to respond to attacks from above. Duncan Smith in his book '*Spitfire into Battle*' explained how he had found a way of keeping the initiative by tempting enemy pilots to attack his formation. By watching FW190s closing the range, they were able to spot the start of their attack and break towards them to evade their fire, and use their superior turning qualities in a classic dogfight, while the top section of Spitfires climbed steeply while the German pilots were preoccupied, ready to circle round behind them and bounce them from above.

Fortunately for less experienced pilots, the faster and more powerful Mark IX Spitfire was about to enter the fray. In the longer term though, the nature of the battle was changing. The use of medium bombers with heavy fighter escorts to force the Germans to accept combat had not worked too well, with increasing RAF losses exacerbated by the new opponents. But the arrival of large US bomber forces delivering massed formation attacks on high priority targets inside Germany would impose a much tougher challenge on Luftwaffe fighters.

At first they would have the best of the fight as the B17s and B24s of the Eighth Air Force reached further beyond the range of escorts like Spitfires and P47

Massive B17 formation over Germany. *(USAF)*

B24s pounding German targets. *(USAF)*

Thunderbolts. Increasing American losses underlined the need for a long-range fighter to accompany the bombers to distant targets, with the performance to shoot down defending fighters when they got there. Ironically the solution was already available in an aircraft designed for RAF requirements, but downgraded to ground attack and reconnaissance missions because of engine limitations.

In the first months of the war, when the British and the French were signing contracts with American manufacturers for desperately needed fighters to meet the growing German threat, the only American fighter potentially capable of matching the latest German warplanes was the Curtiss P40. This was closely based on the older P36 Hawk, with its radial engine replaced by an in-line Allison. In fact Curtiss-Wright had already drafted a design for a successor, the P46. This had a large cooling radiator under the centre section on the XP40 prototype, though for production P40s this was shifted to the nose of the aircraft more on the grounds of appearance than aerodynamics.

When production of the XP46 prototype received US Government backing on 29 September 1939, Curtiss engineers tried to persuade the authorities to switch a huge order placed for P40s to the new fighter instead. However, pressure to build up US fighter strength as quickly as possible kept the vast contract in place and P40s remained in production until December 1944, serving by this time in twenty-eight different air forces.

Then British and French buyers arrived looking for manufacturers with spare capacity. Sir Henry Self, head of the Purchasing Bureau's New York office, had a close working relationship with North American Aviation, makers of the T-6 Texan advanced training aircraft serving the RAF as the Harvard. He suggested North American might start making P40s for European customers. Only when Curtiss suggested they might pass on the work done so far on the XP46 to North American for them to complete the development and build the aircraft for the British and French, did a different possibility arise.

Curtiss could concentrate on the huge P40 contract, while North American went ahead with their own version of the P46. The deal was soon finalized. Curtiss passed to North American data from wind tunnel tests, cooling tests and performance predictions for $56,000, and on 23 May 1940 contracts were signed between the British, the French and North American for 400 new NA-73 (P46) fighters. Since the Curtiss information lacked wing data, the North American team looked at studies from the National Advisory Committee for Aeronautics (NACA) at Langley in Virginia on new aerofoil shapes for laminar flow wings in their state of the art wind tunnel.

The laminar flow wing reduces parasitic drag to the minimum with a surface contour smooth enough to cut through the air with minimal disturbance to the layers of airflow passing over it. This depended on a perfectly smooth surface, free from dents, rivets, dirt or edges of hatches or inspection panels. Nevertheless, the Langley tests produced a highly efficient aerofoil with maximum thickness set 60 percent back from the leading edge, much further back than conventional aerofoils. This delayed the onset of turbulence and produced an extremely low drag coefficient of only 0.003, while providing more interior space than a traditional aerofoil of the same thickness.

Fitting the new wing to the XP46 fuselage was complicated, so the designers simplified it by choosing a straight-edge tapered wing not unlike that of the Bf109. The fin, rudder and tailplane had similar straight edges, and time was saved by using hydraulic and electrical systems from the Harvard for flaps, brakes and undercarriage. By taking these short cuts and working cripplingly long hours, the design team finished the prototype NA-73X in record time. It was powered by the best American engine available, the in-line liquid-cooled Allison V1710, and made its first flight on 26 October 1940. Progress had been so good that a month before that first flight, the British Purchasing Commission ordered another 300, and named the fighter the P51 Mustang.

Such a radical new design made teething troubles inevitable, but its genuine virtues were highly appreciated. There was room in the wing for an inward-retracting undercarriage with a wide enough track for easy landings and take offs. There was enough space for large self-sealing fuel tanks, and more room in the deep, slab-sided fuselage for additional fuel, though the crucial importance of this feature was not realized at first.

Because the Allison engine warmed up slowly compared with other RAF fighters, and supercharger limitations restricted its operating altitude to 25,000 feet, it was assigned to Army Co-operation Command as a tactical reconnaissance aircraft and ground-attack machine where these problems mattered less. Yet it could reach targets deep into occupied Europe and Germany far beyond the reach of the Hurricane and Spitfire. Speed was another advantage, and lack of manoeuvrability mattered less in a plane operating at low altitude.

It would finally be transformed in a process which began on 29 April 1942. Rolls-Royce test pilot Ronald Harker was visiting the Air Fighting Development Unit at RAF Duxford in Cambridgeshire when the CO, Wing Commander Ian Campbell-Orde, invited him to try out one of two Mustang Mark 1s currently under test. Harker spent half-an-hour in the air and was amazed at the aircraft's speed and responses. On returning to his office, he asked his engineers to calculate the effects of replacing the Allison engine with the latest Rolls-Royce Merlin. The results were staggering, promising a much faster climb and a high altitude top speed of more than 440 mph, competitive with any existing German fighter.

Harker proposed to Rolls-Royce management that they should fit one of the latest Merlin 61 engines to the Mustang airframe to see whether tests confirmed the theoretical figures. To his surprise, his suggestion was vetoed on the grounds the Air Ministry would refuse to see engines diverted from Spitfire production, and that the Mustang was a relatively unproven design. Harker spoke to Lord Hives, general manager of the Derby engine plant, who persuaded the Air Ministry to agree three Merlins could be tested on Mustangs. North American were told, and decided to try a Packard Merlin, built under licence in the USA, and fitted to one of their machines.

North American A36 Apache ground attack machine, forerunner of the Mustang. *(USAF)*

With continuing pressure on production, the first engines fitted to Mustangs in Britain were all Merlin 66s, intended for low-altitude versions of the Spitfire Mark VIII and IX. Nevertheless, when an American technical officer was invited to try a test flight in one of the prototypes in November 1942, he applied full climb power after taking off and retracting the undercarriage, and found the aircraft carrying out a snap roll without warning. Having righted the machine, he continued climbing to a height of 33,000 feet where it performed aerobatics far beyond the capabilities of the original version.

Tests carried out in December 1942 at Duxford's AFDU against a Spitfire IX with the same mark of Merlin showed both aircraft were surprisingly evenly matched in all respects bar one. The Mustang was slightly faster than the Spitfire at high altitudes, though slower in the climb to those heights, and faster in the dive. The Spitfire could turn more tightly but the Mustang rolled more quickly. There were problems with rudder control and directional stability, and with limited visibility, but production changes eliminated these; fin extensions increased directional stability and a bulged canopy gave better visibility. But the enormous fuel tank capacity, almost double that of the Spitfire at 154 gallons compared with eighty-five, needed no improvement at all.

American daylight bombers over Germany were suffering heavy casualties. P38 Lightnings and P47 Thunderbolts with long-range tanks could escort them part way to the furthest targets and meet them again on their return journey, but

The Packard Merlin produced in the US under licence transformed the performance of the P51 Mustang. *(USAF)*

The North American P-51D became the first truly successful long-range fighter. *(Will Owen photo)*

the Luftwaffe simply held back its main fighter strength to hit the bombers once their escorts left. Fortunately, thanks to its spectacularly low drag, the P51's fuel consumption was equally low. With a higher cruising speed than the P47, it used less than half its fuel. If its range could be stretched to cover the bombers throughout their flight, this fatal gap could yet be plugged.

This meant either increasing internal fuel stowage, or fitting external fuel tanks, which could be dropped if attacked by German fighters. The easiest internal solution was an extra seventy-gallon self-sealing fuel tank at the rear of the fuselage, but this inflicted an extra weight penalty well aft of the plane's centre of gravity, impairing performance and stability, particularly longitudinal stability, as well.

Fitting extra tanks in the wings would keep extra weight closer to the centre of gravity. Inevitably, this meant using the shallow outer wing sections, where the thickness of the self-sealing layer would limit their additional capacity, and where the extra weight would slow down the P51's roll responses. They finally compromised by fitting the seventy-gallon rear fuselage tank and adding twin under-wing auxiliary tanks for specially long-range missions. Each one held sixty-three gallons, making total capacity 340 gallons, enough to reach the most distant targets, fight over the target and then return to base afterwards.

This gave pilots some difficult decisions when flying over enemy territory. Flying with a full fuselage tank and full underwing tanks degraded combat survivability, so most pilots preferred to use the fuel in the fuselage tank first and then switch

The capacious fuselage and the smooth laminar flow wing produced an unbeatable combination of speed and endurance. *(Will Owen photo)*

to auxiliary tanks. Once the Germans realized why the P51s could fly so deep into *Reich* airspace, they attacked escort groups early on to force them to drop their auxiliary tanks, which kept them from escorting bombers over the target. Later trials showed the P51 could fight well with auxiliary tanks in place, thanks to being faster than both the Bf109G and the FW190, and capable of out-turning the 109, but a full rear fuselage tank caused serious instability, so pilots used this first.

Some of the P51's teething troubles had been lethal. Engine mounting bolts insufficiently heat-treated caused engines to tear loose in flight, losing both fighters and pilots. Wings folded up in aerobatics and tail structures failed after violent manoeuvres, especially with the rear tank full of fuel. Guns jamming proved one of the most persistent bugs of all, but in time they too were eliminated. On tactics, P51 pilots faced the same dilemma as German pilots over England in 1940. Hard-pressed bomber crews wanted fighters close for greater protection, but this made them vulnerable to enemy bounces from above.

Fortunately, Lieutenant-General James Doolittle, taking over the US 8th Air Force in January 1944, ordered escort fighters to assume their first priority was shooting down German fighters, so pilots could use the P51 to the full. By high-speed cruising above the bombers, they could fly straight to wherever they were needed as P38s and P47s could protect a formation in the earlier and later stages. It also gave American fighters a height advantage over their German opponents, who had to stay at the same altitude as the bombers to shoot them down.

So far as the Spitfire and 109 were concerned, these radical new designs did not replace them in front-line service as might have been predicted. So remarkable were the original designs, they remained competitive combat machines, even against the furious pace of wartime development. Already the curtain was about to rise on new variants powered by new versions of existing engines: the DB605 powered 109G and 109K, the Merlin 61 powered Spitfire IX and the even more formidable marks powered by the Rolls-Royce Griffon, a direct descendant of the exotic racing engines of Mitchell's Schneider Trophy seaplanes. But the future would soon prove increasingly bleak for the beleaguered Luftwaffe, facing the formidable American mass-production system turning out heavy bombers by the thousand in massive new factories.

Chapter Twenty-Three

The Fight Continues

New designs promised to make front-line combat survival for the Spitfire and Bf109 more difficult. The FW190 was a severe challenge to the Spitfire Mark V while the P51 could escort large US bomber formations across Europe to distant targets, so Luftwaffe defenders would have to overcome these new fighters before having a chance to attack the bombers. Yet this was the only way to deflect the increasingly threatening future facing Hitler's *Reich* as the war entered its most crucial phase.

It therefore seemed clear the Spitfire and the 109 would be damned by obsolescence. How could it be otherwise when the pace of development made yesterdays' most potent machines too dated for combat? At first, designs like the Bristol Blenheim and the Heinkel He111 had left fighters like the Bristol Bulldog or the Heinkel He51 unable to catch them. Since then these new bombers had changed from state-of-the-art to aeronautical liability. He111s could not survive in daylight over southern England without close-escort Bf109s and the casualty rate of Blenheims almost erased Bomber Command's 2 Group from the RAF Order of

Me309 model by Stahlkocher. *(Via Wikipedia)*

Battle by 1941. Now it was surely time for the fighters which defeated them to bow out and hand over to their successors.

Yet events failed to follow the script. Both designs remained in front-line service thanks to careful development to improve performance, combat capability, range, power and weaponry beyond the most optimistic predictions at the start of their careers. Their outstanding original designs meant updated versions showed little obvious change. The contours of the *Friedrich* would appear again in the Bf109G or *Gustav*. The principal change would be the DB605, a larger capacity version of the DB601 delivering increased power. This was just as well, since the new airframe would be heavier than before.

The reason for keeping a more powerful version of the 109 in front-line service was that there was nothing good enough to replace it altogether. At least two uprated derivatives of the 109 had been designed and developed, only to fall by the wayside, leaving a gap in Luftwaffe fighter strength which only the *Gustav* could fill. The first harked back to the propaganda games played with Willy Messerschmitt's almost lethally unflyable Me209 racer. It was initially numbered the 309 as the next step in the German tradition of renumbering designs. In this case the sequence jumped forwards and then backwards as 209 had already been used for the racing machine's false identity. In any case, designing the next fighter while *Friedrich* development continued, had to fight the official assumption that the 109F did not need replacing anyway.

The 309 tried to eliminate the 109's vicious ground handling and poor runway visibility, due to its high ground angle. It had a tricycle undercarriage with main wheels retracting inwards into the wing and a nose-wheel leg, which turned and folded away under the engine. A central retractable radiator behind the nose wheel let the fighter sit in a more level attitude on the ground, giving a clearer view of the runway during taxiing, take-off and landing, with a reverse-pitch propeller to shorten the landing run and stop quickly.

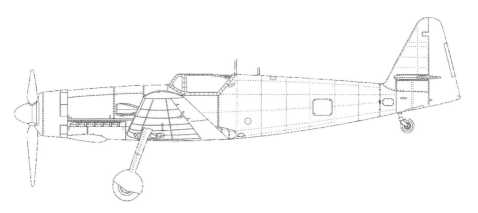

Me209 drawing by Bjorn Huber. *(Released through Wikipedia)*

Sadly, these improvements proved unsuccessful. When test pilot Karl Baur first flew the prototype on 18 July 1942, a chronic lack of directional stability showed up on the ground and in the air, and there were problems with the new radiator. On a later flight in September, the reverse-pitch propeller worked too well, stopping the aircraft so violently the fragile nose-wheel collapsed and the flaps were damaged in the resulting nose-over. Two months later, the first Me309 resumed the test programme with a larger fin and rudder to improve stability. It also suffered the same engine shortages as the *Friedrich*, and instead of the DB605, had to make do with the DB603, largest of the DB600 series of inverted V12s. The cylinders were widened from 150mm to 162mm and the stroke lengthened from 160mm to 180mm, which increased capacity from the DB601's 33.93 litres to 44.52 litres. The engine was originally intended for a Mercedes project to break the world land speed record on a specially prepared stretch of the Berlin-Leipzig autobahn which was cancelled on the outbreak of war, and the engine was reassigned to the aircraft industry.

It was flown in mock combat with a Bf109G, and raised the designers' spirits when it proved faster, but the 109G turned inside the Me309, with fatal implications in a dogfight. The project stumbled on against growing indifference until the third Me309 prototype emerged in March 1943, this time with the DB605. There was little extra performance to compensate for poor manoeuvrability. The fourth prototype followed four months later, carrying heavier armament to shoot down bombers, now the most urgent priority for any German fighter. Half-a-dozen guns of varying calibres wrecked any hopes of performance and when one American raid destroyed the prototype, everyone breathed a sigh of relief and switched to less demanding projects. An official German Air Ministry report said the Me309 was too difficult for the average pilot. In the light of the 109's handling, this was truly devastating criticism.

There was just time for one more 109 replacement attempt, which revived the old Me209 designation for a very different purpose. The new Me209 looked much more like the 109 and used as many 109 parts as possible. It was compromised from the start. It too tried to eliminate the 109 weaknesses of weak undercarriage and inadequate armament, but these meant major alterations to the 109 wing and added extra weight to eliminate airframe failures.

This aircraft too was powered by the DB603, which produced 1874 hp at sea level and 1539 hp of combat power at medium altitude. Once again a powerful engine with wide-track undercarriage produced longitudinal instability, so once again a taller fin was fitted. The prototype first flew on 3 November 1943, with the DB603 hidden behind an annular radiator, looking like an FW190. The second prototype had a Junkers Jumo 213E, another inverted V12 of 150mm bore, 165mm stroke and 35 litre capacity, delivering 1726 hp at take-off power, boosted to 2022 hp with methanol-water injection. It had a two-speed, two-stage centrifugal supercharger with an intercooler between the two stages to increase altitude performance. A third prototype had a greatly extended wing, and was powered by a DB628 two-stage supercharged DB605 or a turbocharged DB603, or even a DB627, a two-stage supercharged DB603.

None of this aero-engine alphabet soup actually materialized. Most were cancelled when the Me209 II project was mercifully shelved in early summer 1944. Some aspects were promising, but making it into a viable operational fighter needed time and effort, which late-war, Germany could not afford. Mass production in dispersed factories was running so efficiently that any disruption from switching to the Me209 was uneconomical for such a poor performance boost. The death knell was finally sounded when tests showed the prototype was 30 mph slower than the already successful 'long-nose' FW190D. Instead they had to accept the inherent limitations of the *Friedrich's* successor. The engineers increased engine power with truly modest revisions. The 109G used the DB605, with cylinder bores widened by a mere 4mm, and stroke left unchanged. Capacity rose slightly from 33.93 litres to 35.7 litres, but changes to valve timing improved engine efficiency and raised its maximum speed from 2600 to 2800 rpm.

The DB605 was originally intended for reasonably plentiful 87-octane fuel. When 100-octane C3 grade fuel became available, the DB605 was modified to run on this. With compression raised from 6.9-to-1 to 7.5-to-1 on 87-octane fuel and 8.5-to-1 with 100-octane fuel, power increased from 1332 to 1455 hp, but weight increased from 700 to 756 kg. Take-off power increased to 1775 hp with MW50 methanol-water injection and the final versions of the DB605 would deliver 2000 hp at take-off.

German engines used a different coolant mixture from the Allies with 50 percent glycol, mixed with 47 percent water and 3 percent anti-corrosion oil. Their fuel injection came at the price of increased complexity. Where a Merlin carburettor contained 433 parts, a Junkers fuel injection pump used no less than 1576 parts. Sir Stanley Hooker of Rolls-Royce felt that 'The Germans paid a large penalty for their fuel injection. When the fuel is fed before the supercharger, as on the Merlin, it evaporates and cools the air by 25 degrees C. This cooling enhances the performance of the supercharger, and increases the power of the engine, with a corresponding increase in aircraft speed, particularly at high altitude'.[1] Former test pilot Bill Gunston said the DB605 'Was fine in the twin-engined aircraft [Bf110] which was nice to fly, but the 109G – made in greater numbers than all other 109s combined – was in my opinion a retrograde step and an aircraft which would never have got through Boscombe Down.'[2]

Ball bearings were replaced by sleeve bearings due to shortages from Allied bombing. Poor quality lubricating oil caused engine fires, so pilots were ordered not

1. Hooker, *Not Much of an Engineer*, p. 62.
2. Gunston, *The Development of Piston Aero Engines*, p. 158.

Late model Bf109G with Erla Haube clear view cockpit canopy. *(USAF)*

to use emergency power, a severe handicap in combat. Modifications to bearings and lubrication produced improvements, but the Ministry (and the pilots!) treated the engine with suspicion.

Otherwise the *Gustav* changed little. Extra engine weight meant stiffer bearers and tougher undercarriage, adding still more weight. Greater power generated even more heat, so a larger and heavier oil cooler was needed. Combat on the Western Front emphasized the value of height so the *Gustav* had a pressurized cockpit, isolated by sealing all holes in sides, floor and front and rear bulkheads with canopy and windscreen openings blocked by rubber strips. These changes allowed cabin pressure to be raised by 4 lbs/sq.inch. Stronger cockpit glazing bars coped with extra stress, and two small scoops were fitted on each side of the fuselage just behind the spinner to maintain pressure. Weight remained a problem. Boosting high altitude performance meant fitting the GM-1 nitrous oxide injection system which added still more weight, though the twenty-five gallon capacity meant pilots could enjoy extra engine power for almost an hour.

This version of the *Gustav* was the Bf109G-1, produced from late spring 1942 in parallel with the G-2 lower altitude version, without the pressurized cockpit or GM-1. The pressurized versions were used by *JG2* and *JG26* in France against Fighter Command while most unpressurized machines went to the Eastern Front. Wider section tyres to strengthen the undercarriage needed bulges to the upper surface of Messerschmitt's uncompromising wing as internal space was so limited.

The Bf109G-6 replaced the cowling guns by 13mm weapons to make up for nose cannon failures. These also needed bulges in the engine cowling for the guns' breechblocks. Either or both changes earned the nickname of '*Beule*' or 'bulge'. The '*Kanonenboote*' or 'Gunboat' 109Gs carried extra cannon in underwing gondolas or the 88lb, 210mm *Werfgranate* (RPG) tube-launched air-to-air rockets in the specialized *Pulk-Zerstörer* ('Formation Destroyer') variant.

How did the *Gustav* affect aerial combat against the Allies? Improvements like replacing the 20mm nose cannon by the 60-round 30mm MK108 cannon transformed its combat power, as a single hit could bring down any enemy fighter. In combat against Spitfire Vs the new 109 was faster both climbing and diving, so if a German pilot resisted the temptation to enter a turning dogfight, at which Mitchell's fighter was still the best, he retained the advantages of the *Friedrich*. He could still enter a dive with a negative-g bunt thanks to fuel injection. However, the more powerful engine and armament marred the 109's agility and handling.

Change was also on the way for British fighters through another clever stopgap to match the 109G's improved high-altitude performance. Spitfires needed more powerful engines to reach these high-level attackers and the Mark III Spitfire was fitted with the Rolls-Royce Merlin XX with two-speed supercharger delivering 1350 hp. This proved a dead-end as all Merlin XXs were used for Mark II Hurricanes, so the Mark III was shelved and the more powerful Merlin 45 transformed Mark I and Mark II Spitfires into Mark V machines.

By summer 1942, RAF Spitfires were losing air superiority to the Luftwaffe's new machines. Casualties had risen so sharply for no effect on the enemy that on 13 June Sholto Douglas was thankfully, and finally, ordered to stop his fighter sweeps which had cost 335 Spitfires in a mere four months. It was long overdue. With more intelligent tactics the casualty rate might be lowered and command of the air retrieved, but only with a new fighter with dramatically improved performance.

In February 1941 the Air Ministry wanted a pressurized Spitfire to take on higher-flying Luftwaffe raiders. They needed a pressure difference of 1 lb/sq.inch at 40,000 feet, using an external compressor and sealing the cockpit like the high-altitude *Gustavs* to reduce pilot stress at extreme altitude. The side door was replaced by sealed metal skin panelling and the sliding cockpit canopy by a fixed hood. However, sealing the canopy before take-off and only releasing it when back on the ground did little for pilot morale.

The new Spitfire Mark VI had a Mark V airframe and a Merlin 47, a Merlin 45 with its supercharger modified for better high-altitude performance and fitted with a four-bladed propeller. It first flew in June 1941 and several had extended wingtips to increase the wingspan from 36 feet 10 inches to 40 feet 2 inches for better high altitude handling. An extra fuel tank in the rear fuselage extended range by one-third and the first production machines went to 616 Squadron in April 1942, but overall it was a disappointment.

Limited engine power and the extra weight of the pressurized cockpit cut top speed at 20,000 feet to 356 mph, falling to 264 mph at its 38,000 feet operating altitude. In many cases modified Spitfire Vs could perform almost as well and the pressurized cockpit was too warm at lower altitudes. As increased incursions by high-flying German raiders dwindled, production of Mark VI Spitfires was limited to 100 machines for four RAF squadrons.

Unlike the changes behind the DB605, each new Merlin retained the same bore, stroke and capacity as the original with power raised by supercharger improvements. It was an astonishingly versatile design and the most important British aero-engine of the war, because 'The engineers at Derby relentlessly strove to wring from it the same level of power as the enemy obtained from engines half as large again.'[3] Though it originally suffered reliability problems, once basic questions of strength, efficiency and breathing had been solved, it delivered twice the power of comparable engines and held together reliably while doing it.

The Merlin switched from water cooling to pure ethylene glycol for early Hurricanes and the disastrous Fairey Battle light bombers, but the searching effect of the hot liquid let it seep through allegedly leak proof joints to set the engine on fire. In August 1938 they found a mixture of water and 30 percent glycol at a constant pressure of 18lbs/sq.inch at all altitudes eliminated leakage and reached 135 degrees C without boiling. Between 1937 and 1940 the weight of a Merlin radiator and coolant was halved, and 100-octane fuel from US suppliers allowed boost to be doubled to 12 lbs/sq.inch, raising sea-level power from 1030 to 1310 hp, a vital improvement against the appreciably lighter Bf109.

In 1938, Rolls-Royce hired a young aerodynamicist, Dr Stanley Hooker. He justified his appointment by redesigning the Merlin's air intake and supercharger to raise the pressure ratio and efficiency of the blower, and with it the peak power and full throttle height without any increase in the size or weight of the engine. In 1941, with the Battle of Britain won, a new engine was needed for Spitfires to meet 109Fs on more equal terms. The initial proposal was based on a high-level pressurized version of the Wellington bomber, the Mark VI. Carrying a pilot and navigator/bomb-aimer with no defensive armament, it needed good high-altitude performance to remain invulnerable to German fighters. This meant much greater boost. Rolls-Royce felt an exhaust driven turbocharger would spoil the radiator aerodynamics and reduce the thrust they contributed. Hooker suggested combining two superchargers in series, with an intercooler to increase the charge density entering the engine, as the Germans had done. For a slight increase in weight, the performance improvement was spectacular. Nine pounds of boost could be maintained to 30,000 feet, where maximum power increased from 500 to 1000 hp.

---

3. Gunston, *op. cit* , p. 154.

In a Spitfire this would raise the combat ceiling by 10,000 feet and top speed by 70 mph.

All engineers know there is no such thing as a free lunch and this radical solution came at a price. The problem was twofold. The extra supercharger would have to be much larger and the fuel-air charge fed to the engine would be far too hot. This might make the fuel detonate prematurely in the cylinders and reduce engine power to less than its predecessors!

The solution proved surprisingly simple. Rolls-Royce had developed a much larger engine, the Vulture, from two V12 Peregrines (a Kestrel development produced in small numbers for the Westland Whirlwind) joined together with a common crankcase and crankshaft to make a 42.47 litre X24 delivering 1780 hp. This was cancelled as unreliable, but its two-stage, single-speed supercharger was ideal for the first stage of the high altitude Merlin. This would deliver a fuel-air charge at a blistering 205 degrees C entering the second stage supercharger, which would raise the temperature still further. Hooker placed the intercooler, a water-fed heat exchanger, between superchargers and engine to reduce charge temperature. Surplus heat would be removed through liquid coolant, but there was a trade-off between the amount of intercooling and additional radiator area, which would add extra drag.

Tests showed increasing intercooling raised peak power with it up to a maximum of 35 percent intercooling. Finally, 40 percent intercooling was chosen as a compromise. The new Merlin 60 performed well on the bench and the engineers calculated its full-throttle height would be 30,000 feet, a massive performance boost.

Eventually, an experimental Wellington was fitted with two Merlin 60s and flight tests confirmed the predictions. At this point, Works Manager Ernest Hives suggested the engine might fit a Spitfire, a much more complex objective. Because double superchargers lengthened the Merlin by nine inches, the Spitfire nose had to be redesigned with revised engine bearers and mountings. Intercooler water was led through a second radiator to remove excess heat, and the oil radiator under the starboard wing was made to cool both intercooler and engine oil in a similar installation to the radiator under the port wing. This counterbalanced the extra weight of the new Merlin, which was fitted with two-piece cylinder blocks and modified supercharger gear ratios.

The new split cylinder blocks were designed in 1938, but because of production pressure only appeared in 1941 on the licence-built US Packard Merlin, and a year later on the UK Merlin 61. They spread loads more evenly and stiffened the engine structure to cope with increased power, and eliminated persistent coolant leaks. Packard also pioneered water injection, boosting peak power of the V-1650-11 of 1944 to a staggering 2270 hp. In late 1943 the maximum boost of the Merlin 66 was raised to 25 lbs/sq.inch for up to five minutes at a time, increasing sea-level power to 2000 hp, enough to let the new Mark IX Spitfires catch and shoot down German V-1 'flying bombs' [actually cruise missiles] launched against Britain from summer 1944.

The new Merlins were mass produced in a factory on Manchester's Trafford Park industrial estate converted by Ford into an engine plant. By 1943 this turned out 200 Merlins a week to Ford's highly exacting tolerances, considerably tighter than those of Rolls-Royce! By the end of the war, the production time for each engine had fallen from 10,000 hours to 2,727 and the unit cost from £6,540 to £1,180, while the overall rejection rate remained at zero, a staggering achievement.

A two-stage Merlin was fitted into the Mark III Spitfire prototype and flown on 20 September 1941. The top speed reached 414mph at 27,500 feet and 354mph at 40,000ft, 90 mph faster than the Spitfire Mark VI! The extra power needed a four-bladed propeller, while the Merlin 66 version powered the new generation Spitfires and also (in Packard-Merlin 266 form), the long-range P51 Mustang. These engines promised to snatch back control of the air war, but the biggest problem would be bringing them into service as quickly as possible.

The original plan was to use a Merlin 61 in an uprated Mark VI pressurized cockpit Spitfire, the Mark VII, together with an unpressurized Mark VIII stablemate. However both versions needed detail modifications like wing-root fuel tanks to increase range by 50 percent, four wing cannon, a stronger fuselage for the heavier engines and revised ailerons to improve manoeuvrability.

Bitter experience with earlier Spitfires showed changes would delay the fighter's complex production system, but with engines already in production, could a simple shortcut speed things up? Another quick stopgap was needed, so two Spitfire VCs (with the stronger 'universal' wing) were taken off the production line. By strengthening the fuselage longerons and modifying the nose to fit Merlin 61s, they were transformed into Mark IX Spitfires. This also eliminated the fuel starvation problem when entering a dive, as the Merlin 61 and its successors had Bendix-Stromberg anti-gravity carburettors to eliminate the problem.

The first Mark IXs emerged from Castle Bromwich in June 1942. In July, one was tested against a captured FW190 by the Air Fighting Development Unit at RAF Duxford. They found both aircraft closely comparable with slight differences in speed and rate of climb. The Spitfire IX could climb faster than the FW190 above 25,000 feet, but the tables were turned when they tried diving, though the difference was less than between the 190 and the Spitfire V. 'The FW190 is more manoeuvrable than the Spitfire except in turning circles, where it is out-turned without difficulty. The superior rate of roll of the FW190 enabled it to avoid the Spitfire IX when in a turn, by flicking over into a diving turn in the opposite direction and, as with the Spitfire VB, the Spitfire IX had great difficulty in following this manoeuvre.

Both aircraft 'bounced' one another in order to ascertain the best evasive tactics to adopt. The Spitfire IX could not be caught when 'bounced' if it was cruising at high speed and saw the FW190 when well out of range. When the Spitfire IX was cruising at low speed, its inferiority in acceleration gave the FW190 a reasonable chance of catching it up and the same applied if the position was reversed and the FW190 was 'bounced' by the Spitfire IX, except that overtaking took a little

Mark IX Spitfire from 402 Squadron RCAF. *(Via Alfred Price)*

longer.... The general impression gained by the pilots taking part in the trials is that the Spitfire IX compares favourably with the FW190 and that provided the Spitfire has the initiative, it has undoubtedly a good chance of shooting it down.'[4]

The first operational machines went to 64 Squadron at Hornchurch, followed by 611 at Kenley in July and two Canadian squadrons in August, then 401 at Biggin Hill and 402 at Kenley, followed by 133 Eagle Squadron at Biggin Hill in September. After training on the new machines, pilots found combat transformed. In particular, the Mark IX's rate of climb let them gain height and remain in a commanding position over approaching enemy formations. Time after time they bounced enemy 109Gs and FW190s to turn the tables altogether.

Group Captain Duncan Smith, commanding 66 Squadron, found Mark IXs could cruise in battle formation at 43,000 feet knowing they were invulnerable to any German attack. Time and again his pilots bounced Luftwaffe fighters and shot them down with relative ease. More and more pilots found the Mark IXs gave them a new immunity against the previously deadly FW190s. Against 109Gs it was the same story. Time and again the Spitfires would use their high altitude to make lethal diving attacks on German fighters and catch them up whenever the combat became a stern chase.

---

4. AFDU report, quoted in Price, *The Spitfire Story, Appendix D*, p. 169.

Wing Commander Al Deere, now commanding the Biggin Hill Wing on a bomber escort mission, found his Mark IX almost too fast when 'One FW190 which was positioning itself to attack the bombers, was jumped and a four second burst was fired, opening at 300 yards and closing to 100 yards. The overtaking speed was so great that I was forced to break to port. As I completed my turn I saw an FW190 in flames and a pilot coming down by parachute. I therefore claim an FW190 destroyed.'[5]

Commandant Rene Mouchotte of 341 Free French 'Alsace' Squadron, was flying over his homeland when he met an assortment of German fighters. 'I was leading 341 (FF) Squadron, and had just crossed the French coast at Cabourg at 23,000 ft. and was about halfway from the coast to Caen aerodrome, when an FW190 dived on us from slightly astern, passing just in front of me. I fired a short burst, but without any results as I must have used too much deflection. Shortly afterwards I saw four Me109's in line astern coming from the south east, who fired on my Yellow Section slightly below me; immediately I dived on the leader, ignoring the others who were by now at a disadvantage in relation to my Yellow Section, and he broke away in a dive, turning to port. I followed and fired from 200 yards dead astern, closing to 100 yards. Here he throttled back still turning; repeating the same tactics, I closed to within 50 yards, still firing, when he took no more evasive action, but dived sharply, crashing into the ground without any trace of smoke or flames.'[6]

Fighter Command pilots revelled in their changed fortunes, but better still was to come. Though the Mark IX was mainly intended for high altitude combat, reports showed more dogfights were happening around 20,000 feet where the Mark IX's engines and superchargers were barely reaching peak performance. So the Mark IX range was widened. The original became the Mark IXA, powered by a Merlin 61 or 63 and fitted with the C-type 'universal' wing for different combinations of armament. This was later reclassified the F (for Fighter) IXC (for 'C' wing) and was joined by the HFIXC fitted with the 'C' wing and the Merlin 70, tuned to deliver 1655 hp at 3,000 rpm with 18lbs of boost at 10,000 feet. Finally, the original Mark IXB was redesignated the LF IXC, which implied a low-altitude fighter similar to the Mark V. This was misleading as its Merlin 66 reached its peak at 22,000 feet rather than the 28,000 feet of the first production machines, with supercharger gear ratios amended to give a wider margin of superiority over the enemy.

Other improvements included new gyro gun sights for more accurate deflection shooting. A 1944 trial showed gunfire effectiveness had almost doubled.

On 29 December 1944, a Canadian pilot of 411 Squadron was flying over the Luftwaffe fighter field of Rheine/Hopsten near Osnabruck where he first met the

---

5. Combat report, Wing Commander A. C. Deere, 4/5/1943.
6. Combat report, Commdt R. Mouchotte, 15/5/1943.

enemy. Four years earlier, his survival chances would have been poor. In this case he saw four Me109s and eight FW190s, and his squadron plunged into the attack.

His first target was 200 yards away and 20 degrees to starboard when he fired. He reported 'At 10,000 feet I opened fire and saw strikes all over the fuselage and..., wing roots. The 109 burst into flames on the starboard side of the fuselage only, and trailed intense black smoke. I then broke off the attack.' He turned his fire on to the nearest FW190, and once again his fire was deadly accurate. 'I saw strikes over cockpit and to the rear of the fuselage. It burst into flames from the engine back, and as I passed very close over the top of it I saw the pilot slumped over in the cockpit which was also in flames.' Seconds later, he shot down another 109 and saw the fighter 'hit and smash into many pieces on the ground.' He managed to shoot down another 190 then on the tail of a comrade, followed by yet another 190 making a head-on attack. Both fighters opened fire, but the Canadian's aim was accurate enough to send his blazing target diving into the ground.[7]

RAF fighters were now using a new formation called the 'Fluid Six' from experience in both Europe and North Africa, and described as 'The best fighter formation of the war.' It was based on the 'Finger Four' of the Battle of Britain, but moved on from the leader-wingman combination used by both sides, to use three pairs of Spitfires for mutual cover. One pair flew higher and up-sun of the leading pair while the third pair flew lower. Any enemy aircraft attacking the formation, from whichever height or direction, would find themselves surrounded by their RAF adversaries. Furthermore, a six-fighter element was better suited to larger squadron formations, now appearing in greater numbers.

The stronger 'E' wing solved the problem of distortion under tight combat turns compromising the accuracy of the machine-guns. Moving the cannon to the outboard positions and replacing the four .303 machine guns with two heavier and more accurate .5-inch guns greatly magnified the Spitfire's weaponry. A larger rudder improved take-off control when the new engine's extra power increased the drift to port. Finally, later Mark IXs had cut-down rear fuselages and streamlined 'bubble' cockpit canopies to transform the pilot's rear vision.[8] This was first tried on a Mark VIII, and sorting out a tendency to mist up and stiffness when opening and closing the hood, made it very popular with pilots.

When Mark IXs went to squadrons in Malta and North Africa, results were similar. Squadron Leader R.W. 'Bobby' Oxspring commanded 72 Squadron over Tunisia when his comrades dealt with 109s with surprising ease. 'Covering a squadron of Hurricanes bombing close support targets, 72 intercepted some Me 109s which tried to interfere. We had the jump of them as Danny took his flight

---

7. Canadian Ministry of Defence, *The RCAF Overseas: The Sixth Year*. Technically, this made him an ace after 52 sorties when he met no enemy aircraft at all!
8. Price, *op.cit.*, p. 155.

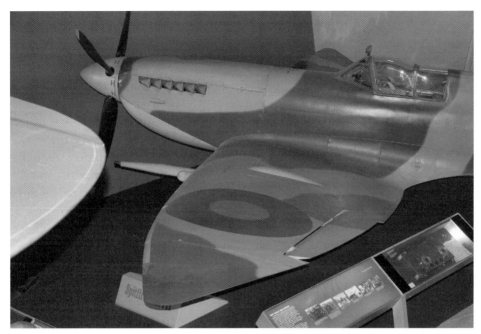

Extended wingtips of high-altitude Spitfire Mark VII. *(Author photo from Smithsonian)*

High-altitude Spitfire Mark VII being test flown at Langley, VA, USA. *(US official photo)*

Restored Spitfire Mark VIII taxiing at a Duxford flying display. *(Author photo)*

down on a formation of four. He and George Malan moved in to clobber a couple which crashed nearby. As the rest of the squadron gave cover, another half-a-dozen 109s appeared beneath and I led the formation down in a copybook bounce. It was a dream set up as we closed range and blasted our targets. Mine shed some bits and hit the deck east of Medjez, as Alan Gear worked over another which spun away. The sudden impact split the rest of the enemy, and in the ensuing mill the squadron damaged three more.'[9]

In the meantime, the Spitfire Mark VII, essentially a re-engined Mark VI, enjoyed a similar performance boost. Its cockpit hood could be slid backwards in the normal way, but preserved pressurization when fully closed. Most had a normal span wing, since the extended-span version reduced the roll rate by 40 percent. Extra leading-edge fuel tanks and larger fuselage tanks held another forty gallons to boost range to 660 miles, though additional forty-five gallon belly, tanks which could be jettisoned on meeting the enemy, were also used on specific missions.

However, the Mark VII's performance had already been overtaken by the high-altitude Mark IX. The first appeared in August 1942 and one per week was produced over almost eight months. The final production series in early 1944 used the Merlin 71 which delivered peak power at 10,000 feet and gave a top speed of 424 mph at 30,000 feet. By then its role had passed to the unpressurized Mark VIII as German high-altitude attacks had ceased. Total production amounted to 140 aircraft, serving with seven squadrons.

---

9. Combat report, Squadron Leader R.W.Oxspring, 11/4/1943.

They proved unexpectedly useful as greater fuel tankage let them fly as top-cover escorts on occasional daylight raids by RAF bombers. On 11 August 1944, 131 Squadron's Mark VIIs defended Lancasters bombing U-boat pens at La Pallice. The target was at extreme range so the Mark VIIs stayed well above the bombers, cruising on lean mixture and low throttle to conserve fuel, in the hope that the Luftwaffe would be deterred by their presence above. Probably because of the shift in combat power resulting from the Mark IXs, the trick worked and the mission was successful.

After the Mark IX stopgap, production switched back to the Mark VIII. This had similar improvements and almost identical performance, but detail modifications made it particularly pleasant to fly. It entered production at different Supermarine plants, rather than the IX's Castle Bromwich. Though never as successful numerically, more than 1600 Mark VIIIs served in North Africa, the Mediterranean, Italy, the Far East and Australia, where they had the same dramatic effect on enemy raiders.

Extra variants were soon added. With the Normandy invasion of June 1944, more detailed photo coverage of the invasion coast was needed and with Bomber Command's more destructive raids on German cities, damage assessment pictures were also needed. To produce machines for high-speed photoreconnaissance missions, guns were removed and vertical cameras behind the pilots' seats were

Spitfire Mk XIV of the Commemorative Air Force (USA). *(Photo released via Wikipedia)*

fitted to convert fifteen Mark IXs into PR IXs. These were followed by PR Mark XIs, some with shorter focal-length cameras under the wings for low-altitude missions and better coverage in cloudy conditions. Removing weapons cut drag and weight, giving a top speed of more than 400 mph at heights from 24,000 up to 32,000 feet. Two used for fast diving trials reached speeds of around 600 mph or almost 90 percent of the speed of sound, an extraordinary achievement for a Second World War piston-engined fighter, and testimony to Mitchell's inspirational design. In one case the supercharger burst and in the other the reduction gear failed, but the airframe coped with the huge stresses, an achievement almost certainly beyond the strength of the Bf109!

Two more reconnaissance Spitfires used Merlin engines. Sixteen Mark Xs, essentially pressurized Mark XIs were issued to two squadrons. Finally twenty-six earlier-mark Spitfires were fitted with Merlin 32s with cropped supercharger impeller blades for faster low-altitude performance, to make high-speed low level passes over the invasion beaches.

Finally, a Mark IX seaplane was built in Spring 1944 for use against the Japanese. This had the same cantilever supported floats as the Mark V seaplanes and an extended engine filter to prevent it ingesting water. Like the Mark Vs, its fin extension below the tail improved longitudinal stability, and its performance with a Merlin 61 was faster than a standard Hurricane, but the project was soon dropped. A two-seat trainer with the instructor sitting in a second cockpit, was built originally for the Soviet Air Force and supplied to Eire and India after the war.

The Mark IX and its Mark XVI American stablemates proved highly successful, but more power and performance would soon have to re-emphasize RAF superiority over the enemy. Unlike Rolls-Royce's previous policy, this would need a larger engine rather than a simple change of supercharger, so they would step back to the mighty Rolls-Royce 'R' racing engine from the 1920s to power the fastest and most successful fighting Spitfires of all.

Chapter Twenty-Four

The *Kurfürst* and the Griffon Spitfires

By autumn 1944 the Luftwaffe faced two almost insuperable problems. The Allies now had more formidable fighters and better trained pilots. On the German side a chaotic jumble of different versions of the Bf109G made replacements and repairs all but impossible, compounded by attacks on production plants, the transport network and the squadrons themselves. Two years of *Gustav* production had seen eighty different variations on the original 109G design, while production co-ordinator Werner Göttel estimated around a thousand detail changes had been made in the first year of production alone.[1]

The final production version of Messerschmitt's fighter, the Bf109K *Kurfürst*, was intended to simplify the system with three basic variants. The Bf109K-1 used the Daimler-Benz DB605DM engine with MW-50 boost, delivering 2000 hp at take-off and 1800 hp at 16,000ft. Three Bf109K-2 prototypes had the DB605DB without methanol and water injection, and the third version would be the 109K-3 reconnaissance fighter, but even this simple breakdown was too ambitious at this late stage.

The revision process was begun early in 1943 at the Wiener Neustadt factory in Austria by 140 technical specialists. Unfortunately, on 13 August 1943, the plant was heavily bombed by the Americans and some 500 workers were killed. The project was rescued by retrieving vital production drawings and switching output to the Messerschmitt plant at Regensburg, where the first production machines finally emerged. These were 109K-4s with the DB605DM engine and a reshaped cowling to accommodate a deeper oil cooler. Wider section tyres tamed the ground handling, but with large wing bulges for the landing gear to retract. To reduce the ground angle and give the pilot a better view, the tail wheel had a taller retractable leg to raise the tail and shorten the take-off run, and also reduce drag when airborne.

The old 7.7mm cowling guns were replaced by heavier 12.7mm weapons which meant still more bulges, and the central 20mm cannon switched to the massive 30mm MK108. The engines used larger blowers from the DB603 bomber engine. The DB605DM and DB605DB had MW50 methanol and water injection on

---

1. Quoted in Ebert, Kaiser and Peters, *op.cit.*, p. 137.

Tight fit: Griffon engine in the nose of a Spitfire 24. *(Author photo)*

87-octane fuel, but the DB605DC production version ran on 100-octane. The propeller had three wide-chord blades like the later *Gustavs*.

Other features included a taller, fabric-covered wooden fin and rudder for greater longitudinal stability. Early 109Ks had undercarriage doors to cover the lower part of the wheels when retracted, but these were usually removed by the units operating them. In all, some 1600 machines were produced at Regensburg up to the end of March 1945 after which records are incomplete. An additional 109K-6 version was tested in late 1944 at the *Erprobungstelle* at Tarnewitz, fitted with two additional 30mm cannon mounted inside the wings, rather than under them in gondolas, reviving the arrangement used for the 20mm wing cannon in the *Emil*. Though the MK108 packed a heavier punch, its rate of fire and muzzle velocity were lower than the 20mm weapons. It was easier to manufacture from a relatively small number of stamped and machined parts, but was heavier, by a factor of 2.5, and sixty rounds of heavier ammunition made things even worse. Pilots also criticized its tendency to jam.

No production figures for this version survive. A few 109K-10 and K-14s were said to have been made, the latter a high-altitude variant with a top speed of 452 mph at 38,000 feet thanks to a DB605L high altitude engine, fitted with a two-stage supercharger. Finally the Bf109K-12 was a two-seat trainer similar to the existing variations on the *Gustav*. The original DB605A produced 1475 hp with 87-octane fuel and no power boost, or 1770 hp with 100-octane fuel and MW50 boost in the DB605AM. When fitted with the DB603 bomber supercharger, it became the DB605AS (109G-5, G-6 and G-14) and with MW power boost it became the DB605ASM. With higher compression ratio and improved boost regulation, upgraded supercharger hydraulic drive and a higher pressure oil system, the engines became the DB605D, or with MW50 power boost, the DB605DM.

The reshaped cowling, retractable tailwheel and wheel well doors increased top speed by 22 mph over later *Gustavs*. Non-strategic materials such as steel sheet or wood were considered to ease shortages and the wing was tested to find the best way to incorporate them. Wooden wings were fitted with integral MK108 cannon, but after tests in August 1944 with a Bf109G-6, the idea was dropped. Instead metal wings were used, but with ailerons made entirely from duralumin instead of being fabric covered. However the leading-edge slats were made from steel sheet. A mechanical undercarriage indicator was located on rib three of both wings, replacing the traditional indicator sited below the starter switch on the port side of the cockpit. Fin, rudder, tailplane and elevator were all made from wood, while the rudder now had a ground-adjustable *Flettner* trim-tab set midway down its trailing edge.

The revised cockpit layout included a full blind flying instrument panel with a variometer, or rate of climb or descent gauge. In addition the pilot's seat bucket, the instrument and electrical panels, and the wheels for adjusting flaps and trim were made from moulded plywood. Oxygen bottles were shifted from the port wing to the starboard to leave space for the compressed air bottles for the MK108 cannon.[2]

As with most late war 109Gs, Ks were delivered without antenna masts, which were fitted by the squadrons. However, the supply chaos meant many Ks departed from specification. Short fixed tailwheel legs were fitted rather than the longer retractable version, and metal tail units without the rudder trim tab were often fitted instead of the wooden version because of shortages or interruption of supplies. None of the Ks carried the WGr.21 rocket tubes for use against bombers. Both BFW and the German Air Ministry referred to the *Erla-Haube* [Erla hood], also known as the Galland hood or the *Vollsichtkabine* (High-visibility cabin), as limited to the 109K-4, but it was also widely used on the G-6, G-8 and G-14.

In the final analysis, the most remarkable truth about both the Spitfire and the 109 is the similarity of their performance after five years of war. In spite of newer and more promising designs like the FW190 on the German side and the Mustang, Typhoon and Tempest on the British, they were still fast enough, powerful enough and heavily enough armed to remain in the front line of the air war. But the background to the fight was tilting irretrievably away from the Luftwaffe.

Increasingly heavy American daylight raids, backed by Mustang long range escorts, were taking a terrible toll of German fighters, a problem compounded by

---

2. Fernandez-Sommerau, *op.cit.*, p. 76.

their tours of duty. Because Germany relied so heavily on high-scoring aces, they had no time to pass on their matchless skills to young men following them through training school and on to fighter units. Worse, there was a shortfall of instructors to give new pilots any inkling of what they would face on their first combat missions.

The problems which worried Dowding and Park in the Battle of Britain, finding sufficient replacement pilots and helping them to survive meeting the enemy, were now crushing their opponents. Though fighter production figures reached new heights, wastage increased too. Rising combat losses and increasing bomb damage to assembly lines, and smaller plants producing essential parts, and transport links between the two, suggested the future for later 109Ks and their pilots was growing ever darker.

Ironically, at the very point when Luftwaffe fighters faced their greatest challenge, a new mark of Spitfire was about to appear through a far more radical transformation than between the *Gustav* and the *Kurfürst*. First, it involved the replacement of the doughty Merlin with the mighty Rolls-Royce 'R' of Supermarine's racing seaplanes, in the form of the Griffon engine.

This promised a huge performance boost. With bores up from 5.6 to 6 inches, and stroke up from 6 to 6.6 inches, the capacity increased from 27 to 36.7 litres, a jump of more than a third, to reach close to the DB605's 35.7 litres. Yet it was almost the same size as the Merlin. Its origins dated back to a Fleet Air Arm request of 1938 for a larger Merlin, reliable and easily maintained, but with good low-altitude performance. Since experience with the Kestrel and the Merlin had simplified the design process, the first prototype of the new engine began bench testing at the end of November 1939. Though still intended for the Fleet Air Arm, an Air Ministry official suggested it might be possible to fit the Griffon into the nose of the Spitfire.

To simplify the task, Supermarine relocated some of the engine auxiliaries to minimize frontal area, cutting it to just 7.9 square feet compared with the Merlin's 7.4 square feet, an increase of just 7 percent. Remarkably, its overall length was just three inches longer than the Merlin, and additional weight was limited to less than 600lbs. This was achieved by connecting drive trains for camshaft and magnetos to the propeller drive reduction gears at the front of the engine rather than extra gear trains at the back. The twin magnetos were replaced by a new dual magneto fitted on top of the propeller reduction gear. Shortening the gear trains made the engine more efficient and more reliable, and lubricating main and big end bearings through the hollow crankshaft distributed the oil more evenly to each bearing.

Ironically, the promising new engine languished on the orders of Lord Beaverbrook, Minister for Aircraft Production, while all priorities were directed towards improving Merlin performance, particularly high altitude power, to match changed Luftwaffe tactics from the close of the Battle of Britain. The result was that the newer Merlins were already delivering more power than the Griffons, but the larger engine promised more for the future.

Dawn of a new age: Messerschmitt 262 swept-wing twin-jet fighter. *(USAF)*

Its fortunes were finally transformed by another change in German tactics. From early 1942, high level attacks by Luftwaffe fighters were backed by low-level raids by fighter-bombers on British coastal targets. These gave little radar warning and by the time defending fighters had responded, the Bf109s and FW190s were on their way back to bases in France. What was urgently needed was a much faster low-altitude Spitfire, capable of catching and deterring these damaging nuisances.

Yet another successful stopgap did the trick. The first Spitfire with the larger engine used a standard airframe fitted with steel main longerons running along the fuselage sides from the engine mountings towards the tail. Stronger engine mountings coped with the extra weight, and the first of two Mark IV prototypes, DP845, was flown by Supermarine test pilot Jeffrey Quill on 27 November 1941. He was delighted by its low level performance, transformed by the new engine and it was one of several candidates he judged to be his favourite Spitfire.

Even at this stage the Spitfire was still threatened with replacement by inadequate designs. To help reach a decision, they staged a high-speed comparison of a Spitfire, a captured FW190 and the latest potential Spitfire replacement, the Napier Sabre powered Hawker Typhoon. This would take place at Farnborough where Jeffrey Quill was ordered to supply and fly a Spitfire. After a brief conversation with Joe Smith, they agreed he should take DP845.

The pilots would fly from Farnborough to the airfield at Odiham some ten miles to the west. They would then form in line abreast at 1,000 feet, heading back towards their starting point and at an agreed signal they would apply full power and race back to the finish line. In Jeffrey Quill's words, 'All went according to plan until we were about halfway between Odiham and Farnborough, and going flat out. I was beginning to overhaul the FW190 and the Typhoon. Suddenly I saw sparks and black smoke coming from the FW190's exhaust and at that moment Willy [Wilson, pilot of the FW190] also saw them and throttled back his BMW engine. I shot past him and never saw him again. The Typhoon was also easily left behind. The eventual finishing order was the Spitfire in first place with the Typhoon second and the FW190 third. This was precisely the opposite result to that expected, or indeed intended.'[3]

True or not, the Spitfire's runaway win boosted its prospects. Immediately Quill landed at Farnborough, Sholto Douglas insisted Fighter Command pilots try the Griffon Spitfire and report back to him on performance and handling. An Air Ministry meeting the following week discussed the new Spitfire's performance, particularly its ability to outclimb the FW190 up to 12,000 feet. Plans already envisaged fitting the same twin superchargers and intercooler as the 60 series Merlins, but for the time being the simpler, single carburettor version of the engine gave the Spitfire the speed to tackle the Luftwaffe's low-level nuisance raiders.

So was born the Spitfire Mark XII, basically a strengthened Mark V airframe powered by the IIB version of the Griffon with a single two-stage supercharger providing 1730 hp at just 750 feet, ideal for catching incoming raiders. With clipped wings and the standard float-type carburettor, its top speed of 355 mph at 5,700ft rose to 397 mph at 17,800ft. It was armed (after discussing heavier options like six 20mm cannon or twelve .303 machine guns – the inspiration of Sholto Douglas) with two cannon and four .303 machine guns. An order was placed for 100 Mark XIIs for issue to 41 Squadron at Hawkinge in April 1943 and 91 Squadron a month later.

This made the Spitfire XII one of the fastest low-level fighters in the world, but exploiting this performance took time. First the Griffon rotated opposite to the Merlin, so that as the pilot took off the fighter would drift to the right instead of the left, demanding instant rudder correction. Even when pilots learned to handle their new machines confidently enough to use their performance, bringing the enemy to battle was far from straightforward. Spitfire XII units had to fly standing patrols over the coast, ready to dive on the enemy when they spotted them, but the Luftwaffe refused to co-operate. Finally, six 91 Squadron Spitfires attacked a larger force of FW190s attacking Folkestone. Bouncing the enemy, the Spitfires achieved almost total surprise and claimed six FW190s for no loss.

This was promising, but the impossibility of predicting attacks made a repeat unlikely. The possibility of tackling enemy fighters on a regular sweep over northern

---

3. Quill, op. cit., p. 234.

Hawker Tempest V. *(Author photo)*

France found the Germans refusing a turning dogfight where the Spitfires held all the advantages. Furthermore, the Spitfire XIIs might have to climb to higher altitudes where they no longer had a performance advantage.

One particular action showed what a formidable prospect this stopgap fighter could present to the enemy. Both squadrons joined a fighter sweep over northern France on 20 October 1943. Near Rouen the equivalent of two squadrons of Bf109Gs and FW190s manoeuvred into position above them with the sun at their backs in an apparently perfect 'bounce'. So often this led to carnage as the diving attackers hacked down their opponents before they reacted. This time they spotted the German fighters in time to turn inwards to meet their attackers head-on and force them to steepen their dives and lose their targets. Normally the enemy would dive out of danger, but this time they began a turning dogfight. Here Spitfires had no equal, and the fight ended with British pilots claiming to have shot down eight opponents. Whatever the truth of those claims, no British fighters were lost.

Afterwards, the Luftwaffe learned its lesson and left the Mark XIIs no chance of repeating their success. Instead they were switched to different high-speed, low-level attacks, where their targets could take no evasive action. These were the V1 flying bombs or 'Doodlebugs', primitive pulse-jet powered cruise missiles following a pre-set course, diving at the end of a timed flight to detonate almost a ton of explosive somewhere in London or its suburbs. Their cruising speed was around 350 mph between 3,000 and 4,000 feet, offering a real challenge to fighters. The

Replacing the BMW radial engine of the FW190 with a liquid-cooled inline power unit produced the fast and formidable FW190D '*Dora*'. *(USAF)*

first-line defence along the coast and Thames estuary massed batteries of radar-aimed heavy anti-aircraft guns to put up box barrages of proximity-fuzed shells and destroy as many missiles as possible.

Next came interceptor fighters, fast machines like Griffon-powered Spitfires, Hawker Tempests and de Havilland Mosquitoes directed to their targets by radio from the Observer Corps. With enough warning and a height advantage, they could accelerate in a shallow dive and get close enough to their targets to detonate their warheads, but not so close as to fall victim to the explosion, or to bring it down on people and property on the ground.

Launches went on day and night, and night time chases relied on the flames of the missile's exhausts making an unmissable target. On at least three occasions, skilled and brave pilots flew alongside the missiles to bring their nearest wingtip to within six inches of that of the missile. Then the airflow over the fighter's wing would tilt the missile far enough for its gyros to topple and send it plunging into the ground.

The V1 campaign began in June 1944 a week after D-Day and lasted until early September when advancing Allied troops pushed the missile launchers out of range. Thousands of V1s hit London, causing a sharp drop in morale. Nevertheless, fighters brought down almost 1700 of the missiles and barrage balloons, anti-aircraft guns, bad weather and mechanical malfunctions meant that only some 2,500 reached their objective. Finally, the Spitfire XIIs of 41 and 91 Squadrons were stood down and it was time for their role to be taken over by yet another brilliant compromise.

So far the Griffon had used a single, two-speed supercharger, the effective equivalent of the Merlin 45 in the long outclassed Spitfire Mark V. The next step was

What the Luftwaffe was up against: A Huge B17 production facility. *(USAF)*

to replace the single blower with a twin blower and intercooler combination like the Merlin 61 series. This had entered the script for future marks of the fighter, with a revised larger-area wing profile to eventually appear as the Mark 21 in the autumn of 1942. This proved unrealistically optimistic. Production delays with the two-stage Griffon 61 meant the first engines would be available from autumn 1943, so the Mark 21 sank to a longer-term priority. Then engine development accelerated, so the first six pre-production power units would be ready by the end of 1942, creating an opportunity and a problem. They needed fitting into an airframe as soon as possible to eliminate the inevitable bugs. Waiting for production lines to tool up for the Mark 21 at Castle Bromwich would take too long, so what else could fill the gap?

The answer was the delayed Mark VIII intended for the Merlin 61, diverted to the Mark IX and now in production at Supermarine. Why not use these straight away? The strengthened airframe posed no problems and the Griffon 61 promised a bigger performance leap than the Merlin 61. Where the single-stage Griffon turned out 1730 hp at low level, the two-stage Griffon 61 delivered 2035 hp at 7,000 feet and 1820 hp at 21,000 feet. So much extra power needed a five-blade propeller and detail changes included placing the oil pressure pumps inside the sump and eliminating most outside pipework. Moving the supercharger drive from the front of the engine to the back cut the length of the drive shaft and made a more robust connection.

Why the Luftwaffe was doomed to defeat: an equally huge B24 production line. *(USAF)*

The result was the Spitfire Mark XIV, which Jeffrey Quill first flew on 20 January 1943. 'It had a spectacular performance, doing 445 mph at 25,000 feet with a sea-level rate of climb of over 5,000 feet per minute, and I remember being greatly delighted with this aircraft. It seemed to me that from this relatively simple conversion, carried out with a minimum of fuss and bother, had come something quite outstanding.'[4] Yet another remarkable stopgap, this would make life even more depressing for Bf109 pilots. The skies over the *Reich* were darkening with chickens coming home to roost: increasing combat losses against better trained Allied pilots, the breakdown of the training system to prepare new recruits to replace those casualties, and the draining away of fuel reserves for operations to be continued.

Like every aircraft ever made, the Spitfire XIV had shortcomings, though these were trivial compared with its splendid performance, and relatively easy to cure. A lack of directional stability meant pilots had to re-trim after every change in throttle setting, while the elevator needed re-trimming more often than with the Mark IX. The heavier engine also meant more care was needed in ground handling and in the landing approach, where it tended to sink more quickly than the Mark IX. It

4. *Ibid.*, p. 239.

burned fuel 25 percent faster than its predecessor, but as it was 35 mph faster at most heights, its range remained similar, though its endurance time was reduced. Its rate of climb was better, its dive speed was higher and its manoeuvrability much the same as the Mark IX, as were armament and armour protection.

When compared with the other front-line RAF fighter of the time, the Hawker Tempest Mark V, the Spitfire XIV was better in most respects above 20,000 feet, a staggering result for an aircraft which began its front-line career so much earlier. The Mustang held the prize for range, and dived faster than the Spitfire XIV, but the Spitfire climbed faster and was more manoeuvrable. All these comparisons, while encouraging for those who flew Mitchell's fighter as developed by Joe Smith, were less important than how it compared with its Luftwaffe opponents.

Compared with the FW190, the Mark XIV opened up a wider gap than the Mark IX had done. In the two narrow gaps, from ground level to 5,000 feet and between 15,000 and 20,000 feet, the Mark XIV was 20 mph faster than its adversary. At all other heights it was up to 60 mph faster. It could climb faster than the 190, dive faster and turn more tightly. Only in its unbeatable rate of roll did the German aircraft do better, and even then the Spitfire could get the best of it in combat. Against the formidable new 'long-nose' 190D, with its BMW radial replaced by the Jumo 213 in-line inverted V12 to boost power and performance, the Mark XIV could more than hold its own. Against the Bf109 *Gustav*, the Mark XIV could outclimb it, catch it up in a dive, roll faster, turn tighter and fly faster. Apart from a narrow height band around 16,000 feet where the Spitfire was only 10 mph faster, the 109G lagged by 40 mph at all other altitudes. Overall, this left RAF fighter pilots in an unusually secure situation against formerly formidable adversaries, and for their opponents, conditions could only get worse.

The first production Spitfire XIVs emerged from the factories in late October 1943. In January 1944, 610 (County of Chester) Squadron began operating with the new fighter, followed by 91 Squadron from the end of February. More squadrons followed in quick succession, in time for the Normandy invasion, and in September 1944 they began escorting US bombers on the first stage of their flights to German targets. As the Allied bridgehead expanded, squadrons moved to European bases in Holland and Belgium. Later squadrons flew the Fighter Reconnaissance version (FRXIV) with cameras as well as cannon and machine guns.

In some ways the Mark XIV was actually a little too good. With superb high-altitude performance, pilots found they were operating in an environment where Luftwaffe fighters were only found much lower down. This meant switching to ground attack missions, attacking German trains and airfields. When these provoked their opponents to react, the parlous condition of German pilot training and aircraft maintenance were shown up in pitiless clarity.

The reports of two pilots from 402 Squadron showed how much conditions had changed by 6 October 1944, just four months after D-day. Flight Lieutenant J.B. Lawrence was leading the squadron. 'We were scrambled after Huns coming in from Venlo – Wesel area. I was flying Red 1 and led the squadron south.... We climbed under a loose gaggle of 15 plus 109's when I sighted one 109 alone crossing in front of me. I turned into line astern and closed quickly. I fired one very short burst and the 109 went into a diving turn to starboard. I turned inside him and at about 20 degrees off 200 yards range I fired another burst of about 2 seconds. Strikes were observed on cockpit and engine. Pieces flew off, and white and black smoke poured out. The enemy a/c turned into a steep spiral to port. He dived into the ground two or three miles S. of Nijmegen. I saw no parachute. I claim one Me 109 destroyed.'[5]

Flying Officer W.H. Whittaker flew as his No 2 in the same action. 'When at approximately 17,000 ft. a Me 109 passed above and in front of my No 1 who turned towards the e/a while in a steep climbing turn. Red 1 however was not able to get a shot at the e/a which then passed directly in front of me turning and climbing steeply. I fired a two second burst from about 200 yards at 40°–50° angle off. The a/c seemed to shudder and stall, and went down in an almost vertical spin. The e/a then disappeared under my nose as I climbed to regain Red 1. F/Lt. Lawrence however saw the action and saw the strikes near the cockpit. He saw the e/a go down in the spin and hit the ground 5 miles south of Nijmegen near a small wood. I claim this e/a destroyed.'[6]

By this time the Luftwaffe was using nitrous oxide and methanol injection systems to boost engine power and fighter performance. Both the Bf109G-6 and the G-14 had these systems from the summer of 1944 to increase low-altitude performance, but they had their work cut out to match the Mark IX and Mark XIV Spitfires. By then, the Griffon 65 engines powering the Mark XIVs had been limited to 21lbs of boost pressure, because of main bearing problems, but this was soon increased to 25lbs boost.

Bf109 pilots faced even tougher restrictions. In the extreme danger of combat, pilots wanted the lifesaver of selecting emergency combat power, but experience showed that extra engine speed and power caused overheating, so they were forbidden to use extra power without the cooling effect of MW-50 injection. Since severe shortages of necessary tanks and chemicals meant injection was not always available, combat options for German pilots were severely limited. Even with the 109K, attempts to increase boost pressure caused frequent engine failures. In addition, by October 1944 when 402 Squadron met their Messerschmitts over Holland, the 109s were no longer in good condition. Hans Knickrehm of *I/JG 3* recalled the condition of new Me 109 G-14/ASs received by his group in October, 1944. 'The machines that were delivered were technically obsolete and of considerably lowered quality. The engines proved prone to trouble after much too short a time because the factories

---

5. Combat report, F/L J. B. Lawrence, 402 Squadron, 6 October 1944.
6. Combat report, F/O W. H. Whittaker, 402 Squadron, 6 October 1944.

had had to sharply curtail test runs for lack of fuel. The surface finish of the outer skin also left much to be desired. The sprayed-on camouflage finish was rough and uneven. The result was a further reduction in speed. We often discovered clear cases of sabotage during our acceptance checks. Cables or wires were not secured, were improperly attached, scratched or had even been visibly cut.'[7] At one time, sorting out these problems could have been left to capable ground crews, but too many had been transferred from the Luftwaffe to Army infantry units, to feed the insatiable appetite of the ground fighting as the Allies pushed back the German lines. Had the men still been available, the increasing intensity of bombing and strafing attacks on fighter bases made it virtually impossible to maintain repair schedules.

From November 1944, the Allies increased their attacks on the German synthetic fuel industry. Indications showed that at last they hit on what 'Bomber' Harris, Commander-in-Chief of Bomber Command, scornfully dismissed as 'panacea' targets, but which forced the Luftwaffe to defend them as the price of continuing the fight. American bombers stepped up the weight dropped on fuel targets from one-eighth to one-third, and even Harris allowed Bomber Command to increase their percentage of bombs devoted to the industry from 5.9 percent to 25 percent.

The effects of the onslaught took time to make themselves felt. Although the Germans still possessed large numbers of fighters, a combination of bad weather, restricted training and inexperienced pilots kept losses high. For example, I Fighter Corps claimed they had shot down 155 Allied aircraft during the course of the month. Their own losses, from a smaller defending force, amounted to 244 day-fighter pilots killed or missing by the month's end. December proved even worse. In vile weather, better trained Allied pilots could operate more consistently than their opponents, and were able to extend their attacks to rear area airfields, and rail transport. Losses to aircraft and pilots mounted. For example, Flying Officer Harry Walmsley of 130 Squadron was with eight other Spitfires over Burgsteinfurt near Rheine on the afternoon of 8 December 1944 when they spotted a train. 'We saw a locomotive with about ten trucks. We had made one attack on the loco and two on the trucks and were preparing to make another when about a dozen aircraft appeared from the east and they dived straight past us as if they were joining in the attack on the train. These aircraft had cigar-shaped drop tanks slung under the centre of the fuselage and I thought at first they were American aircraft. I then saw the crosses on the wings and I could see that they were Me 109s and FW 190s. A dogfight started with everyone milling round. After about five minutes I found myself alone. I saw another train pulled up in a station, so I went down and had a squirt at it and saw strikes on the locomotive. When I pulled up I saw a Spitfire in trouble. It was smoking and the undercarriage partly down. I joined up with it to protect it. There

---

7. Jochen Prien & Gerhard Stemmer, *Jagdgeschwader 3 "Udet" in World War II, Vol 1.*, (Schiffer Publishing Ltd., Atglen, PA, 2002), p. 365.

were five Spitfires there. I do not know what happened to the damaged Spitfire for suddenly six e/a, probably some of the ones I had first seen, came diving down out of cloud. They had obviously climbed and reformed after the initial attack. This second attack made from 10/10th cloud at 1,500 feet was obviously directed against the damaged Spitfire. Some of the others in the Squadron chased them off. I went for two which were making an attack. I made a quarter attack on one of them, an Me 109, closing to 300 yards and giving a two second burst with all guns. I saw strikes behind the cockpit; the e/a dived straight into the ground. I found I was being fired at by two e/a so I used full evasive tactics for about five minutes and finally got away into cloud. I landed at Heesch [RAF airstrip in Holland] as I was short of petrol and made my claim to the Intelligence Officer there. I then returned to base. I claim this Me 109 destroyed.'[8]

December saw the balance tilt still further in favour of the Allies. More and more attacks on German airfields destroyed many fighters on the ground, and as German-held territory shrank under Allied advances, even training flights with instructors and pupil pilots came under attack. Further diminution of fuel reserves restricted the retaliation German fighters could inflict, and RAF Spitfires found their opposite numbers easier and easier prey.

Increasing desperation finally forced the Luftwaffe to plan a massive low-level attack with around a thousand fighters to hit as many Allied airfields in North-Western Europe as could be reached in a single operation. The result, tagged *Unternehmen Bodenplatte* (Operation Baseplate) was launched on the first day of 1945, and achieved a high degree of surprise, but failed to reach its objectives. Losses were unsustainably heavy and many pilots were killed by heavy anti-aircraft fire, not only from guns protecting the Allied airfields, but also from German batteries protecting V2 rocket launch sites which had not been warned of the operation. Overall the Luftwaffe lost some 300 fighters, and the lives of more than 150 pilots. Allied losses were certainly larger, involving around 500 aircraft, but since most of these were destroyed on the ground, pilot casualties amounted to just over thirty. With fighter production reaching staggering levels, replacing the aircraft took between one and two weeks. For the Luftwaffe however, replacing their most experienced pilots would prove impossible, and much worse was about to happen to their hopes of survival.

In the middle of January 1945 the massive Russian attack on the River Oder began, which would finally see Soviet soldiers raise the Red Flag on the ruins of the *Reichstag* in Hitler's capital. More immediately it had the effect of diverting what was left of the Luftwaffe eastwards to help stem the tide and protect the defending troops. Just four *Jagdgeschwader* – *JG2*, *JG26*, *JG27* and *JG53* were left to guard the Western Front, and most attacks on the massed ranks of heavy bombers

---

8. Combat report, F/O Walmsley, 130 Squadron, 8 December 1944.

were abandoned. Priority for remaining fuel stocks also went to the Eastern Front squadrons, leaving Allied pilots with less and less to do.

Flight Lieutenant L.J. Packwood of 2 Squadron was flying with another Spitfire on a tactical reconnaissance mission east of Amersfoort in Holland on 1 January 1945. 'We sighted 2 Ju88s escorted by 30+ mixed Me109s and FW190s flying west. We turned up sun of the formation and attacked the last section. I attacked an Me109 from dead astern and above, the enemy a/c took no evasive action. I gave it a 5 sec. burst with cannon and machine guns, closing from 400 to 150 yards. I observed strikes on the cockpit and fuselage, the enemy a/c disintegrated, the starboard wing broke off and [it] flicked over on its back and hit the ground in flames at E.3596. This was also observed by F/Lt. YOUNG. I pulled vertically upwards and broke off the engagement. I claim 1 Me. 109 destroyed.'[9]

For Flight Lieutenant J.R. MacElwain of the same squadron on a similar mission on 10 February 1945, it was much the same story against a Bf109 pilot apparently unable to take skilled evasive action. 'I was flying as No 1 with F/O Jefferies on Tac/R [Tactical Reconnaissance] in ARNHEM and had pulled up to 8,000ft. when I saw 2 unidentified a/c at 200ft. flying S.W. along the Canal towards LOCHEM. I went down, followed by my No 2, in order to identify the a/c and recognized them as ME.109Gs with L.R.T. [Long Range tanks] They were camouflaged dark and light green with the dark crosses showing clearly. As I positioned for attack the section broke to port – one very sharply and the other more slowly, dropping their tanks. The latter was the one I attacked. As we did a series of turns I gave him several short bursts and he finally flew straight with full throttle, climbing slightly. I opened up and held him quite easily and from slightly on his starboard allowed ½ ring deflection, seeing strikes on his starboard wing; I closed to about 150 yards dead astern and saw further strikes on tail and fuselage. The E/A pulled up to 400 feet, jettisoned his hood and bailed out. The a/c went straight in and was seen blazing on the ground by No 2. The pilot's chute failed to open. I claim 1 Me109G destroyed.'[10]

By this time the German squadrons left in the west were running short of aircraft. So bad had conditions become that some were issued with older Bf109G-6s, probably requisitioned from non-operational training units. In the first week of April, signals revealed production of Bf109s had finally halted and the fight was all but over. Still the survivors battled on in conditions which beggared belief. To save the last drops of fuel, fighters were pushed to and from dispersal to the runways by exhausted ground crew or pulled by teams of oxen. By 12 April, the Spitfires of 2nd Tactical Air Force faced a mere forty-one Bf109Gs from the remnants of *JG27* and from *JG 26*'s sixty

---

9. Combat report, F/L Packwood, 2 Squadron, 1 January 1945.
10. Combat report, F/L MacElwain, 2 Squadron, 10 February 1945.

remaining FW190s, though many of these were the more formidable long-nosed FW190D *Doras*, powered by the Jumo 213 in-line V12s.

On 28 March 1945, three pilots from 130 Squadron tangled with FW190Ds and shot down all three. Flight Sergeant Clay reported, 'At about 1645 I saw aircraft approaching us as we were flying north. They were about 3,000 ft and they approached us to port. Red 1 called up and said "Watch these" and we broke round after the aircraft which were FW 190's of the long-nosed variety. I chose one and he started to turn going down and I went through cloud after him right down to about 200 feet. I got in behind him and opened fire and I saw strikes to the engine. I closed right into him, firing all the time. I broke away to one side and then I saw the pilot jettison his hood. He climbed to 500 ft and then he rolled over on his back and he bailed out, but the parachute did not open.'[11]

Flight Sergeant B.W. Woodman saw 'Aircraft, which I identified as long-nosed 190's passed us on the port side travelling in the opposite direction. We broke round into them and I found there were three of them in front of me. I picked on the last one of the three and climbed after him. I opened fire from about 800 yards and I saw strikes on the wing. The e/a rolled away.... I then got on to the second one and from 30 degree angle off, opened fire again from about 800 yards. I closed in to about 400 yards and fired again from dead astern whilst climbing. I saw strikes on the wing roots on the port side. There was a burst of yellow flame from where I had seen the strikes. The aircraft went over on its back and it went down out of control and I saw a trail of smoke right down until the aircraft hit the ground. It finished up in a field where it continued to burn and there was just a mass of red embers.'[12]

To make things even worse for the beleaguered Germans, they were being attacked even in the course of falling back to new airfields further away from the advancing Allies lines. *JG27* was relocating from Achmer, north of Osnabruck, to its new base at Störmede near Lippstadt on 19 March 1945 when it was hit, first by American fighters and secondly by the Spitfire XIVs of 130 Squadron. Flying Officer G. Lord, 'Saw a number of e/a circling the aerodrome at Rheine at about 1,000 ft. I went in behind one Me109 and closed very fast. The e/a took no evasive action and I opened fire with all guns from dead astern from about 200 yards closing to 50 yards. I saw strikes behind the cockpit. I overshot this e/a and I saw him crash land on the aerodrome.... After this I pulled round and saw another 109, but as I closed in I overshot him. The e/a was trying to turn so I pulled round on to him a second time and got behind him. The e/a was trying to do a tight turn. I turned inside him and fired from 200 yards. I saw strikes behind the cockpit and the

---

11. Combat report, F/Sgt. Clay, 130 Squadron, 28 March 1945.
12. Combat report, F/Sgt. Woodman, 130 Squadron, 28 March 1945.

machine blew up in the air. The pilot was able to bail out and I saw the parachute go down and finish up in a tree about a quarter mile to the east of the aerodrome.'[13]

Flight Sergeant G. Hudson 'Was flying Red 3 with my Squadron sweeping to Rheine-Osnabruck-Munster. We were at 12,000 ft when Red 2 (W/O Edwards) reported e/a orbiting the aerodrome at Rheine. I saw the e/a about 10,000 ft below and went down with my No 2 (W/O Miller.) We were the first in and we went for six 109s which were orbiting the aerodrome. I picked out one e/a and attacked from almost dead astern; opening fire from about 200 yards. I saw strikes on the jet tank and on the underside of the fuselage. There was a terrific burst of flame and the e/a went straight into the aerodrome and crashed.'[14]

Wing Commander George Keefer found, 'When we were at 12,000 ft between the aerodromes at Rheine and Hopsten e/a were reported below. I led the Squadron down and a dogfight began at deck level near Rheine aerodrome from which there was intense light flak. I found two Me109s going round in a turn. Eventually one straightened out and flew due east. I gave him a quick squirt from dead astern and saw strikes on the starboard wing. Closing in further I fired again and this time there were strikes on the top of the cockpit and I saw that the hood was dragging. The e/a slowed, pulled up and he stalled in from about 20 feet. I saw the e/a crash into a field.'[15]

By April, the one-sided struggle was all but over with Spitfires hunting down the last of their familiar adversaries over the countryside of north-western Germany. Experienced enemy pilots were rarely encountered and Luftwaffe losses were heavy. Flying Officer A.G. Ratcliffe of 402 Squadron 'Was flying Red 6 on patrol over LINGEN/RHEINE area when twelve plus e/a Me109s and FW190s were sighted flying westerly course at about 2,000 ft. We broke starboard into them. Red 5 went for the last one in a section of three and I closed in on the leader as they circled anti-clockwise. I opened fire at about 400 yards about 20 degrees starboard and held fire until 50 yards. I saw tracers going into cockpit and watched him go down. The plane on hitting the deck burst into flames. It was a Me109, did not see the pilot get out.'[16]

By now, RAF Spitfires were operating further east than ever before. Flight Lieutenant C.J. Samouelle of 130 Squadron was patrolling near Wittstock, northwest of Berlin on 20 April, 1945 when 'I heard Red 3 report two aircraft at 12 o'clock. I saw these two aircraft at about 6000 ft going in the same direction as ourselves. They were Me109s and they began to climb immediately. I opened up and gained height rapidly. I caught one of the e/a at 8000 ft and closed in and opened fire at 300 yards from astern and I saw strikes all round the cockpit and on the back of the e/a. There

13. Combat report, F/O Lord, 130 Squadron, 19 March 1945.
14. Combat report, F/Sgt Hudson, 130 Squadron, 19 March 1945.
15. Combat report, W/C Keefer, 19 March 1945.
16. Combat report, F/O Ratcliffe, 402 Squadron, 5 April 1945.

was a big red flash, white smoke came out and I found myself flying through debris. I had to pull up sharply to avoid hitting the e/a. When I was able to look again the e/a was in a flat spin and at 4000 ft the pilot bailed out. I saw the aircraft go down and crash in a wood.'[17]

Wing Commander George Keefer DSO DFC was leading the section 'When we spotted two Me109s slightly above us at 12 o'clock to us and going the same way. We opened up and caught up with the e/a. I picked one and opened fire at dead astern. I saw strikes on the fuselage. I gave him another burst at closer range whereupon the e/a caught fire, crashed into a field and exploded.'[18]

On 25 April 1945, Flight Lieutenant Ponsford of 130 Squadron flew over the airfield at Rechlin where ironically, the Bf109 prototype had first been demonstrated to the Luftwaffe a decade before. 'Yellow 4 (F/Lt Bruce) reported a/c taking off. I told him to go down and saw him attack an e/a without results. I then went down with my No 2 (W/O Coverdale) and I saw a 109G with its wheels down making a slight turn to port. I closed to about 50 yards and started firing at 20 degrees off. I saw strikes all round the cockpit, engine and wings. The e/a began to pour white and black smoke, and it rolled over slowly onto its back, crashed into some woods and exploded as it hit the ground.'[19]

One of the last combats between a Spitfire and a Bf109 took place on the final day of April 1945, when Squadron Leader Shephard of 41 Squadron was patrolling the Elbe bridgehead. 'I climbed to 6,000 ft and patrolled between two layers of cloud; AA fire started to burst a short distance away and an Me109 appeared through cloud. I chased the e/a and opened fire from approx. 400 yards, obtaining strikes and causing a thick trail of Glycol smoke to pour from underneath the e/a. He immediately jettisoned hood and dived for the deck; I followed giving him about three more short bursts and getting strikes each time. He finally crash landed heavily in a field a few miles N.W. of Ratzeburg Lake. I gave another burst on the ground causing wreckage to catch fire. No one appeared to get out of the wreck.'[20]

At this last gasp of the once mighty Luftwaffe, the only aircraft able to out-perform Spitfire XIVs of the RAF fighter squadrons were the latest products from Willy Messerschmitt's design team; the twin-engine, swept-wing Me262 which ushered in the jet age too late to make a real difference. Even here, experienced pilots could still catch and even shoot down these formidable machines.

As far back as 14 February 1945, Flight Lieutenant F.A.O. Gaze of 610 (County of Chester) Squadron, the first unit to be issued with the Mark XIV, had tried several times to catch an Arado 234 jet bomber without success, 'Then I did an

---

17. Combat report, F/L Samouelle, 130 Squadron, 20 April 1945.
18. Combat report, W/C Keefer, 130 Squadron, 20 April 1945.
19. Combat report, F'L Ponsford, 130 Squadron, 25 April 1945.
20. Combat report, S/L Shephard, 41 Squadron, 30 April 1945.

orbit at 13,000 ft. to clear off the ice on the windscreen and sighted 3 Me262s in Vic formation passing below me at cloud top level. I dived down behind them and closed in, crossing behind the formation and attacked the port aircraft which was lagging slightly. I could not see my sight properly as we were flying straight into the sun, but fired from dead astern, at a range of 350 yards, hitting it in the starboard jet with the second burst; at which the other 2 aircraft immediately dived into cloud. It pulled up slowly and turned to starboard, and I fired, obtaining more strikes on fuselage and jet which caught fire. The enemy rolled over on to its back and dived through cloud. I turned 180 [degrees] and dived after it, calling on the R/T to warn my No 2; on breaking cloud I saw an aircraft hit the ground and explode about a mile ahead of me....'[21]

On 13 March Flying Officer Howard Nicholson of 402 Squadron 'Was flying Yellow 3 on a fighter sweep in the Gladbach area when I sighted a Me262 at about 5,000 feet flying south west. He did not appear to see me. I broke and fired a 3 second burst from 250 yards line astern into his starboard wing and the base of the fuselage. Smoke poured out and pieces flew off the starboard wing. I kept firing, observing many hits and the aircraft tended to fall out of control, regaining slowly. At 2000 ft. he went into a sharp dive to port, but owing to the extremely heavy flak from Gladbach, I broke to starboard. I did not see him crash, but this is confirmed by the C.O. of 402 Squadron.'[22]

So at last the epic fight smouldered to its end, leaving Spitfires in possession of the field. Even with the Luftwaffe out of the fight, there still remained the other enemy still fighting on in South-East Asia, and more and more Mark VIII and Mark XIV Spitfires were sent there to end the air war as speedily as possible. By June 1945, they equipped three squadrons and more units were being sent to Germany as part of the forces of occupation.

---

21. Combat report, F/L Gaze, 610 Squadron, 14 February 1945.
22. Combat report, F/O Nicholson, 402 Squadron, 13 March 1945.

Chapter Twenty-Five

After the War was Over

The end of the air war and the conflict for which the Spitfire and Bf109 were developed, left them bereft of purpose. Yet it was not the end of their story. Though the final versions of the 109, the *Gustav* and the *Kurfürst*, represented a halt in the development of Germany's premier wartime fighter, the march of innovation was diverted into less successful avenues. Granted, the 109 would have no more powerful engines to boost performance. Speed, armament, range and other qualities remained frozen at the levels of 1943. Instead the 109's Luftwaffe replacement was a spectacular leap into the future. The Me262 was from a different age; a streamlined shark-like form with a triangular-section fuselage, swept wings and two Junkers Jumo axial-flow turbo-jets. With four 30mm MK108 cannon in the nose, a top speed of more than 500 mph from sea level up to 30,000 feet, tricycle undercarriage and a superb view from its bubble canopy, it was a stupendously effective machine.

The streamlined nose of the Me262 held a battery of four 30mm cannon. *(Will Owen photo)*

However, the engines were temperamental and under-powered, and needed very careful treatment to work properly, including a full overhaul after ten hours' running, and replacement after twenty-five hours. A 'flame-out' would leave only one engine delivering power, making the aircraft uncontrollable unless the pilot reacted quickly and accurately. Take-offs were exciting as the aircraft had to reach 180 mph before it could cope with engine failure. Likewise the landing involved lowering full flaps at 155 mph and crossing the runway threshold at 125 mph for a long and anxious touchdown.

The 262 made its first flight under jet power on the morning of 18 July 1942, early enough to have become a terrible threat to RAF Spitfires. It failed to do so because Hitler, excited by an aircraft too fast for Allied fighters to catch, insisted it be built as a bomber, wasting its stupendous performance advantage. There were two-seat trainer and night-fighter versions, and later examples carried a dozen R4M air-to-air rockets on wooden racks under each wing. Rockets were also used to boost take-off and climb to altitude, to minimize exposure to enemy fighters.

When it finally emerged as a fighter in late 1944, it appeared in small numbers and under such difficult conditions that it could make little difference to the air war. Despite low-drag aerodynamics, it could edge no closer to the speed of sound in controlled dives than Mach 0.86 (Mach 1.0 being the speed of sound). Though this seemed impressive, piston-engined Spitfires had dived to Mach 0.88, and RAF

Hawker Tempest Mark V of 486 RNZAF Squadron at RAF Castle Camps, a satellite airfield of RAF Debden in April 1944. *(Official photo via Wikipedia)*

and USAAF fighter squadrons soon developed tactics to take on the jets at their most vulnerable; when making the long and careful landing approach at the end of a sortie, giving the pilot little chance to spot outside threats.

High-speed fighters like Spitfire XIVs, Tempest Mark Vs and Mustangs would patrol around known jet fighter airfields. When warned by radio of 262s returning to base, they would climb to gain height over their opponents. As the jets slowed on nearing the runway, Allied fighters could bounce them with a devastating burst of gunfire to send them plunging to destruction. Lacking an ejector seat, the unfortunate pilot would almost certainly die in the ensuing crash.

Beside the dramatic switch to jet power, the sole option available for the Luftwaffe fighter force was to improve the performance and weaponry of existing fighters. To attack massed US bomber formations, special units of Bf109s and FW190s were fitted with extra armour for the pilots to press home attacks through withering crossfire. Unfortunately, the armour's weight reduced performance, making it harder to catch the bombers and to evade their fighter escorts. Wartime material shortages forced Göring to suggest using wood like the Mosquito, which he greatly admired, and steel which was more plentiful than aluminium to build airframes. Messerschmitt suggested wings with spar flanges made from steel, but ribs, skin and tail surfaces of light alloy or wood. The first arrangement would cut aluminium consumption by a third and the second by two thirds. Wooden wings would exploit spare capacity in the woodworking industry, and cut by half the man-hours needed for normal wings. In the end, though Bf109Gs and Ks used wooden tails, the wooden wing was never fitted.

Another idea involved increasing the 109's bomb load by mounting it on top of a pilotless bomber like a Ju88, packed with explosive. This *Mistel* ['Mistletoe'] project involved the 109 pilot flying the combination to its target before releasing the bomber for it to crash into its objective while the pilot flew the 109 back to base. Operational versions of the *Mistel* soon switched to the FW190 as the control plane, which separated from the bomber by firing explosive bolts. The Ju88's crew compartment was replaced by a shaped-charge warhead weighing more than three tons, meant for major warships or massive strategic targets like dams, power stations or bridges. Several *Mistel* flights were completed with varying degrees of success, but the fast-moving land war soon eliminated their options.

Finally, after five years of struggle during which first one contender then the other had held the advantage, the fight was over. The Luftwaffe's defeat was less a result of any failing of Willy Messerschmitt's fighter than of the service's wider defects combined with the wickedness and incompetence of the regime for which it fought. Herculean efforts were made to boost production to the end, under crushing Allied fighter superiority and bombing raids which destroyed the transport infrastructure and the synthetic fuel industry, to keep the German war machine running.

Though fighter aircraft were still plentiful, there was insufficient fuel to power them and too few pilots to fly them. The final wartime variant of the 109 reflected

Stupendously powerful but hideously unreliable and temperamental: one of the two Jumo 004 gas-turbines of the Me262. *(Author photo)*

the desperate situation: a stripped down G or K model armed with a single machine gun, to climb above US bombers, and ram the B17s and B24s in a semi-suicidal vertical dive. By then defeat was inevitable, and even the advanced German jet fighters and bombers failed to change the outcome.

The Allies lacked that ultimate level of desperation. Instead, the successful Spitfire Mark XIV spawned later versions using the Griffon's superb performance for specialized purposes. First was the Mark 18 (Arabic numerals replaced Roman ones for subsequent variations) which looked like a copy of a later Mark XIV with the clear view bubble canopy. With the same Griffon 65 its performance was almost identical, but the changes were under the skin. The Mitchell wing spar's complex arrangement of interlinking tubes was replaced by solid spar booms to increase stiffness and the fighter version had two thirty-one gallon fuselage tanks to increase range. The FR (fighter reconnaissance) Mark 18 had one tank replaced by three cameras, two vertical and one oblique, in the rear fuselage.

The Mark 18 appeared in June 1945, a month after the end of the European air war, with numbers sharply reduced by post-war RAF cuts. Mark XIV production totalled 957 aircraft up to the end of 1945, but only 300 Mark 18s emerged before production ended in early 1946. They equipped four squadrons in the Far East, and two Middle East-based squadrons became involved in complex air fighting after the establishment of Israel in 1948, when Spitfires fought other Spitfires, and even stopgap modifications of the Bf109, as a final round in a new theatre of war.

One more Spitfire XIV variation appeared as early as May 1944, a month before D-Day, but took no part in combat. The PR Mark 19 carried the same cameras as the Mark XI, but the Griffon engine transformed performance. With extra fuel tanks holding a total of 256 gallons for the longest range missions, it could cruise at 370 mph at 40,000 feet, immune to all late-war German fighters – including the jets!

The Mark 19's greatest quality was its prodigious altitude. The official service ceiling of the pressurized version was 42,600 feet, but on UK exercises, pilots could reach 49,000 feet given sufficient time. Tropical conditions let pilots fly even higher and one Hong Kong-based machine broke all records on a weather reconnaissance flight on 5 February 1952. Flown by Flight Lieutenant Ted Powles of 81 Squadron, it reached 51,550 feet or 9.76 miles above the earth before the decompression warning light lit up on the instrument panel. Powles throttled back and eased the stick forward, coarsening the propeller pitch to prevent over-speeding. Chaos followed. The dive steepened with violent buffeting and the controls apparently locked solid, with a violent yawing motion. Through the canopy, Powles saw the wings covered in a layer of mist caused by shock waves in the transonic region. Eventually he regained control and returned to his airfield severely shaken. The flight records showed the Spitfire reached a maximum dive speed at 15,000 feet of Mach 0.94, a piston engine record of 690 mph. Yet engineering checks showed no lasting damage, tribute to the soundness of the design and the toughness of the airframe – hard to imagine the Bf109 standing up to such ill-treatment!

There remained one more twist to the Spitfire story. One problem for aircraft designers under the hugely accelerated pace of wartime development was keeping up with changes in combat and tactics. For example, the Spitfire XIV was developed to produce its best performance at the kind of heights adopted by Luftwaffe hit-and-run attacks, but it appeared in reasonable numbers when most missions were flown at much lower heights, where most enemy aircraft were found and ground-attack targets were plentiful. Before this had been realized, Joe Smith and the Supermarine development team had begun on the next version of the Spitfire, the Mark 21. Their first priority was to correct one of the few limitations of the excellent Mark XIV; a lack of directional control at speeds over 470 mph, more than 100 mph faster than the Spitfire prototype of almost a decade earlier! The cause was aileron reversal and upfloat at these higher speeds. Larger ailerons would impose higher loads on the wing, which needed stiffening, and detail changes to provide room for four 20mm cannon and extra fuel tanks. The undercarriage also needed strengthening with longer legs to match the large five-blade propeller needed for the extra engine power.

The first Mark 21 prototype had a modified wing with extended wingtips like the Mark VII and VIII high-altitude variants. Supermarine test pilot Jeffrey Quill flew it on 1 December 1942 and reported its performance was good, but described its directional and longitudinal handling as 'appalling', whereupon the aircraft proved the problem by crash-landing. Test flights were being made from Hartford Bridge airfield near Camberley because of problems at Worthy Down near Winchester, the base for Supermarine development.

In May 1943, Quill flew the prototype back to Worthy Down and the station workshops. He landed just short of a ridge in the runway surface, which overstressed the undercarriage. The starboard leg collapsed, the wingtip touched the ground, and the plane spun around and skidded backwards, completely wrecking itself. Fortunately, the second prototype with the new wing was almost ready and this first flew on 24 July 1943.

Revised ailerons with piano-type hinges improved the handling slightly and top speed was 10 mph faster than the Mark XIV, but control stability was poor. The real solution was a larger fin, rudder and tailplane, but incorporating them into the production line would cause unacceptable delays. Even when production machines emerged, handling problems remained. When LA201, the fifteenth production Mark 21 machine, was tested by the Air Fighting Development Unit at Boscombe Down, they 'Recommended that the Spitfire 21 be withdrawn from operations until the instability in the yawing plane has been removed and that it be replaced by the Spitfire XIV or Tempest V until this can be done.' It went on to suggest that 'If it is not possible then it must be emphasized that although the Spitfire 21 is not a dangerous aircraft to fly, pilots must be warned of its handling qualities and in its present state it is not likely to prove a satisfactory fighter.' Finally, and most damningly of all it concluded that 'No further attempts should be made to perpetuate the Spitfire family.'[1]

There were three reasons why this opinion held weight. It was confirmed by trials carried out by the A&AEE. In addition, the Spitfire 21 represented such a radical reshaping of wings and fuselage that it had been suggested that it amounted to a new aircraft, and should have been renamed the 'Victor'. Thirdly, it was now clear to all that the demands of the air war had changed, and high altitude performance was no longer a trump card. Nonetheless, Joe Smith and Jeffrey Quill and their colleagues were convinced the Spitfire 21 deserved better than a new name and premature retirement, and were confident the problems could be solved.

First to go were the extended wingtips, which worsened the control problems. A revised tip shape closer to Mitchell's beautiful elliptical wing increased the wing area, but left the span just an inch greater than the prototype at 36 feet, 11 inches. The Spitfire's elevators had always been over sensitive, and replacing them with metal-skinned surfaces and lower-geared trim tabs reduced this tendency while revised rudder trim cut yaw instability.

Detail revisions to reduce over-control eased the situation. The AFDU tested one aircraft in March 1945 and agreed the problems reported on the Spitfire 21 were 'Largely eliminated by the modifications carried out to this aircraft. Its

---

1. AFDU report on LA201, quoted in Price, *op.cit.*, p. 247.

handling qualities have benefitted to a corresponding extent and are now considered suitable both for instrument flying and formation flying. It is considered that the modifications to the Spitfire 21 make it a satisfactory combat aircraft for the average pilot…, [and] that the modifications carried out on the Spitfire 21 tested be incorporated immediately in all production models, including the present Squadron equipped, and that the aircraft is then cleared for operational flying.'[2]

This was just as well as 91 Squadron had already been issued with Spitfire 21s. Based at a former Royal Navy airfield at Ludham in East Anglia, it searched for German midget submarines attacking Allied shipping. When joined by No. 1 Squadron's Spitfire 21s providing air cover for liberating the Channel Islands, the European air war was over, and peace descended over Germany and the occupied countries.

More than 3000 Mark 21s had been ordered, but orders were cut back to just 120. Nonetheless, Spitfire development still continued, although Allied jets would soon follow the trail blazed by Messerschmitt's 262. The Mark 21s went to reserve squadrons and during the final months of production were joined on the assembly lines by a similar version with shorter wingtips, an extra rear fuselage fuel tank and a bubble hood. This was the Mark 22, intended mainly for the Far East, and 627 were ordered, with the first appearing in March 1945. This proved another success, though the rear fuselage tank caused instability, so was not used in practice. It was replaced by a fifty-gallon cylindrical drop tank slung below the fuselage and an extra 170-gallon drop tank for ferry flights.

After the Japanese surrender, orders were cut to 200. Two improvements were tried. The torque reaction on take-off from increasingly powerful Griffons was eliminated on one Mark 21 by harnessing the engine to a pair of three-bladed contra-rotating propellers. This cancelled the swing force altogether, though a larger rudder had to be fitted to counterbalance the Griffon 83 contra-prop system.

The stability problem was cured by fitting a tailplane almost double the area of its predecessor, with correspondingly larger elevators. A larger fin and rudder was also provided, thanks to an ultimately disappointing Spitfire development, too late for combat. The Supermarine Spiteful was a Mark 21 airframe with a specially designed laminar-flow wing like the P51 Mustang to reduce drag and postpone compressibility effects at near sonic speeds. It also had an inward-retracting undercarriage for better ground handling.

Sadly, the Spiteful and the naval Seafang never lived up to this promise, thanks to persistent aileron jamming, instability and unpleasant stall characteristics. Improving Mitchell's incomparable wing would take more time, work and expense than predicted. Wind-tunnel tests showed over-sensitivity to the smallest imperfections in surface finish so for most purposes the Spitfire wing was still the

---

2. AFDU report on LA215.

Mass take-off of P51s to escort American bombers over Germany. *(USAF)*

best option available, ten years after its original appearance. Finally the Air Ministry lost interest in the project and the only two bright spots in the whole exercise were a staggering top speed, with the Spiteful Mark XVI prototype on 6 February 1946 recording 494 mph at 27,500 feet before its Griffon 101 two-stage, three-speed blower failed, causing a crash-landing.

However, attempts to improve yaw stability had produced an elegantly shaped taller fin and rudder, and this was fitted to the later production examples of the Mark 22 and the final Mark 24. This eliminated stability problems and turned both variants into the most capable and powerful versions of the fighter ever produced. The Mark 24 had extra fuel tank capacity and fittings for rocket launchers. Just twenty-seven were made by modifying late production 22s on the line at Castle Bromwich before the factory's final closure. After that, parts for another fifty-four Mark 22s went to the South Marston factory for completion as Mark 24s, the last of which emerged on 20 February 1948. With jets equipping front-line squadrons, most 24s went straight to the reserves, except for sixteen machines of 80 Squadron in Hong Kong from August 1949 to early 1952, finally replaced by the twin-engined, single-seat de Havilland Hornet, a formidable development of the brilliant Mosquito.

One of the 80 Squadron pilots, Gordon Bowtle, described the Mark 24 as 'A superb fighting machine' and 'An exhilarating aircraft and a challenge to the pilot's skill.' He also highlighted the 'Immediate' spin recovery and the 'Ample stall warning' by 'Airframe buffeting and…, aileron snatching.' He found it stable and easy to fly in the circuit, though it needed care on landing with a threshold speed of ninety knots, while the larger ailerons made rolling a delight, with 'Very slow rolls and hesitation rolls being performed with precision.' He felt the credit should go to Joe Smith, who had taken Mitchell's design and made the name Spitfire 'Synonymous with "fighter" for two generations. With a performance in some respects superior to jet-propelled fighters then in service, the Spitfire F.24 remains at the pinnacle of propeller-driven fighter development.'[3]

Supermarine test pilot Jeffrey Quill emphasized in his autobiography that he felt 'Extremely proud of the Spitfires Marks 22 and 24' and was sad that most of his RAF colleagues never had the chance to experience their qualities for themselves. After an interruption in his flying career for several years on medical grounds, he had recovered his flying licence in time to fly reconditioned Mark 22s from England to Cairo for the Royal Egyptian Air Force in 1954. By then he had flown jet aircraft, but remained convinced, 'What a magnificent aircraft the Mark 22 was.' On one of those delivery flights he recalled landing at Luqa airfield on Malta on a dark, cloudy night. 'I remember setting the trimmers carefully, with a small amount of power on, for a gently curving approach; as I chopped the throttle at the threshold of the runway, the aircraft virtually landed itself. And the thought occurred to me as it had so many years before, after my first flight in the prototype – this aeroplane was a real lady. By then a much more powerful, noisy, tough and aggressive lady, certainly, but a lady just the same.'[4]

Perhaps the most surprising quality of the Spitfire was the way in which Mitchell's original design was developed, stretched and tweaked, and boosted to transform its qualities far beyond what would normally be possible even under the pressure of a continuing air war. Dr Alfred Price summed it up in his splendid book, *The Spitfire Story*[5] that 'Compared with the prototype in its initial form, the Mark 24 was more than one-third faster, had its rate of climb almost doubled and its firepower increased by a factor of five. At its maximum permissible take-off weight, the Mark 24 tipped the scales at just over 6790 lbs more than the prototype; this is a weight equivalent to 30 passengers each with 40lbs of baggage on board a modern airliner, and it shows just how far the design had been pushed (the final Seafire 47 had a maximum permissible take-off weight slightly higher still).'

---

3. The Ultimate Spitfires, by Harry Robinson, *The Aeroplane*, February 1980, pp. 66.
4. Quill, *op.cit.*, pp. 294–5.
5. Price, *op.cit.*, p. 250.

So the story ended, except for one or two loose ends. With the Luftwaffe dead and the RAF becoming an all-jet fighter force, many countries still wanted these superb machines. Customers included Holland, Belgium, Norway, Sweden, Egypt, Thailand, Burma and the Irish Republic. Others were resold by their new owners, the Russians to the Communist Chinese, the Rhodesians to the Syrians and the Czechs' Spitfires to the Israelis, then striving to build up their defences against imminent attack from their Arab neighbours. In doing so they created the most poignant twist of the entire Spitfire saga.

The Czech National Air Guard continued to operate the Bf109G, still in production at a satellite plant set up by the Germans and left working at the end of the war. Two 109G-12 trainers and twenty 109G-14 fighters were assembled after checking for Resistance-inspired sabotage, and more fighters were planned for a revived Czech Air Force. Unfortunately, the country's entire stock of DB605 engines was stored in a former sugar refinery with stocks of ammunition, all of which were destroyed in a catastrophic fire in September 1945.

Faced with engineless fighters, the Czechs found stocks of Junkers Jumo 211s meant for Heinkel He111H bombers also turned out under German rule. These

To continue building 109s after the supply of DB605s dried up, the Spanish were forced to switch to first, Hispano engines and then Rolls-Royce Merlins, as in this Buchon ('Pigeon'). *(Diego Dabrio via Wikipedia)*

fitted the *Gustav* airframes well enough, but were totally unsuitable for fighters with excess torque but inadequate power. The resulting hybrid showed Messerschmitt ancestry in appearance, but not behaviour. Designated the S199, it first flew on 25 March 1947. Later a two-seat tandem advanced trainer, the CS199, proved essential to cope with the vicious handling of the single-seat machine, which Czech pilots dubbed the *Mezec* or Mule.

With war approaching, the Israelis were desperate for modern fighters, but most nations refused to supply them, so that even this limited hybrid was better than nothing. With many war veterans available, all they needed was to pay the Czech asking price and have twenty-five *Mezecs* smuggled aboard a C54 four-engined transport. When reassembled and tested, the fighters did the job, though with a high attrition rate. They tended to ground loop on landing and turn over on take-off even more readily than their German parent. But on 19 May 1948, they attacked an Egyptian armoured formation and on 3 June they shot down two Egyptian Dakotas modified to serve as bombers and about to attack Tel Aviv. Finally and most remarkably, one of these substitute fighters showed a flash of the old Messerschmitt spirit when on 18 July 1948, it shot down a Spitfire IX of the Royal Egyptian Air Force!

Worse was to follow. The Egyptians attacked RAF Spitfires on the ground at the Israeli airfield of Ramat David in Northern Israel on 22 May 1948, just four days after the base was handed over to the Israeli Defence Forces. When they returned, the RAF Mark 18s were scrambled and shot down two Egyptian Spitfires while another fell to ground anti-aircraft fire. The Czechs raised more hard currency by selling Spitfire IXs to the Israelis, some of which then attacked a formation of RAF Mark 18s on patrol over the Suez Canal Zone. Three RAF machines were shot down and two pilots killed, a sad coda to a story which had seemed to end with total triumph over the Axis.

At last comes the inescapable question in a story covering the two most successful wartime fighters; which in the end was the better combat machine? The answer is complex, with historians and former combat pilots on both sides asserting the superiority of one or the other, and citing examples to back up their claims. Even combat loss figures allow different assessments and in most cases the normal and entirely comprehensible human tendency to place the aircraft each individual flew in battle at the top of the tree influences their opinion. Former Luftwaffe pilots and allies like Italians, Hungarians, Rumanians and (especially) the Finns often paid sincere tributes to the Spitfire in their reminiscences, but few ever felt the British fighter could match the Bf109. Similarly, RAF pilots were equally convinced of the Spitfire's virtues and combat superiority as were those of the Free French, Dutch, Belgian, Norwegian, Polish, Czech, Canadian, Australian, New Zealand and American pilots who flew them against 109s.

In some ways, the verdict is as hard to prove as controversies over different cars in the field of motor sport, where objective comparisons are impossible without

The Czechs had to turn to Junkers bomber engines for their 109s, producing the less than inspiring Avia 199 *Mezec* ('Mule'). *(Alf van Beem via Wikipedia)*

assessing the skills of individual drivers. When closely matched fighters clashed, the outcome depended on the skill and experience of the pilots and the tactics they used. Over Dunkirk and in the Battle of Britain, Luftwaffe pilots did well, and it seemed even to British analysts that perhaps the *Emil* had the edge over the Spitfire Mark I and Mark II. But when British pilots gained experience and improved their tactics, the German pilots had a much rougher time of it and the Spitfire held its own. Later the balance depended more on new engines and improvements in performance and armament between the Spitfire Mark V and the Bf109 *Friedrich*, and between the Spitfire IX and the *Gustav*. Only at the end, with the development of the two-stage Merlin and the hugely powerful Griffon did the scales come down on the Spitfire side. Even then, another Messerschmitt, the twin-jet 262, could have upset the balance once more had the fighting continued longer.

Perhaps the nearest possible approach to an objective opinion is to ask a pilot who has flown several marks of both aircraft and who is also a qualified and experienced test pilot. The person best fitting this description from a very select number is retired Royal Navy Captain Eric 'Winkle' Brown, CBE DSC AFC. He wrote, 'A pilot's first impression of an aeroplane is made in the cockpit, and although both these fighters are a tight fit, the Me109 is so much so as to be claustrophobic. Efforts have been made to improve the design of the cockpit canopies, but with [the] main success being in favour of the Spitfire. What was being sought was a canopy that would improve the streamlining of the fuselage and provide the pilot with greater all-round vision, particularly rearwards which was the sector from which the danger of attack was most likely to come.

'Externally the two fighters look quite different, the Spitfire with its beautiful aerodynamic lines looking sleek and feminine, while the 109 looks rather square-jawed and ruggedly male by comparison. Also the German aircraft looks taller by virtue of its inverted-V engine which brings the propeller clearer off the ground, and this impression is heightened by the slight splaying of its longer main landing gear.

'Preparation for flight involves cockpit checks, and therefore layout of the instruments and ancillary controls becomes an important factor, especially so in alarm scrambles as were prevalent in the Battle of Britain. In this respect the Spitfire was narrowly better than its German counterpart, largely due to the introduction of the standardized six instrument blind-flying panel into British aircraft. My main criticism of the 109's cockpit was of the flaps control, which was a manually-operated handwheel, making any movement a slow process.

'Starting the engine was by a different system in each aircraft. The Rolls-Royce Merlin used electrical power (internal or external battery) and the Daimler-Benz 600 series used the inertia type whereby a flywheel was wound up by one of the

Many Avia 199s ended up being smuggled to Israel to equip the fledgling Air Force. *(Via Wikipedia)*

ground crew turning a handle until sufficient revs were obtained for the pilot to engage the starter clutch by a cockpit handle. In the light of the climates in which these fighters were to operate in their careers, the German choice for starting was certainly the more practicable.

'Taxiing was difficult in both aircraft due to the poor view over the long nose, but I preferred steering on the Spitfire's hand-operated air-pressure brakes to the 109's pedal-operated hydraulic brakes, which gave a suspicion that a little misuse could tip the nose forward for the airscrew to contact the ground.

'There was not much to choose between these aeroplanes in take-off swing or distance, and even though the Me109 used 15 degrees of flap, it was inadvisable to pull the machine off the ground early otherwise wing snatching could occur as the slats opened unevenly. The Spitfire could be settled into the climb more smartly than the 109, but the latter's angle of climb was significantly steeper although there was no great difference in the overall rate of climb.

'Engine response to throttle felt sharper in the German aircraft due to its direct injection fuel system. The Spitfire had a slight level speed advantage but the 109 could leave it behind in a dive. The harmony of control was superb in the Spitfire, but poor in the 109, being particularly heavy on the elevators. Also, although the rudder was light there was no rudder trimmer and this meant it was necessary to apply moderate right rudder during the climb and considerable left rudder in a dive – a bloody nuisance in combat! Another shortcoming was the drastic heaviness of the 109's elevators with increasing speed, till they became almost immovable at 400 mph. In the Battle of Britain, many 109s seeking to avoid Spitfire attack by diving away at low altitude finished up diving straight into the English Channel, unable to pull out at high speed.

'Manoeuvrability is of course the fighter's key to success, and this is where the Spitfire excelled over the Me109, particularly in the matter of stick force per 'g'. Personally, however, I preferred the German aircraft's straight control column for combat flying rather than the spade grip of the Spitfire although this preference changed for transonic test flying to deal with the high Mach number compressibility effects. Another minus to the 109 was that under high 'g' in turns, the slats snatched unevenly and ruined gun sighting on the target being attacked. The obvious combat advantage factor for the Me109 was its more powerful and better positioned armament, with an engine-mounted cannon firing through the propeller hub, and two fuselage-mounted heavy machine guns.

'It was in the final act of landing that the real Achilles heel of the 109 became evident. The approach speed with undercarriage and flaps lowered was 115 mph and was steeper than that of the Spitfire. A substantial change of attitude was called for at the flare or round-out before touchdown at 100 mph, and even after ground contact the lift did not spill rapidly, and ballooning or bouncing could easily be experienced. Once the tailwheel was firmly on the ground, the rudder became blanked by the fuselage and the pilot had to be very alert to the possibility of a swing occurring which could become very vicious due to the narrow track undercarriage. Such accidents occurred at an exceptionally high rate during pilot training and caused great concern.

'After the end of the war in Europe, I interrogated Willy Messerschmitt and tackled him on the matter of the structural integrity of the 109 as there were quite a significant number of cases of aircraft losing their wings under heavy 'g' forces. He reacted rather aggressively to this, but eventually admitted that pilot pressure, especially from Galland, to increase the rate of climb and reduce the wing loading to improve manoeuvrability, led him to shed some structural weight to achieve these goals.'[6]

The Bf109 far outdid the Spitfire in numbers produced (33,000 against 20,000 or so Spitfires) and length of service. During the war, the Germans supplied Franco's Spain with 25 Bf109G-2s and the jigs and tools to produce more under licence. For different reasons they ended up with none of the DB605s to power the aircraft as the Germans slid into defeat. They fitted their own Hispano-Suiza upright V12s in a reshaped nose with a carburettor air intake and oil cooler slung underneath. The Hispano engine rotated in the opposite direction to the DB605, but pilots coped well enough though performance was poor.

More machines were fitted with the fuel-injection French version of the engine until 1953 and the end of production. It was replaced by the Rolls-Royce Merlin with two-stage supercharging, which increased top speed from 382 mph to 419 mph. These Merlin-engined Messerschmitts remained in service until 1967, when their replacement freed them to appear opposite their old adversaries in the '*Battle of Britain*' film of 1969.

RAF Spitfires had been out of operational service for even longer. Their duties ended with attacking terrorist bases and supplies in the Malayan campaign from 1951 to 1954. Their naval opposite numbers lasted slightly longer over Korea and there was one final service the old fighter could provide to the jets which had taken over. In 1963, with Indonesia threatening a 'confrontation' campaign with Malaysia, the RAF wanted to see how the latest RAF fighters, twin-jet supersonic Lightnings could cope with Indonesian P51s. The RAF P51s had long gone, so a PR Mark XIX Spitfire was flown in mock combat against a Lightning.

The result was a foregone conclusion. The jet pilot was warned not to bounce his target from above, but to approach from below and behind in the Spitfire pilot's blind spot. The reason for this changed approach? To avoid being drawn into a turning dogfight, where the Spitfire's marvellous manoeuvrability could well have proved decisive as it had for so many Bf109 pilots who made the same mistake. Not bad for a 28-year-old design!

---

6. Letter to the author, 3 March 2012.

Spitfire 18s of 32 Squadron preparing to take off from Dawson Field in Jordan. *(Via Alfred Price)*

Since then both old adversaries continue to appear in museums and air shows all over the world. Surviving Spitfires and Bf109s are being restored and put back in the air from the most unpromising beginnings. Several 109s have been recovered from crash sites in the wildest parts of Russia, and one Spitfire has been excavated from a beach on the coast of Northern France and put back into the air after years of effort. Slowly but surely, the numbers of these beautiful and iconic old warriors are rising, so that people much too young to have known them in their combat prime can see for themselves what the legends were all about …

Bibliography and Sources

Allen, Wing Commander H. R. 'Dizzy': *Battle for Britain*, Arthur Barker (London), 1973
—— *Fighter Squadron 1940–42*, Granada Publishing (St Albans) 1982.
Arthur, Max: *Last of the Few*, Virgin Books (London) 2010.
—— *Forgotten Voices of the Second World War*, Ebury Press (London), 2005.
—— *Lost Voices of the Royal Air Force*, Hodder (London), 2012.
Austin, A. B., *Fighter Command*, Victor Gollancz (London), 1941.
Baker, David, *Adolf Galland, The Authorised Biography*, Windrow and Greene (London) 1996.
Barnett, Corelli, *The Audit of War*, Pan Books (London) reissued 1996.
Beaver, Paul, *The British Aircraft Carrier*, Patrick Stephens Limited (Wellingborough), 1987.
Bekker, Cajus, *The Luftwaffe War Diaries*, Macdonald, London, 1964.
Bergel, Hugh, *Flying Wartime Aircraft: ATA Ferry Pilots' Handling Notes for seven World War II Aircraft*, David & Charles (Newton Abbot), 1972.
Bishop, Patrick: *Bomber Boys: Fighting Back 1940–1945*, Harper Press (London), 2007.
—— *Fighter Boys, Saving Britain 1940*, Harper Collins (London) 2003.
—— *Wings: the RAF at War, 1912–20*, Atlantic Books (London), 2012.
Blackah, Paul and Louise, *Supermarine Spitfire Restoration Manual*, Haynes Publishing (Sparkford) 2014.
Blackah, Paul and Lowe, Malcolm V, *Messerschmitt Bf109, Owner's Workshop Manual*, Haynes Publishing (Sparkford) 2009.
Böhme, Manfred, *Jagdgeschwader 7*, Motorbuch Verlag (Stuttgart), 1983.
Bowyer, Chas: *Air War over Europe 1939–1945*, Pen & Sword (Barnsley), 2003.
—— *Fighter Command, 1936–1968*, Sphere (London), new edition 1981.
—— *Hurricane at War*, Ian Allan, Shepperton, 1974.
Bowyer, Michael J. F., *Interceptor Fighters for the Royal Air Force 1939–1945*, Patrick Stephens Limited, (Wellingborough), 1984.
Brickhill, Paul, *Reach for the Sky*, William Collins & Sons (London), first published 1954.
Braham, Group Captain J. R. D., *Scramble*, Frederick Muller (London), 1961.
Brown, Captain Eric 'Winkle': *Wings on My Sleeve*, Weidenfeld & Nicolson, (London), reprint 2007.
Wings of the Luftwaffe, Creçy Publications (Manchester), revised and expanded 2010.
—— *Wings of the Navy*, Hikoki publications (Aldershot) reissued 2013.
—— *Wings of the Weird and Wonderful Volume 2*, Airlife (Shrewsbury), 1985.

Brown, Malcolm, *Spitfire Summer: When Britain stood alone*, Carlton Books (London) 2000.

Budiansky, Stephen, *Air Power from Kitty Hawk to Gulf War II*, Penguin (London) 2004.

Bungay, Stephen, *The Most Dangerous Enemy: A History of the Battle of Britain*, Aurum Press (London) 2000.

Burns, Michael G., *Bader: The Man and His Men*, Cassell (London) 1998.

Caidin, Martin, *Me109, Willy Messerschmitt's Peerless Fighter*, Macdonald (London), 1968.

Caldwell, Donald, *JG26 Luftwaffe Fighter Wing War Diary (2 vols.)*, Grub Street, London, 1996.

Carruthers, Bob, *Luftwaffe Combat Reports*, Coda Books (Henley-in-Arden), 2011.

Caygill, Peter: *Flying to the Limit: Testing World War II Single-Engined Fighter Aircraft*, Pen & Sword (Barnsley), 2005.

—— *Combat Legend: Spitfire Marks VI-24*, Airlife (Marlborough), 2004.

Clayton, Tim and Craig, Phil, *Finest Hour*, Hodder Headline (London) 2001.

Clostermann, Pierre: *The Big Show*, Chatto & Windus (London), 1951.

—— *Flames in the Sky*, Chatto & Windus (London), 1952.

Cole, Lance, *Secrets of the Spitfire: The Story of Beverley Shenstone, The Man Who Perfected the Elliptical Wing*, Pen & Sword (Barnsley), 2013.

Crosley, Commander R. 'Mike', *They Gave Me a Seafire*, Airlife (Shrewsbury), 1986.

Darling, Kevin: *Warbird Tech Series, Volume 35, Merlin Powered Spitfires*, Specialty Press (North Branch, MN) 2002.

—— *Volume 32, Griffon Powered Spitfires*, Specialty Press (North Branch, MN) 2002.

Deere, Air Commodore Alan, *Nine Lives*, Hodder & Stoughton (London), 1959.

Deighton, Len: *Fighter: The True Story of the Battle of Britain*, Triad Panther, (St Albans) 1979.

Blitzkrieg: From the Rise of Hitler to the Fall of France, Triad/Granada (London) 1981.

Blood, Tears and Folly: An Objective Look at World War II, William Collins & Sons (London), reissued 2014.

Dibbs, John and Holmes, Tony, *Spitfire: Flying Legend*, Osprey Publishing (Botley), 1999.

Dierich, Wolfgang, *KG51 'Edelweiss'*, Ian Allan (London) 1973.

Donald, David (ed), *Air Combat Legends (Vol. 1)*, AIRtime Publishing (Norwalk, CT) 2004.

Douglas, Sholto and Wright, Robert: *Years of Command*, Collins, (London), 1966.

—— Excerpt from dispatch to Secretary of State for Air on 29 February 1948 on fighter campaign over France in 1941.

Duke, Neville, *Test Pilot*, Allan Wingate (London), 1953.

Duncan Smith, Group Captain Wilfrid, *Spitfire into Battle*, John Murray (London), 1981.

Ebert, Hans J, Kaiser, Johann B and Peters, Klaus, *The History of German Aviation: Willy Messerschmitt, Pioneer of Aviation Design*, Schiffer Military History, (Atglen, PA), 1999.

Edgerton, David: *Britain's War Machine, Weapons, Resources and Experts in the Second World War*, Penguin (London) 2012.

—— *England and the Aeroplane, Militarism, Modernity and Machines*, Penguin (London), 2013.

Fernandez-Sommerau, Marco, *Messerschmitt Bf109 Recognition Manual, A Guide to Variants, Weapons and Equipment*, Ian Allan Publishing (Hersham), 2004.

Fleming, Peter, *Operation Sea Lion*, Simon & Schuster, (London), 1952.

Flint, Peter, *Dowding and Headquarters Fighter Command*, Airlife, Shrewsbury, 1996.

Franks, Norman: *Dieppe: The Greatest Air Battle*, Grub Street (London), 2000.

—— *Air Battle of Dunkirk*, Grub Street (London), 2000.

Frayn Turner, John: *The Battle of Britain*, Airlife (Marlborough), 1999.

—— *Douglas Bader*, Airlife (Shrewsbury), 1995.

Galland, Adolf, *The First and the Last*, Methuen (London) reprinted 1955.

Gardiner, Juliet, *Wartime, Britain 1939–1945*, Headline (London), 2004.

Gelb, Norman, *Scramble: A Narrative History of the Battle of Britain*, Pan Books (London), 1986.

Glancey, Jonathan, *Spitfire: The Biography*, Atlantic Books (London) 2007.

Grant, R. G., *Flight: 100 Years of Aviation*, Dorling Kindersley (London) 2002.

Green, William: *Messerschmitt Bf 109: The Augsburg Eagle, a Documentary History*, Jane's Publishing Company, London, 1980.

—— *Famous Fighters of the Second World War*, Macdonald and Jane's, London, 1975

Gunston, Bill, *The Development of Piston Aero Engines*, Patrick Stevens Limited, (Sparkford) 1999.

Henshaw, Alex, *Sigh for a Merlin*, Arrow Books, London, 1990.

Hillary, Richard, *The Last Enemy*, Macmillan, London, first published 1942.

Holland, James, *The Battle of Britain: Five Months that changed History*, Corgi Books (London) 2010.

Hooker, Sir Stanley, *Not Much of an Engineer: An Autobiography*, Airlife (Marlborough), 2009.

Irving, David: *The Rise and Fall of the Luftwaffe: The Life of Field-Marshal Erhard Milch*, Focal Point, US, 2002.

—— *Göring*, Focal Point, US, 2010.

—— *Goebbels: Mastermind of the Third Reich*, Focal Point Publications (London) 1999.

—— *Hitler's War*, and *The War Path*, Focal Point Publications (London) 2002.

Isby, David, *The Decisive Duel: Spitfire vs. 109*, Little, Brown (London), 2012.

Jackson, Robert: *Dunkirk: the British Evacuation, 1940*, Cassell (London) 2002.

—— *Jane's Fighting Aircraft of World War I*, Random House Group (London) 2001.

—— *Jane's Fighting Aircraft of World War II*, Random House Group (London) 2001.

Johnson, Air Vice-Marshall J. E. J., *Full Circle: The Story of Air Fighting 1914–1953*, Pan Books, London, 1968.

Jones, Professor R. V., *Most Secret War: British Scientific Intelligence 1939–45*, Hamish Hamilton (London) 1978.

Ketley, Barry and Rolfe, Mark, *Luftwaffe Fledglings, 1935–1945: Luftwaffe Training Units and their Aircraft*, Hikoki Publications, (Aldershot) 1996.

Knokke, Heinz, *I Flew for the Fuhrer*, Cassell (London) (republished), 2007.

Levene, Joshua, *Forgotten Voices of the Blitz and the Battle for Britain*, Ebury Press (London) 2007.

Lucas, Laddie (ed), *Wings of War*, Grafton Books (London), 1989.

Lyall, Gavin (ed.), *Freedom's Battle: The War in the Air*, Arrow Books, London, 1971

McKinstry, Leo, *Spitfire, Portrait of a Legend*, John Murray (London) 2007.

Mason, Francis: *Battle over Britain*, McWhirter Twins, London, 1969.

—— *The Hawker Hurricane*, Creçy Publishing (Manchester), 2005.

Masters, David, *So Few:* Corgi Books (London) paperback edition, 1956.

Michulec, Robert, *Messerschmitt Bf109F*, MMP (Petersfield), 2013.

Mitchell, Gordon, *R. J. Mitchell: Schooldays to Spitfire*, The History Press (Stroud), 2009.

Morgan, Eric B. and Shacklady, Edward, *Spitfire: The History*, Key Books, (Stamford), 2000.

Murray, Williamson: *War in the Air 1914–45*, Cassell (London) 1999.

—— *Luftwaffe, Strategy for Defeat 1933–45*, Grafton Books (London) 1988.

Neil, Tom, *Gun Button to Fire*, William Kimber (London) 1987.

Nesbit, Roy Conyers, *An Illustrated History of the RAF*, Colour Library Books, Godalming) 1990.

Newton Dunn, Bill, *Big Wing, the Biography of Air Chief Marshal Sir Trafford Leigh-Mallory*, Airlife (Shrewsbury), 1992.

Orange, Vincent: *Dowding of Fighter Command: Victor of the Battle of Britain*, Grub Street (London), 2008.

—— *Park: The Biography of Air Chief Marshal Sir Keith Park*, Grub Street (London), 2000.

Overy, R. J., *The Air War 1939–1945*, Europa Publications, London, 1980.

Pilot's Notes: *Messerschmitt Bf109: Technical Data and Handling Notes*, reprint 2000.

—— *Spitfire IIA and IIB Aeroplanes, Merlin XII Engine*, HMSO, 1940.

—— *IX, XI and XVI: Merlin 61, 63, 66, 70 or 266 Engine*, HMSO, 1947.

Price, Dr Alfred: *Instruments of Darkness: the History of Electronic Warfare*, Macdonald and Jane's, London, 1977

—— *Battle of Britain: 18 August 1940: The Hardest Day*, Granada Publishing (St Albans) 1980.

—— *Battle of Britain Day: 15 September 1940*, Greenhill Books (London), 1990.

—— *Battle over the Reich (2 vols.)*, Ian Allan Publishing (Hersham), 2005.

—— *Luftwaffe Handbook*, Ian Allan (Shepperton), 1986.

—— *The Last year of the Luftwaffe, May 1944 to May 1945*, Greenhill Books (London), 2001.

—— *Focke Wulf FW190 in Combat*, The History Press (Stroud), 2009

—— *Targeting the Reich: Allied Photographic Reconnaissance over Europe, 1939–45*, Greenhill Books (London), 2003.

—— *Spitfire at War: Parts 1 and 2*, (Ian Allan, Shepperton), 1974, 1985.

—— *Spitfire: A Documentary History*, Macdonald and Jane's, London, 1977.

—— *The Spitfire Story*, Cassell (London) 2002.

Price, Alfred and Blackah, Paul, *Owner's Workshop Manual, Supermarine Spitfire 1936 onwards, all marks)*, Haynes Publishing (Sparkford) 2007.

Priller, Josef, *Geschichte Eines Jagdgeschwaders*, Wowinkel Verlag (Neckargemund), 1969.

Pritchard, Anthony, *Messerschmitt*, Vantage Books (London), 1975.

Quill, Jeffrey, *Spitfire: A Test Pilot's Story*, Creçy Publishing (Manchester), 2005.

Radinger, Willy and Schick, Walter, *Messerschmitt Bf109A-E: Development, Testing, Production*, Schiffer Military History, (Atglen, PA), 1999.

Ramsey, Winston, *The Battle of Britain: Then and Now*, After the Battle, (London), 1980.

Ray, John: *The Battle of Britain: New Perspectives. Behind the Scenes of the Great Air War*, Arms and Armour Press (London), 1994.

—— *The Night Blitz, 1940–1941*, Arms and Armour Press, London, 1996.

Reschke, Willi, *Jagdgeschwader 301/302*, Schiffer Military History, (Atglen, PA), 2005.

Richards, Dennis and Saunders, Hilary St G., *Royal Air Force 1939–45 (3 vols.)*, HMSO (London), 1953–4.

Richey, Paul, *Fighter Pilot*, Cassell, (London), reissued 2001.

Ring, Hans and Girbig, Werner, *Jagdgeschwader 27*, Motorbuch Verlag (Stuttgart), 1971.

Ritger, Lynn, *The Messerschmitt Bf109, Part 2: 'F' to 'K' Variants*, SAM Publications (Bedford) 2007.

Robinson, Derek, *Invasion 1940: The Explosive Truth about the Battle of Britain*, Constable & Robinson, London, 2005.

Roberts Andrew, *The Storm of War*, Penguin (London), 2010.

Roussel, Mike, *Spitfire's Forgotten Designer: the Career of Supermarine's Joe Smith*, The History Press, Stroud, 2013.

Russell, Cyril Richard: *Spitfire Odyssey*, Kingfisher Railway Productions, 1985.

—— *Spitfire Postscript*, (published privately), 1995.

Sarkar, Dilip: *Last of the Few*, Amberley Publishing (Stroud) 2010.

—— (ed): *Spitfire Manual 1940*, Amberley Publishing (Stroud) 2010.

Schmoll, Peter, *Nest of Eagles: Messerschmitt Production and Flight Testing at Regensburg, 1936–1945*, Ian Allan Publishing (Hersham), 2004.

Scutts, Jerry: *Messerschmitt Bf109: The Operational Record*, Motorbooks International, (Osceola, WI), 1996.

—— *Combat Legend, Messerschmitt Bf109*, Airlife, (Shrewsbury), 2002.

Shacklady, Edward, *Messerschmitt Bf109, Classic WWII Aviation (Vol.2)* Tempus Publishing, Stroud, 2000.

Shirer, William, *The Rise and Fall of the Third Reich: A History of Nazi Germany*, Secker & Warburg, (London) 1961.

Shore, Christopher, and Williams, Clive, *Aces High*, Neville Spearman (London), 1972.

Spick, Mike, *Luftwaffe Fighter Aces: the Jagdflieger and their Combat, Tactics and Techniques*, Greenhill Books (London), 1996.

Steinhilper, Ulrich, *Spitfire on my Tail*, Independent Books, (Bromley), 1989.

Steinhoff, Johannes, *Messerschmitts over Sicily*, Stackpole Books (US), 2002.

Terraine, John, *The Right of the Line*, Wordsworth Editions (Ware, Herts), 1997.

Toliver, Colonel Raymond F. and Constable, Trevor J., *Horrido ! Fighter Aces of the Luftwaffe*, Arthur Barker (London) 1968.

Townsend, Group Captain Peter, *Duel of Eagles*, Weidenfeld & Nicolson, (London), 1990.

Vader, John, *Spitfire*, Macdonald & Co (London), 1970.

Vann, Frank, *Willy Messerschmitt: First Full Biography of an Engineering Genius*, Patrick Stephens, Yeovil, 1993.

Van Ishoven, Armand: *Messerschmitt Bf109 at War*. Ian Allan, (Shepperton), 1977.

—— *Messerschmitt*, Gentry Books Limited (London), 1975.

Vigors, Tim, *Life's Too Short to Cry*, Octopus Books (London) 2010.

Warner, Graham, *The Bristol Blenheim, A Complete History*, Creçy Publications (Manchester), revised 2005.

Wellum, Geoffrey, *First Light*, Penguin (London) 2002.

Williams, Neil, *Airborne*, Airlife (Shrewsbury), 1977.

Winterbotham, Group Captain F. W., *The Ultra Secret*, Weidenfeld & Nicolson (London) 1974.

Winton, John, *Find, Fix and Strike: the Fleet Air Arm at War 1939–45*, Batsford, London, 1980.

Wisniewski, Jason R., *Powering the Luftwaffe: German Aero Engines of World War II*, Friesen Press (Victoria, BC), 2013.

Woods, Derek and Dempster, Derek, *The Narrow Margin*, Hutchinson (London), 1961.

Wright, Robert, *Dowding and the Battle of Britain*, Macdonald and Jane's London, 1972.

Websites:

The Battle of Britain Historical Society at: www.battleofbritain1940.net

The Spitfire Site at http://spitfiresite.com

Aces of the Luftwaffe at: http://www.luftwaffe.cz

Kurfürst: the Bf109 Performance Resource site at: http://kurfurst.org

The Archives of 'Flight' magazine at: http://www.flightglobal.com/pdfarchive/view

Conversations and Correspondence (1970 to 2014):

Air Commodore Peter Brothers, DSO, CBE, DFC

Captain Eric 'Winkle' Brown, CBE, DSC, AFC

Commander R. M. 'Mike' Crosley, DSC

Professor M. R. D. Foot, CBE, TD

Generalleutnant Adolf Galland

Wing Commander Pat Hancock, OBE, DFC

Professor R.V. Jones, CH, CB, CBE, FRS

Dr Alfred Price, FRHistS

Rt Hon Duncan Sandys, CH, PC, MP

Wing Commander James Storrar, DFC

Index